Endurance and War

Endurance and War

The National Sources of Military Cohesion

Jasen J. Castillo

Stanford Security Studies

An Imprint of Stanford University Press

Stanford, California

Stanford University Press
Stanford, California

Printed in the United States of America on acid-free, archival-quality paper

Library of Congress Cataloging-in-Publication Data

Castillo, Jasen, author
Endurance and war: the national sources of military cohesion / Jasen J. Castillo.
pages cm.
Includes bibliographical references and index.
ISBN 978-0-8047-8910-3 (cloth : alk. paper)
1. Unit cohesion (Military science)—Case studies. 2. Sociology, Military. 3. Military history, Modern—20th century. I. Title.

U22.C39 2014
355.3—DC23 2013035081

Typeset at Stanford University Press in 10/14 Minion

Special discounts for bulk quantities of Stanford Security Studies are available to corporations, professional associations, and other organizations. For details and discount information, contact the special sales department of Stanford University Press. Tel: (650) 736-1782, Fax: (650) 736-1784

CONTENTS

Figure, Maps, and Tables

FIGURE

MAPS

TABLES

Acknowledgments

Not only does a book project take a long time—longer than anyone anticipates!—it also requires a great deal of help from family, friends, and colleagues. Both John Mearsheimer and Charles Glaser at the University of Chicago guided me as I started this project. While John forcefully demanded that my book offer a simple answer to a complex question, Charlie reminded me that the world is complicated. They remain important mentors and sources of inspiration. Both encourage their students to use the theories and methods of social science to address important questions in national security policy.

Along the way, a number of scholars commented on portions or all of the manuscript. For their insights, I thank Alexander Wendt, Robert Pape, Stephen Walt, Duncan Snidal, Keir Lieber, Negeen Pegahi, Adam Grissom, Jeffrey Martini, Dan Reiter, Taylor Fravel, Owen Cote, Charles Hermann, Joseph Cerami, Domonic Bearfield, Arnold Vedlitz, Adam Seipp, Larry Napper, Brian Linn, and Valerie Hudson. I also appreciate contributions from participants who attended presentations I made at the Massachusetts Institute of Technology, the U.S. Naval War College, Duke University, and the RAND Corporation. Christopher Layne, Michael Desch, Paul Kapur, and David Edelstein deserve special thanks for reading the entire manuscript and providing sharp but useful criticism. I am particularly indebted to Alexander Downes, who read the manuscript several times, helped me improve the argument significantly, and was always available for discussions. Alex embodies the model colleague: a tough reader but also eager to help me make my argument. I pity the scholar that does not have a friend like him as an ally.

I received generous research support from the Scowcroft Institute of International Affairs as well as from the George H. W. Bush School of Government and Public Service, both at Texas A&M University. For all she does to ensure the smooth running of the International Affairs Program at the Bush School, I am grateful to Janeen Wood. Additionally, an army of research assistants kept this project going day to day. This merry band included Nick Reves, Miguel Gonzalez, Kip Whittington, and Frank Edmiston. I offer special thanks to Joshua Murray, Dante Miller, and William Mayborn for helping me lay the foundations of the current draft. A determined Danielle Menard labored along with me to put the final manuscript together. Leanne Sulewski carefully and quickly created the book's maps. At Stanford University Press, James Holt provided timely editorial assistance. My editor, Geoffrey Burn, gave prompt, patient guidance to an anxious author.

My family proved an important source of staying power. For her constant encouragement, I thank my mother, Estelle Newidouski. In their unique canine way, Henry, Jamaica, Sophie, and Toot bolstered my morale with morning walks, games of fetch, or merely by sleeping next to me as I wrote. My biggest supporter in this endeavor, however, has been my wife, Victoria Roach Castillo. Not only did she endure countless discussions about military cohesion, not only did she suffer through long, hot summers by my side as I finished the manuscript, not only did she read and edit the manuscript twice, but she never stopped believing in my ability to finish. In the dark moments we all confront in our professional lives, she rallied my spirits. For her unconditional loyalty and love, I will be forever grateful. I dedicate this book to her.

Jasen Castillo
College Station, Texas

Preface

Why do the militaries of some countries fight hard when facing defeat, while others collapse? Scholars have advanced three kinds of arguments to explain why some countries can create cohesive militaries and other countries cannot. One view argues that military cohesion depends on the bonds of loyalty that soldiers form within small units, that soldiers fight for their comrades in arms instead of grand political causes. This perspective, however, ignores instances where militaries fought hard without small-unit bonds, such as the Soviet Red Army during World War II. A second view contends that a country's ideology motivates a nation's armed forces. Although ideologies sometime rally a nation's armed forces, at other times they fall on deaf ears. France in 1940 represents such a case. A third view claims that democracies produce militaries with unrivaled effectiveness. Despite the popularity of this argument, the historical record suggests nondemocracies fight with equal, if not more, determination on the battlefield. Nazi Germany fought with great tenacity and staying power even to the end of World War II.

This book presents *cohesion theory* to explain why national militaries differ in their staying power. My argument builds on insights from the existing literature on military effectiveness and collective action. According to my theory, the cohesion of national militaries depends on two variables: the degree of control a regime holds over its citizens and the amount of autonomy the armed forces possess to focus on training for warfighting. With a high degree of regime control, governments instill and enforce norms of unconditional loyalty throughout the population, including within the armed forces. A hard-core group of regime supporters inside the military will fight no matter the strategic circumstances and pressure others to do the same. With a high degree of organizational autonomy, the armed forces can demand and enforce norms of unconditional loyalty as well as trust among its personnel. These norms will motivate most units—even reserve units—to fight with determination and flexibility on the battlefield. Regime control and organizational autonomy interact to produce four distinct types of militaries, each with a different degree of cohesion: *messianic*, *authoritarian*, *professional*, and *apathetic* armed forces. The book's case studies explore each of these types of militaries.

Overall, the goal of the book is twofold: to present a new theory to explain military cohesion while also providing an initial empirical test of the theory's explanatory power. The study, therefore, is one part theory creation and one part theory testing. To demonstrate the plausibility of cohesion theory, as well as to test the proposition that democratic armies fight with the greatest endurance and determination, I present a series of comparative case studies. I test to see if my theory's variables are associated with the predicted degree of military cohesion across different cases.

The question of what motivates a military during war is important to policy-makers and scholars alike. Understanding why armed forces differ in their cohesion should help U.S. military planners better assess the military capabilities of potential adversaries, such as Iran and North Korea. Such estimates might have improved the planning for Operation Iraqi Freedom, where the tenacity of the Baathist regime's most ardent supporters, the Fedayeen Saddam, surprised coalition forces.

The book also makes two contributions to the field of international security. First, it helps explain how countries create military power and win wars, which is a central puzzle in the field of international relations. Second, the book's theory helps identify which national armed forces could fight with enough cohesion to adopt the doctrine and tactics required for success on the contemporary battlefield, what Stephen Biddle calls the "modern system." Although scholars have paid increasing attention to issues of military effectiveness, they have focused on skill, not will. Cohesion represents a crucial determinant of who wins and loses on the battlefield. Militaries might understand what the "modern system" requires for victory but nevertheless lack the determination and discipline to employ it.

Finally, scholars can employ cohesion theory to explain the staying power of groups other than national militaries, including insurgencies, terrorist organizations, and social movements. Cohesion theory bridges two literatures on collective action: rationalist accounts that emphasize the use of incentives to motivate individuals in a group, and constructivist arguments that focus on the ability of norms to shape how strongly members identify with a group's interests.

Endurance and War

1 Introduction

> One might say that the physical seems little more than the wooden hilt, while the moral factors are the precious metal, the real weapon, the finely-honed blade.
>
> Carl von Clausewitz[1]

WHY DO NATIONAL MILITARY ORGANIZATIONS differ in their cohesion? Why do some countries' militaries fight hard when facing defeat, while others collapse? Why are some countries better than others at creating and sustaining cohesive armed forces in war? Consider, for example, the performance of the German military in World War II. Even when defeat appeared inevitable, the Wehrmacht fought tenaciously, prolonging the war and adding to the tremendous suffering it had already inflicted.[2] In contrast, the battlefield determination of French armed forces varied widely in 1940, with some units offering stiff resistance, while others dissolved at crucial moments. As a result, France could never recover from initial German breakthroughs in early May 1940, and its military leaders, fearing an internal insurrection, pressured the civilian politicians to capitulate in late June.[3]

What accounts for the greater cohesion of the German Army? Why did French armed forces disintegrate with their country's survival at stake? Traditional definitions of military power that focus on material capabilities offer only some clues as to why national militaries differ in their performance. Although knowing how many tanks and troops a country can send into battle can occasionally serve as a good predictor of how its military might perform in war, material capabilities alone cannot tell the entire story. Military cohesion, or the capacity of national armed forces to fight with determination on the battlefield, and *to keep fighting* even when a war appears lost, also plays an important role.

The historical record suggests no countries are alike in their military cohesion. Some countries can field armies capable of overcoming deficiencies in military hardware through their determination. Similarly, other countries bring superior military capabilities to the battlefield, but they cannot match their opponent's tenacity. Both France and the United States painfully discovered the importance of military cohesion while fighting in Vietnam.[4] The balance of capabilities suggested that the North Vietnamese stood very little chance in their struggles against the Western Powers. In the end, however, strong military cohesion helped North Vietnam to outlast its opponents and emerge victorious.[5] Battlefield determination and the staying power to endure a long, painful war trumped the balance of forces. Today, a similar dynamic appears at work in Afghanistan. The Taliban have proven a "tenacious and determined" foe hindering the nation-building efforts of the United States and its NATO allies.[6] The pervasive nature of the insurgency has led some analysts to worry that sectarian divides in the newly raised Afghan National Army will undermine its military cohesion.[7] Victory or defeat frequently depends on which side can field the more cohesive force.

In this study, I use the term "military cohesion" to describe two characteristics of a country's armed forces in war: (1) the capacity to fight with determination and flexibility on the battlefield; and (2) the ability to resist internal pressures to collapse as the likelihood of winning a war decreases. Military cohesion, therefore, refers to both battlefield performance as well as the staying power of a country's armed forces to endure difficult strategic circumstances without disintegrating. Traditionally, the literature uses cohesion to discuss the ability of small units, such as squads or platoons, to fight with determination and staying power. I broaden the concept of cohesion to encompass all of a country's military organizations during war.

Scholars have advanced two kinds of arguments to explain why some countries can create and sustain cohesive militaries, while other countries cannot. One view argues that military cohesion grows out of the strong bonds of loyalty that soldiers form inside small units, such as platoons or squads; soldiers fight for their buddies, not for grand political causes.[8] This claim, however, ignores instances where militaries fought hard even when they lacked small-unit bonds, as a result of faulty personnel policies, internal divisions, or high rates of casualties during war. Another view argues that ideology, and not small-unit ties, motivates a country's armed forces to fight hard. By forging strong connections between soldiers and their nations, ideologies create cohesive militaries, capable of enduring long and costly wars. In this view, militaries fight with a high degree of cohesion because of nationalism or devotion to democracy.[9] Of course, sometimes militaries respond to ideological rallying cries, but at other times they ignore them. Nationalism and democracy galvanized France in World War I but failed to do so in World War II. Ideologies could play an important role, but the conditions under which they promote military cohesion remain unclear.

SUMMARY OF THE ARGUMENT

This book presents what I call *cohesion theory* to explain why national militaries differ in their staying power and battlefield determination. My argument builds on insights from both the rationalist and social constructivist literatures on group solidarity.[10] The core idea underpinning my theory is that any large group, including a country's armed forces, motivates members to sustain collective action by both promoting and enforcing norms of unconditional loyalty among its members. Norms define individual member obligations to any group. Groups remain cohesive even under challenging circumstances because these norms motivate some members to remain committed to group goals. At the same time, these hard-core group members pressure and coerce less motivated members to do the same.

According to the theory, the relationship between a government and its military organizations determines the cohesion of its armed forces during war. Specifically, the cohesion of a country's military depends on two explanatory variables: (1) the degree of regime control over a population; and (2) the degree of autonomy the armed forces possess for training. With a high degree of control, the regime can promote an ideology of unconditional loyalty and enforce it through coercion. This level of control requires both a national ideology that demands unconditional loyalty, such as communism, fascism, or nationalism, combined with a highly repressive government with no civil society to oppose it. Regimes that dominate their societies in this fashion bolster military cohesion by creating hard-core supporters in the armed forces who fight with determination and pressure others to display a similar commitment. Some personnel are true believers, with strong ties to the regime, while others feel compelled to fight.

For many regimes, however, obtaining a high degree of control over society is too difficult or simply too repugnant to consider. Consequently, many countries create cohesive armed forces by granting military organizations the freedom to train without interference. With a high degree of autonomy, the armed forces can use training to promote norms of unconditional loyalty and trust inside their ranks. Personnel develop bonds of loyalty and trust to one another as well as their organization. These norms will motivate most units—even reserve units—to fight with determination and flexibility on the battlefield. Units will display a willingness to fight even after suffering the shock of local defeats, encirclement, and intense bombardment. They will also fight with flexibility, capable of performing without tight command and control, as well as avoiding panic. Strong internal ties inspire personnel in battle and sustain them in times of defeat.

To maintain the cohesion of its armed forces in war, then, governments face several choices: they can exert a high degree of control over the country, they can allow military organizations the autonomy to train, they can do both, or they can fail to do either. Taken together, these two variables, regime control and organizational autonomy, produce four ideal types of national armed forces, each capable of fighting with different degrees of cohesion.

A high degree of regime control and organizational autonomy for the armed forces creates *messianic* militaries. I describe them as messianic because the regime's ideology advances a national mission that galvanizes the armed forces in war. Because it believes the military will remain faithful, the regime allows the armed forces autonomy to forge strong internal bonds of their own. Messianic militaries fight with a high degree of cohesion. They exhibit strong staying power: the regime's hard-core supporters fight under all circumstances and pressure others to do the same, even if victory looks impossible. They also display strong battlefield performance: strong internal bonds within the armed forces enable most units of messianic militaries to fight with determination and flexibility on the battlefield.

When a high degree of regime control undermines the autonomy of the armed forces, states field *authoritarian* militaries. These militaries draw their cohesion from the regime's ability to demand and enforce unconditional loyalty from the population. A high degree of regime control bolsters staying power, giving the armed forces the ability to fight even when victory looks impossible. Zealous followers fight and coerce others to do the same. However, they fight with less battlefield performance. Without autonomy, a country's military organizations cannot form strong internal bonds of their own. The state does not trust the military with the autonomy to create loyalty within the ranks. Devoid of these internal military bonds, most units will fight with determination but lack the ability to fight with flexibility on the battlefield.

In contrast, a regime exercising a low degree of control and permitting a high degree of autonomy for the armed forces will create a *professional* military. I label them as professional because with autonomy the armed forces can concentrate on training for warfighting and, thereby, create their own internal bonds of loyalty. Under these conditions they develop a corporate identity centered on their expertise and responsibility to the county. The armed forces serve as trustworthy organizations, with the singular purpose of preparing and fighting the country's wars. With a focus on training, strong organizational bonds develop, enabling the military to show strong battlefield performance: most units fight with high levels of determination and flexibility on the battlefield. Without strong regime control, however, professional militaries will slowly lose their staying power as the probability of victory decreases. A hard-core group of regime supporters will not exist to coerce the armed forces to fight when defeat looks imminent.

Finally, countries with a low degree of regime control as well as a low degree of military autonomy will produce *apathetic* armed forces. These organizations fight with a low degree of cohesion for two reasons. Without military autonomy, the armed forces cannot instill norms of unconditional loyalty and trust across most of its members. As such, they will display weak battlefield performance: only elite units will fight with determination and flexibility in combat. This means that reserve units, critical in long wars and conflicts requiring large forces, will unlikely display tenacity in battle. Similarly, without a high degree of regime control, the government will lack a hard-core group of supporters willing to fight and coerce others to fight even if defeat in war appears inevitable. After a few defeats, apathetic militaries will quickly lose their staying power, as the armed forces voice their opposition to the war and their battlefield units disintegrate.

WHY STUDY THE SOURCES OF MILITARY COHESION?

The question of what motivates a military during war is important for policy-makers and scholars alike. How governments and military organizations can create forces with maximum staying power remains an open question. For governments and military organizations, it is no easy task to convince individuals to risk injury and perhaps death in combat to defend their country. Moreover, the issue of motivation does not pose a problem only for nation states. The fighting units of nonstate actors, insurgency movements and terrorist organizations, for instance, often rely on highly motivated members to carry out very dangerous military operations, or, in some cases, to give their own lives in support of a cause. For example, members of the al Qaeda terrorist network, by their capacity to conduct attacks in several countries in spite of widespread pressure by the international community, by their tenacious fighting ability in places like Afghanistan and Pakistan, and by their willingness to engage in suicide attacks, have demonstrated the dangers a highly motivated adversary can pose to international security.[11]

Understanding why armed forces differ in their cohesion should help U.S. military planners assess the military promise of potential adversaries, such as Iran, North Korea, and Venezuela. Ultimately, planners want to gauge prospects for defeating potential opponents on the battlefield, what analysts commonly describe as net assessments.[12] Analysts performing net assessments of potential adversaries commonly examine material capabilities, but they should also attempt to discern a variety of nonmaterial factors, such as the willingness of an enemy's forces to fight. Taking the cohesion of an opponent's military into account can provide a more detailed picture of an adversary's capabilities.

For example, militaries armed with modern technology operated by well-trained personnel may do well on the battlefield but lack the staying power for long, inconclusive wars. Arguably this describes the pressures democracies face when fighting counterinsurgencies abroad.[13] Alternatively, ethnic or political divisions, for example, might keep an opponent's forces from performing effectively on the battlefield, becoming the Achilles' heel of an otherwise formidable country.[14] Still further, some regimes possess a coercive apparatus strong enough to resist a powerful attacker, even when defeat looks inevitable. Such estimates might have improved the planning for Operation Iraqi Freedom, where the tenacity and coercive power of the Baathist regime's most ardent supporters, the Fedayeen Saddam, surprised coalition forces.[15]

This study on the sources of military cohesion should interest students of security studies, and international relations more generally, for three reasons. First, it offers a broad and systematic definition of military cohesion that encompasses not only battlefield performance but also the ability of military organizations to avoid the pressures of collapse when victory seems unlikely. Recall that military cohesion is the capacity of a country's military units to fight with determination and flexibility in combat while enduring difficult strategic circumstances. In this way, military cohesion represents a national-level phenomenon that applies to all of a country's military organizations. Looking at military cohesion from this vantage point follows a growing trend in the field of international security that examines the domestic and societal sources of military effectiveness. More important, this definition allows scholars to better conceptualize and understand variations in staying power. On the one hand, national militaries rarely fight to the last soldier; even Nazi Germany's armed forces eventually surrendered. On the other hand, national militaries rarely collapse without a fight; even the French Army in 1940 took and inflicted casualties. This study helps explain the range of outcomes in cohesion between these two extremes.

Second, the book's theory helps identify which national armed forces could fight with enough cohesion to adopt the doctrine and tactics required for success on the contemporary battlefield, what Stephen Biddle calls the "modern system."[16] Although scholars have paid increasing attention to issues of military effectiveness, they have focused primarily on why armed forces differ in skill, not will. Cohesion represents a crucial determinant of who wins and loses on the battlefield. Militaries might understand what the "modern system" requires for victory, but nonetheless lack the determination and flexibility to employ it.

Finally, cohesion theory could help scholars explain the success of collective action in several types of groups, not only national militaries but also insurgent and terrorist groups. The challenge of motivating individuals to risk their lives in combat is at the center of the collective-action problem, a subject of interest to a variety of disciplines. Economists, political scientists, social psychologists, and sociologists remain deeply divided over why some groups can achieve collective goals and why others are stymied by the free-riding tendency of individuals when pursuing collective action.[17] The view of individuals as self-interested actors, largely the view in economics and the dominant perspective in political science, contends that individuals require either coercive or pecuniary incentives to persuade them to engage in collective action. When convinced they will gain more from participating than from sitting on the sidelines, individuals will participate in groups to secure collective goals.

Social constructivist scholars, borrowing from sociology and social psychology, offer a different perspective. They argue that individuals can form group identities that facilitate collective action. Instead of seeing the world from the vantage point of self-interested individuals who ask, "What is in it for me?" social constructivists believe that there are instances when people might participate in collective action by asking, "What is in it for the group?"[18] Norms create individual members that strongly identify with a group's interests.[19] Cohesion theory bridges the constructivist and rationalist approaches by explaining how norms and coercion work together to motivate militaries.[20]

THE LITERATURE ON MILITARY EFFECTIVENESS

There is a substantial literature debating why militaries win and lose battles. Studies of military effectiveness try to move beyond explanations of battlefield success based on technology and the balance of forces. Instead, they look to nonmaterial factors to explain combat outcomes. Most of these studies address the sources of military skill, or why some armed forces fight better than their competitors. They describe military skill along the strategic, operational, and tactical levels of war.[21] Stephen Biddle, for example, emphasizes the dominance of firepower on the contemporary battlefield and how skilled militaries will adopt "modern system" tactics to succeed under these conditions.[22]

To explain differences in skill, scholars have pointed to several national characteristics. According to a few analysts, tensions between civilian leaders and military organizations frequently affect battlefield performance. Stephen Biddle and Robert Zierkle examined civil-military relations to describe why some militaries encountered trouble with emerging weapon technologies.[23] Risa Brooks argues that the balance of civilian and military power shapes the effectiveness of the armed forces during war.[24] Sometimes a regime's effort to coup-proof the military reduces the skill of the armed forces on the battlefield. According to James Quinlivan, these civil-military tensions account for the poor performance of many Arab militaries.[25]

Other scholars locate the sources of military skill in a society's domestic political institutions, its political culture, or the organizational culture of its military. For example, Stephen Peter Rosen uses historical evidence from India to illustrate how societal divisions can prevent countries from realizing their full military potential.[26] These same societal cleavages might undermine unity within the ranks of the military. In a similar fashion, Elizabeth Kier argues that domestic political disputes about the role of the military in society can overwhelm strategic considerations when militaries construct their doctrines. She notes that debates about the role of the military in society can also influence the organization's culture and overall combat effectiveness.[27] Some scholars, most notably Kenneth Pollack, have also tried to connect a country's national culture with its battlefield performance.[28]

This book, in contrast, focuses on questions of will. In particular, the central issue here is why militaries differ in their battlefield determination and overall staying power to endure wars where victory looks unlikely, what I define as cohesion. Although skill plays an important role in combat outcomes, will, or cohesion, also determines who wins and loses on the battlefield. For this reason, military planners and scholars have long sought to understand why people risk their lives in combat for a cause, country, or both.[29] Military theorists ranging from Sun Tzu to Clausewitz consider the motivation of armies a crucial facet of warfare.[30] The fact that some militaries collapse and others hold together under stressful combat conditions has puzzled generations of historians, military sociologists, and political scientists.[31] To inspire their soldiers and keep their units intact under difficult circumstances, military organizations have relied on a variety of techniques.[32]

Scholars have focused on four sources of military motivation.[33] The most obvious method of motivating involves the use of positive and negative incentives, or carrots and sticks. For mercenaries, the promise of payment or the prospect of looting the vanquished proved sufficient to keep units fighting.[34] The capacity of military organizations to provide for the physical and psychological needs of their personnel has received attention as a factor in wartime motivation.[35] Coercion and fear also play a role in maintaining cohesion.[36] Some armies threaten their men with painful penalties or death if they attempt to break ranks. Fear of punishment, including execution, encourages reluctant soldiers to fight. To complement threats of punishment, militaries drill their personnel to make battlefield discipline and tactics second nature.[37] The French military theorist Ardant du Picq believed training also facilitated the tight bonds among soldiers that made them more effective in battle.[38] Of course, the problem with positive and negative incentives is identifying the conditions under which they keep militaries fighting and when they do not.

A second line of research emphasizes the importance of small-unit bonds, from the platoon to the battalion level, and the role of training in improving an army's staying power. Awash in data from World War II, American sociologists in the late 1940s began systematic investigations to understand the behavior of soldiers in battle.[39] Samuel Stouffer and his colleagues questioned American infantrymen and bomber crews to learn how they coped with the stress of battle. In a more controversial work, S. L. A. Marshall argued that fear of injury and death prevented significant numbers of American soldiers from firing their weapons. Marshall argued that men too fearful to fire their weapons posed the key obstacle to a unit's combat effectiveness.[40]

Edward Shils and Morris Janowitz interviewed German prisoners of war to ascertain how the Wehrmacht remained an intact and capable fighting force in the face of impending defeat.[41] Their path-breaking research argued that the cohesion of the Wehrmacht reflected strong primary group ties within units. Loyalty to their friends in the unit, and not ideology, kept German soldiers fighting in the face of a losing cause. Little and Janowitz made a similar argument from their studies about the determination of American soldiers in the Korean War.[42] Charles Moskos pointed to the U.S. Army's individual replacement system and poor leadership at all levels of units in the Vietnam War as an obstacle to the formation of primary group ties in American units.[43]

Theories about the formation of primary-group ties illustrate the power of strong bonds within groups, but wrongly disregard the potential for these bonds to form at higher levels, such as within the military organization as a whole and society at large. In addition, some armed forces fail to adopt the personnel and training policies small-unit cohesion requires, but can still display tenacity on the battlefield.[44] Robert Rush has argued that strong bonds between soldiers and the military organization held the U.S. Army together in World War II, while the combination of heavy casualties and an individual replacement system undermined primary group ties. Alternatively, some militaries can fight with great determination, even though high casualty rates and their corresponding high rates of replacement can prevent small-unit bonds from forming.[45] The Iranian military, for example, still fought with great ferocity even though it suffered terrible losses during its long war with Iraq in the 1980s.[46]

A third line of research argues that a country's ideology accounts for an army's determination in war. At odds with the findings of Shils and Janowitz, the historian Omer Bartov claims that Nazi ideology, and not primary group ties, explains the determination of the Wehrmacht as it fought on the Eastern Front.[47] Stephen Fritz arrives at a similar conclusion from his study of the German Army.[48] The importance of ideology as a motivating factor in war has received increasing attention from scholars.[49] Military historians suggest, for example, that ideological commitments probably played a great role for both sides of the American Civil War.[50]

Some scholars argue that the most powerful motivating ideology is nationalism. As evidence, this view points to the effects of the French Revolution not only on battlefield tactics but also on the ability of states to raise mass armies and motivate them in combat.[51] In a case study of post-Meiji Japan, Dan Reiter shows why nationalism increases the willingness of soldiers to die for their country, to employ suicide tactics, and to innovate on the battlefield.[52] Barry Posen illustrates how Germany and France inculcated their populations with nationalism as a way to improve their military power on the eve of World War I.[53]

Nevertheless, the historical evidence provides mixed support for the nationalism argument. Sometimes nationalism rallies a country and its forces; sometimes it does not. As they stand today, arguments about nationalism fail to describe the conditions under which these appeals work. Compare, for instance, the case of France in 1914 with France in 1940. Next to its Russian ally, the French Army of World War I paid the second highest blood price of any combatant to defend its homeland, but it displayed substantially less staying power in World War II.[54] The defeat of France stands in stark contrast to the determination shown by Imperial Japan, Nazi Germany, or the Soviet Union when protecting their homelands in World War II.[55] There seems little doubt that nationalism can motivate a country's armed forces. The key lies in understanding the conditions under which ideologies like nationalism create staying power, and when they do not. The nationalism of Wilhelmine Germany, for instance, differed from the nationalism of Nazi Germany, both in its message and how it motivated their respective armed forces.[56]

Consistent with claims about ideology, other analysts argue that regime type, or a country's form of government, accounts for differences in military effectiveness, including battlefield determination and staying power.[57] Allan Stam and Dan Reiter hypothesize that democracies win wars because they choose the right wars and their personnel display more skill than soldiers from nondemocracies. Democracies field armies with high morale, they assert, because soldiers are fighting for a popular form of government.[58] Reiter and Stam, however, find little empirical support for their hypothesis about democracies and morale, since militaries with strong staying power are not unique to democracies. Their findings seem reasonable given notable examples of nondemocracies capable of fielding tenacious armed forces. For example, the undemocratic government of Japan emboldened its citizens and soldiers to fight World War II with great determination.[59] The Soviet Union likewise rebounded from the invasion of Germany in World War II by relying on a combination of communist ideology and nationalism to rally its citizens.[60]

A fourth line of research claims that a country's military fights hard when it has more at stake in the conflict than its adversary. In particular, countries fight with great tenacity when they are defending their own territory.[61] Few other causes can seem greater than defending the fatherland or motherland. These explanations, nevertheless, cannot account for the occasions when even armies defending their homeland, such as the French and Italian armies in World War II, the Russian Army in World War I, and Iraqi forces in both Persian Gulf wars, eventually lost their will to fight.

The argument I present here, cohesion theory, borrows from each of these arguments. Claims about small-unit cohesion are important because they illustrate the power of strong bonds within groups, but they are wrong to ignore the potential for these bonds to form at higher levels, such as in the larger military organization. Similarly, arguments about ideology—for example nationalism—correctly identify the power of ideas to motivate individuals, but they do not explain the conditions under which these concepts galvanize groups.

PLAN OF THE BOOK: TESTING COHESION THEORY'S PLAUSIBILITY AND CASE SELECTION

In the following eight chapters, I present the hypotheses and causal logics of cohesion theory, test the plausibility of the theory's explanatory power, and then offer some implications of my argument for scholars and policy-makers. Overall, the book pursues two goals: to present a new theory to account for military cohesion while also providing an initial empirical test of the theory's explanatory power. The study, therefore, is one part theory creation and one part theory testing.[62]

In Chapter 2, I unpack the book's theoretical argument connecting regime control and organizational autonomy to military cohesion. First, I define military cohesion, the book's dependent variable. Second, I outline the theory's hypotheses, in the form of four ideal types of national armed forces (messianic, authoritarian, professional, and apathetic), along with the associated causal logics explaining why different combinations of regime control and organizational autonomy produce varying levels of military cohesion. Next I deduce three alternative explanations to cohesion theory from the existing literature: nationalism, small-group ties, and democracy. Although popular, most of these arguments remain vague about how and why they explain military cohesion. This omission forces me to sketch some causal logics connecting each of these alternative hypotheses to military cohesion.

The bulk of the book provides an empirical test of my argument about the sources of military cohesion. To demonstrate the plausibility of cohesion theory, I present a series of comparative case studies.[63] Using the comparative method, I test to see if my theory's variables correspond with the predicted degree of military cohesion across different cases. This approach also permits a test of my theory's causal logic through process tracing, a unique advantage of case studies.[64] Case studies give the researcher leverage on difficult-to-measure concepts, such as determination and flexibility on the battlefield.[65]

Within each case, I assess two aspects of a country's military cohesion: staying power and battlefield performance. For staying power, I assess the ability of national leaders to keep the armed forces fighting as the probability of victory begins to fall and the pressures to quit the war rise. To evaluate battlefield performance, I select key battles within the case to assess whether most units fight with determination on the battlefield. At the end of each case study, I briefly compare cohesion theory with the three aforementioned existing alternative explanations of why militaries fight hard: the strength of small-unit ties, nationalist ideologies, and democracy. I use these comparisons at the end of each chapter as another method for gauging the plausibility of cohesion theory. Again, my aim is not an exhaustive test of cohesion theory against these alternatives, since they remain largely undeveloped. Instead, I show that my argument sheds more light than these alternatives do on the selected historical cases.

The first set of case studies come from World War II in Europe. Chapters 3, 4, and 5 examine, respectively, the messianic German military from 1944 to 1945, the apathetic French armed forces in 1940, and the authoritarian Soviet Red Army in 1941. In each of these initial cases, a national military fights for its country's survival. The first two, and longer chapters, on the German and French militaries, assess extreme values on my theory's independent variables (regime control and military autonomy) and the study's dependent variable (the degree of military cohesion). By comparing these extreme values, I can better illustrate the causal logics of my theory connecting regime control and organizational autonomy to military cohesion.[66] These chapters also briefly compare the cohesion of national armed forces during World War II with their cohesion in World War I, when Germany and France both raised professional militaries. Through these within the case comparisons, I demonstrate my theory's ability to explain how and why military cohesion can change from one war to the next. The shorter chapter on the Soviet Union contrasts with the German and French militaries by describing the cohesion of an authoritarian military. These first three historical chapters depict each of my theory's ideal type of military organization: messianic, authoritarian, professional, and apathetic. In the next two chapters, the book's focus turns to the Vietnam War, comparing the armed forces of North Vietnam (Chapter 6) with those of the United States

(Chapter 7). These two chapters test the plausibility of cohesion theory outside of Europe during World War II.

My case selection reflects four criteria. I choose cases covering a wide range of variation in regime control and organizational autonomy, along with variation in military cohesion during war. This first criterion allows me to test the causal logic of cohesion theory across different values of its independent variables. Additionally, I select data-rich cases that provide a chance to explore the details of the causal processes my theory predicts. These cases have the virtue of also being familiar and important to students of military effectiveness.[67] To demonstrate cohesion theory's breadth, I choose cases from different regions and time periods. By selecting cases from Europe and Asia, I avoid arguments that defy generalization by relying on a particular culture to explain cohesion.[68] Furthermore, to compare cohesion theory with the democracy argument, I also choose cases from wars where democracies fight nondemocracies. This criterion allows me to compare the relative explanatory power of my argument against the democracy argument within the same conflict. Each case study compares the performance of democratic militaries versus their less democratic opponents.

Finally, I select cases to control for several confounding variables (see Table 1.1). The cases consist of national militaries fighting on the defense where the probability of victory appears at best uncertain, and at worst increasingly unlikely. This focus provides a way to keep the operational and strategic circumstances comparable across cases. I also select cases where countries are defending their own territory to control for defending one's homeland as a possible explanation of an army's determination. Most of my cases involve countries fighting to protect their native soil.

The case studies provide three types of evidence to evaluate cohesion theory. First, for each case, I try to determine if variations in the strength of regime and organizational ties are congruent with the predicted degree of military cohesion. I then compare congruence *across* these difference cases. Second, the case studies also assess national militaries over a period of time, with multiple battles, each viewed at the time as critically important to a campaign, as well as different decision points where military leaders worried whether they could continue the war. This element of my research design also allows me to test *within* case congruence between independent and dependent variables. Lastly, the case studies provide opportunities, through process tracing, to test the causal logic connecting my theory's variables to military cohesion. Through

Table 1.1: The Cases

The Cases: Country and Year	Major Battles within the Cases
Chapter 3: Messianic German Military, 1944–45	Retreat from France West Wall Ardennes Offensive Defense of the Rhine
Chapter 3: Professional German Military, 1917–18	Cambrai Amiens
Chapter 4: Apathetic French Military, 1940	Dinant Sedan Montcornet, Laon, and Arras Weygand Line
Chapter 4: Professional French Military, 1914–18	First Marne Second Marne
Chapter 5: Authoritarian Soviet Military, 1941	Bialystock-Minsk Smolensk Kiev Moscow
Chapter 6: Messianic North Vietnamese Military, 1965–73	Ia Drang Operation Dewey Canyon
Chapter 7: Professional U.S. Military, 1968–72	Tet and Hue Khe Sanh

process tracing I can evaluate whether or not the predicted behaviors of my theory are occurring for the reasons it anticipates.

In the last chapter, I summarize the book's findings. The conclusion then offers some suggestions for future research and the implications for national security policy.

2 Cohesion Theory

Explaining a Military's Will to Fight

WHY DO NATIONAL MILITARIES differ in their cohesion or will to fight? In this chapter, I introduce cohesion theory to explain why national militaries differ in two important ways: (1) the ability of most units to fight hard on the battlefield and, (2) the capacity of the armed forces as a whole to avoid disintegration when facing the possibility of losing a war. The first section defines military cohesion, the study's dependent variable. In the subsequent section, I outline the collective action problem associated with motivating groups to sustain cooperation even though individual members risk suffering horrible costs to achieve a common goal. Fighting in a war represents an extreme example of collective action. Here, I lay the groundwork for my argument about why group cohesion depends on the ability of a group's leaders to promote as well as to enforce norms of unconditional loyalty. This core idea applies to groups of any size and function, from social movements to insurgent organizations, but I apply it to a country's armed forces.

Next, I outline my argument explaining why military cohesion depends on the degree of control a regime holds over its citizens and the autonomy of its armed forces. By combining these two explanatory variables, regime control and organizational autonomy, I create four ideal types of militaries. Each of these national militaries represents a hypothesis connecting the degree of regime control and the autonomy of the armed forces to cohesion. The chapter then lays out three alternative arguments to cohesion theory. These explanations point to small-group bonds, nationalism, and democracy, as sources of military cohesion. In the conclusion, I set the stage for the case studies found in the following chapters, where I test cohesion theory against these alternative perspectives.

WHAT IS MILITARY COHESION?

Because wars are bloody and costly affairs, laden with uncertainty, most countries avoid them as their primary means to obtain their national objectives. When confronted with war, countries mobilize their resources in an effort to bring the conflict to a rapid and, it is hoped, victorious conclusion. The odds of a country winning, let alone securing a quick and decisive victory on the battlefield, depend on a number of factors, not the least of which is luck.[1] To estimate a country's chance of success, analysts typically compare the balance of forces between the combatants. These assessments examine the types and quantity of forces a country can deploy for battle. By examining standard measures such as manpower and military technology, analysts try to predict who might emerge the victor in any given conflict. Many analysts rely on extremely sophisticated computer simulations capable of modeling combat at different levels of detail while also controlling for differences in military technology as well as fighting effectiveness.[2]

In addition to these material measures, military planners also consider the combat effectiveness of a country's forces, the nonmaterial or human factors that contribute to a state's ability to win wars.[3] One element of combat effectiveness receiving increasing attention from scholars as well as policy-makers is skill, or how well a state's armed forces make use of weaponry in battle.[4] A second and more familiar component of combat effectiveness is will, or how hard a country's soldiers fight on the battlefield in pursuit of their country's objectives and their ability to remain together under difficult strategic circumstances. This book focuses on the domestic determinants of this second facet of combat effectiveness, the cohesion of a country's military forces. Although analysts use computer simulations that account for will, they need to know what value to assign to it. In this way, the book's argument should supplement traditional methods of net assessment and operations research.

Cohesion has always played a crucial role in determining which side prevails in battle. Conquering fear of death and injury remains a critical obstacle to an army's successful military performance.[5] In the view of many military theorists, inspired troops can frequently triumph over well-armed but less driven soldiers. The Greek military leader Xenophon (434–355 B.C.), for example, recognized the importance of will in war when he wrote, "I am sure that not numbers or strength bring victory in war, but whichever army goes into battle stronger in soul, their enemies cannot withstand them."[6] Endorsing this belief, Napoleon noted, "The moral is to the physical as three to one."[7] The French military theorist Ardant du Piq echoes a similar theme in his famous *Battle Studies*, "The art of war is subjected to many modifications by industrial and scientific progress. But one thing does not change, the heart of man. In the last analysis, success in battle is a matter of morale."[8] Scholars have used a variety of concepts and terms when describing an army's will to fight.[9]

Although military professionals and scholars universally agree that these "moral factors" play a role in war, they have not arrived at a consensus about how to define them. Most emphasize the significance of morale, the ebb and flow of fighting spirit during the course of battles. Some explore the ability of militaries to fight under difficult conditions, such as recovering from significant defeats. Even though most descriptions focus exclusively on the operational performance of military units, a few recognize the need to incorporate the "spirit and other moral qualities of an army, a general, or a government, the temper of the population of the theater of war, the moral effects of victory or defeat—all of these vary greatly."[10]

My definition of cohesion integrates these different conceptions.[11] Traditionally, the literature uses cohesion to refer to the determination of small units, such as squads or platoons.[12] The concept of cohesion I adopt here broadens the idea to include all of a country's military organizations during war. Specifically, I define military cohesion as the capacity of a country's armed forces to fight with determination and flexibility on the battlefield, while also resisting the internal pressure to collapse when the likelihood of winning a war diminishes. Military cohesion, then, is a dependent variable with two dimensions: battlefield performance and staying power. Cohesion reflects, not only tenacity in battle but also the ability of a nation's military to hold together as a coherent, fighting force as the chances of victory in a war decrease.

This definition improves our understanding of military cohesion in two ways. First, it increases leverage on the concept of staying power, or what military organizations can endure before they begin to unravel. On the one hand, national militaries rarely fight to the last soldier. On the other hand, national militaries rarely collapse without a fight. Historical cases lie somewhere in between these common caricatures. This study helps explain the range of outcomes in cohesion between these two extremes.

Similarly, the definition incorporates the actions of units on the battlefield as they perform assigned missions. Scholars sometime describe this behavior as "task cohesion."[13] The distinction made here highlights how hard subordinate units fight on the battlefield. For this reason, I add the concept of battlefield performance to my definition of cohesion. Taken as a whole, then, military cohesion is one part battlefield performance and one part staying power.

The First Dimension of Military Cohesion: Staying Power

Staying power is the ability of a country's armed forces to hold together and to fight even as the odds of winning a war diminish. In particular, it refers to a national military's ability to remain disciplined and capable of conducting operations as the probability of victory decreases. Staying power describes the capability of military organizations to prevent and to resist disintegration from within their ranks when a country faces the prospect of defeat. These internal sources of collapse arise as military personnel become increasingly unwilling to expend blood and treasure for a war they believe appears increasingly unwinnable. The prospect of losing a war erodes staying power because members of the armed forces begin to conclude that they are fighting and dying for an objective out of reach.

Staying Power: The Sources of Erosion. In general, the longer a war continues the greater the costs of fighting. With this calculus in mind, political and military leaders prefer to initiate wars when they possess a strategy for quick and decisive victory.[14] Long and costly wars strain a country's population, with military organizations often bearing the brunt of this strain. Both attrition and time undermine the ability of national armed forces to fight by consuming the resources for fighting, fraying lines of supply, and exacerbating pre-existing divisions within the armed forces.[15] When the probability of victory begins to fall, the pressure for military disintegration begins to emerge. Members of the armed forces will question the reasons for war. As the costs of war rise, and the benefits of fighting fall, soldiers confront a stark set of choices: they can remain committed to the cause, they can abandon the cause, or they can demand a different course of action.[16] Shortages of materiel and problems with supplies will further undermine support for the war within the armed forces.

Additionally, when opposition to a war surfaces at home, this dissension will eventually reach the front lines to further erode military staying power. The connection between dissent at home and discontent in the armed forces will emerge more strongly in militaries forced by high rates of casualties to recruit a similarly large number of conscripts as replacements. Newly inducted members bring fresh memories of public dissatisfaction with the war to the front lines. As these new recruits reach the front, they will bring with them the home front's dissatisfaction with a long and costly war.

TABLE 2.1: Indications of Staying Power

Strong: National military collapses only when an adversary possesses crushing material superiority	Moderate: National military collapses slowly as the probability of victory decreases	Weak: National military collapses quickly as the probability of victory decreases
Top military officials obey regime	Gradual disobedience by top military officials	Quick disobedience by top military officials
Few large-unit mutinies, strikes, or organized surrenders	Gradual increase in large-unit mutinies, strikes, or organized surrenders	Quick increase in large-unit mutinies, strikes, or organized surrenders
Low rates of desertion, combat refusal, and crime	Slowly rising rates of desertion, combat refusal, and crime	Quickly rising rates of desertion, combat refusal, and crime

The erosion of military staying power could take different forms within the armed forces. Generally, a reduction in overall discipline serves as the first sign that a military organization has begun to lose its stamina. Rates of crime, combat refusal, and desertion will increase across the armed forces. Because they face the immediate pressure of combat, personnel in the rear or in noncombat support positions typically show the first signs of disorder. As staying power decays, more extreme paths to military collapse will follow. Field officers could begin to organize or to support acts of disobedience, such as mutinies or organized surrenders. They could decide that fighting only prolongs a losing war, requiring greater sacrifice from soldiers with whom they likely share strong bonds of loyalty and trust. Similarly, top-level military officers could also conclude that they need to compel the regime to end the war. Such acts could include the refusal to execute the regime's strategy for war, or even a military coup.[17]

As the first dimension of military cohesion, staying power varies from strong, to moderate, to weak. When staying power is strong, the armed forces can avoid the pressure to collapse regardless of the probability of victory. With strong staying power, a country's military organizations will retain their fighting capacity under all circumstances, carrying out large-unit operations, retreating in good order, obeying regime leadership, and overcoming internal dissension. In contrast, when staying power is moderate, pressure to quit slowly builds inside the armed forces the more victory seems impossible. Lastly, when staying power is weak, the capacity of political-military leaders to keep the armed forces fighting will dissipate quickly as the probability of victory diminishes (see Table 2.1).

The Second Dimension of Military Cohesion: Battlefield Performance

The second dimension of military cohesion is battlefield performance, which I define as the willingness of combat units to fight with determination and flexibility. Battlefield performance here refers to larger units, ranging from the army to battalion level.[18] Determined units fight after experiencing setbacks on the battlefield. Indications of determination include defending when attacked, fighting after suffering from local defeats, carrying out operations after experiencing prolonged periods of aerial or artillery bombardment, and resisting after becoming encircled by enemy forces. Units with determination can maintain high levels of morale even when fighting losing campaigns. They will, on average, prefer fighting to surrendering. Determined armies will undertake great amounts of self-sacrifice by engaging in last-ditch efforts at combat. Rather than offer passive resistance, by simply not firing their weapons, or by actively surrendering, determined units will risk injury or death in order to perform their assigned missions.[19] Even when their original strength diminishes, units from a determined army can still operate and inflict damage on opposing forces.

As a result, some armies can compensate for their material inferiority by showing greater determination to win than their enemies. For example, from late 1981 through the middle of the following year, the Iranian Army showed great determination as it drove the Iraqi Army out of the areas near the cities of Khorramshahr and Abadan. Displaying greater tenacity, the Iranians relied on human wave attacks to defeat the better-equipped Iraqi forces, a tactic that relied on sheer manpower to compensate for their deficiencies in armor. By the summer of 1982, the Iranians had made impressive gains at the cost of nearly 200,000 of their own soldiers, stunning their Iraqi opponents. Rumors still persist that the initial assaults so shocked the Iraqis that many officers and soldiers "refused to mow down the masses of people deliriously careening towards them."[20]

Flexibility describes the willingness of units to overcome panic when surprised or confronted with difficult conditions. When units fight with flexibility, officers show a readiness to take risks. They deviate from script when previously conceived strategy and doctrine fail on the battlefield. Flexible units resist panic during retreats and can also form new formations from shattered ones. When displaying flexibility, units will adjust the tactics they learned through

TABLE 2.2: Indications of Battlefield Performance

Strong Battlefield Performance: Most units fight with determination and flexibility	Moderate Battlefield Performance: Most units fight with determination but lack flexibility	Weak Battlefield Performance: Only the best units fight with determination and flexibility
Most Units Determined	Most Units Determined	Only the Best Units Determined
Most units attack and defend when ordered	Most units attack and defend when ordered	Only the best units attack and defend when ordered
Most units fight despite high percentage losses	Most units fight despite high percentage losses	Only the best units fight despite high percentage losses
Most damaged units combine to form new units	Most damaged units combine to form new units	Only the best units can combine to form new units
Most Units Flexible	Most Units Inflexible	Only the Best Units Flexible
Most units resist panic	Most units cannot resist panic	Only the best units resist panic
Most units can change tactics and planning	Most units cannot change tactics and planning	Only the best units can change tactics and planning
Most units fight when command and control disrupted	Most units cannot fight when command and control disrupted	Only the best units fight when command and control disrupted

training to fit conditions on the battlefield. Lastly, when units fight with flexibility, they perform missions when command and control links fail.[21]

The level of battlefield capacity varies from one national military to another. If battlefield capacity is strong, most units, including reserve and recently created formations, fight with determination and flexibility. When battlefield capacity is moderate, most units fight with determination but lack flexibility; they fight with vigor but often succumb to panic. If battlefield capacity is weak, only the best units fight with determination and flexibility. By best units I mean those elite formations with the most experience, resources, and training. The term "best units," therefore, excludes reserve units and recently raised units (see Table 2.2).

Cohesion Theory: The Sources of a Military's Will to Fight

The central question of the book is why national militaries differ in their cohesion. To answer this question, I step back to first address a more basic issue: a variant of the collective action problem. Specifically, under what conditions can groups maintain collective action? As collective action becomes costly, and members begin to see reasons to abandon the cause, how do groups stick together? This is an especially thorny problem for states trying to persuade their people to risk their lives fighting long and costly wars. In this section, I first sketch a variety of existing solutions to this problem. One set of explanations focuses on appealing to individuals' self-interest to inspire them to participate and to sustain collective action. Another set of answers focuses on how to appeal to group interests. I discuss why cohesion theory provides one promising solution to the collective action problem in general, and to the difficulty of motivating national armed forces in particular. Overall, this discussion lays the groundwork for my own argument about the importance of regime control and organizational autonomy in determining a military's will to fight.

Collective Action: The Rationalist View

From a rationalist perspective, one that sees people as driven primarily by self-interest, getting individuals to participate in organized violence on behalf of their country should be much more difficult than it usually is. Indeed, two obstacles should discourage individuals from risking their lives in combat for the good of the state—or any group for that matter. First, the costs of war, which include the possibility of death, should discourage, if not outright frighten, any person from participating in a war. Second, since the aims of war usually entail fighting for some collective objective rather than personal gain, individuals should face huge incentives to let someone else take their place on the battlefield.

Individuals deciding whether or not to fight for their country or a group, therefore, face a classic collective action problem: individuals should prefer to free ride rather than risk their lives in combat. This tendency should stymie collective action.[22] At least that is what should happen in a world of individuals sensitive to the costs and benefits of war. The frequency of domestic and international conflict, however, suggests otherwise. What, then, are the factors that might motivate self-interested individuals to fight their country's wars? How might governments overcome this collective-action problem? From the egoistic perspective, there are at least three types of incentives self-interested individuals might find compelling enough to enlist in the armed forces and to fight for their country.

Negative incentives, such as the fear of punishment, are one set of reasons people march, albeit grudgingly, off to war. Relying on old-fashioned coercion, some governments compel citizens to serve in the military by offering a very nasty choice between joining the armed forces and joining the ranks of the imprisoned, or even worse punishment, such as torture or death. Coercion almost certainly plays a role once soldiers reach the battlefield. Getting people into uniform is much easier than actually getting them to fire their weapon in combat.[23] Military organizations might draw up a long list of penalties, such as execution, to instill a modicum of wartime discipline. Preventing individuals from breaking ranks and running from the battlefield might also call for unsubtle forms of coercion. During World War II, the Soviet Union resolved this problem by deploying blocking formations that would shoot deserters.[24] Fear alone, however, cannot keep soldiers fighting. Battlefield situations reach points where soldiers will begin to fear the enemy more than the blocking formation pushing them forward.

In the interest of appearing more benevolent and avoiding insurrection, governments could also rely on positive incentives to motivate its citizens to fight.[25] Although coercion emphasizes negative sanctions to induce compliance, strategies predicated on self-interest require the promise of some personal reward for participating in combat. Instead of avoiding punishment, individuals join the military and freely participate in the fighting because their governments offer to pay them. Indeed, before the French Revolution, governments often relied on mercenaries to fight their wars.[26] Another way self-interest can play a role is that citizens might decide that defending their nation is better than letting it fall victim to some aggressor. Facing an adversary bent on exterminating or enslaving a country's population might encourage an otherwise disinterested citizenry to take up arms in defense of the state. Once again, however, egoistic individuals will reach a point where they will find it more in their self-interest to free ride than to continue the fight.

A third solution scholars have proposed for motivating individuals to take collective action entails the creation of institutions that create combinations of rewards and punishments that motivate people and promote group solidarity.[27] Institutions are a set of commonly accepted expectations of how people will act under certain conditions. These norms of behavior can either manifest in laws or simply represent common practices. Michael Hechter, for instance, argues that group solidarity embodies a principal-agent problem.[28] The principal, group leaders in this instance, must create control mechanisms to monitor agents, or group members, in order to punish those that shirk their responsibilities. In addition to this "control capacity," Hechter also contends that norms of obligation drive individuals to participate in groups. The problem, however, is that Hechter never explains how groups form and where these norms of obligation originate.

Russell Hardin suggests that individuals form certain identities and norms based on self-interest. Once in place, these norms engender group solidarity by creating membership rules for who can and cannot belong to a particular group.[29] Unfortunately, Hardin puts the cart before the horse, failing to spell out why individuals see themselves as groups in the first place. He does not, moreover, explain why individuals hold varying commitments to the group to which they belong. In this story of collective action and group formation, individuals can simply select from a pre-existing set of groups, or identities, but it remains unclear why individuals consider some group affiliations as more salient than others.[30]

Governments might also rely on institutions to distribute fairly any obligations of military service that it might need to impose on its citizens. When a population feels that a government is trustworthy, individuals are more likely to undertake military service. In the absence of fair rules about the requirements of entering the military, citizens are not likely to consent to becoming soldiers, a situation that undermines a military's morale. According to Margaret Levi, democratic governments vary in their standards of fairness and thus motivate their populations at differing rates of success.[31] Although the argument might shed light on patriotism in democracies, it does not explain how less democratic governments motivate their citizens in times of war. Nor does it account for the instances when democratic armies show little staying power in battle.

Collective Action: The Constructivist View and the Importance of Cultivating Group Interests

Many disciplines have begun to recognize the limitations of the egoistic perspective to explain collective action, especially the collective act of risking one's life in war. This shift from the assumption that individuals are self-interested actors reflects a growing trend in sociobiology contending that people do not necessarily inherit selfishness because of natural selection. Some researchers now believe that natural selection takes place at the group level, a highly controversial proposition in the past.[32] According to David Sloan Wilson, "Despite its widespread acceptance, the case for individualism as a general prediction that emerges from evolutionary theory, or as a general principle to explain human behavior, actually is very frail."[33] In nature, animals as well as humans form groups to enhance their prospects for survival. For social science, Sloan suggests, "a notion of group-interest must be added to the notion of self-interest, to the extent that group selection is important in nature."[34]

Even economists have recently begun to move away from the egoist assumption to account for more variation in their models of human behavior.[35] This new line of research seeks to explain why individuals sometimes conduct themselves in nonselfish ways. In the past, some economists resolved this anomaly by incorporating altruism into the utility functions of individuals. The change assumes that people could act in altruistic ways because it maximizes their self-interest.[36] Still further, some have argued that concepts like trust and social capital facilitate cooperation among individuals in a variety of institutions.[37] In contrast to these more traditional views, other economists have proposed a radical solution that alters the assumption that individuals are interested in more than maximizing their own conceptions of utility to account for the same altruistic actions.[38] Rather than individuals who regard only their own interests, these economists explain more behavior by assuming that people can view their interests from the perspective of a group or a team.[39]

Social psychologists have always questioned the value of the egoist assumption. As Henri Tajfel explains, "It is a crude oversimplification to conceive of social motives as being capable of direct derivation from a hedonic algebra of self-interest—real or fictitious—based on a few universal human drives, whatever the choice may be."[40] Although people might behave in selfish ways, they display other kinds of behavior that confound explanations based on mere self-interest. Take, for example, the numerous experiments in which strangers form cohesive groups to win competitions.[41] Scholars in this research tradition maintain that individuals possess not only material needs but social needs as well. That is to say, people have a "hierarchy of needs" beginning with food, shelter, and security, but including the desire to belong to groups.[42] Individuals are social creatures that crave membership in a group for not only their physical security but also for a sense of belonging. From membership in a group, people satisfy their needs for human contact, self-esteem, and the ability to develop and grow.[43]

Thus there are good reasons to believe that individuals might identify their interests in terms of a group's interests. Humans, we know from sociobiology, have always formed groups to improve their chances of survival in nature, a trait still common in individuals today. Individuals, according to social psychology, join groups to satisfy their social needs. Individuals are group-oriented beings requiring human interaction. Lastly, research from the field of social psychology shows that individuals readily separate themselves into groups when forced into competitive situations.

How and why, then, do individuals think as a group, or equate their personal interests with the interests of a larger group? Groups are defined along a variety of characteristics (or commonalities), such as class, ethnicity, language, religion, and even like tastes in music.[44] Once groups are formed, how much individuals identify with them can vary. For some people, they can belong to a particular group but not see their interests as defined by it. Other individuals view their own interests as the group's interests. These people make decisions based on what is best for the collective to which they have membership. In other words, they have team preferences. When individuals think in terms of the group, then, they do not consider whether they prefer x to y, but they think in terms of what their team or group prefers.[45]

If individuals can possess "team preferences," or they can see their interests in terms of a group's interests, then under what conditions do they take collective action? Scholars interested in the study of war have long recognized the virtues of relaxing the assumption that individuals are self-interested egoists to explain group solidarity in combat. This is why military sociologists have focused on how the cohesion of small military units, companies or below, can motivate soldiers to fight under adverse conditions. Common experiences, such as training and fighting together, forge strong bonds within units, commonly known as primary group ties.[46] In this view, soldiers do not fight for abstract causes, but instead fight for their fellow soldiers. Solutions that emphasize small-unit cohesion, however, cannot account for why some armies keep fighting when long wars and the casualties associated with them undermine their primary group ties. Still further, these arguments do not explain the depth of commitment among group members, nor the ideas that unify them in the first place.

Other researchers have emphasized the importance of ideas, such as a country's ideology, as the key to building cohesive societies of individuals willing to risk their lives in defense of their nation.[47] For some analysts the crucial element in forging group preferences is social stability or agreement about the role of the military in society.[48] Still others see nationalism as the key factor in inspiring a society to take up arms to fight for the homeland.[49] A more recent variant of these arguments contends that the regime type is key to understanding military cohesion. Specifically, some scholars claim that democratic armies fight harder than the armies of nondemocracies. However, like the arguments about small-unit cohesion, claims about how and why ideology creates military cohesion remains vague. Namely, they do not explain the conditions under which groups embrace certain ideologies, nor do they explain why ideas vary in their power to motivate groups.

Cohesion Theory: Combining the Two Views to Explain a Military's Will to Fight

By incorporating insights from both rationalist and constructivist accounts of collective action, I derive what I call cohesion theory to explain the staying power of national militaries. Contrary to existing arguments, cohesion theory argues that strong bonds can exist outside of small units, motivating members of the armed forces to fight for their country and their fellow soldiers with varying degrees of intensity. The central idea underlying my theory is that large groups, like states, motivate members to sustain collective action not only by promoting norms of loyalty but also by enforcing them. Members take action because some are motivated by these norms, and they pressure others to follow suit.

Two key variables determine the cohesion of a nation's armed forces during war: the degree of regime control over society, and the degree of autonomy the armed forces possess to train for warfighting. With a high degree of control, the state promotes an ideology of unconditional loyalty and can enforce it through coercion. Regime control bolsters military cohesion. Hard-core supporters of the regime fight with determination and pressure others to do the same. With a high degree of autonomy, the country's military organizations can promote norms of unconditional loyalty and trust among officers and enlisted personnel through training. Personnel develop bonds of loyalty and trust that enable them to fight with determination and flexibility on the battlefield. Below, I define regime control and organizational autonomy. Then I explain why these two explanatory variables interact to produce four ideal types of militaries capable of fighting with differing levels of cohesion in war.

Regime Control. Regime control defines the amount of loyalty a state can demand and enforce from the populace. Control serves as a mechanism a regime can use to motivate both the general population as well as the military for collective acts, such as national defense. Using a mix of ideological appeals and coercion, a high degree of regime control motivates citizens and soldiers to fight for their country. A high degree of regime control rests on two conditions: a national ideology demanding unconditional loyalty, and the coercive power to enforce these norms. In contrast, a regime exercises a low degree of control when its ideology fails to demand unconditional loyalty from its citizens, or it lacks the ability to compel individuals to embrace these ideological demands, or both.

The degree of regime control, then, depends firstly on a state's ideology. In simplest terms, "An ideology is a normative theory of action. Ideologies 'explain' prevailing social conditions and provide individuals with guidelines for how to react to them."[50] Ideologies "energize groups" into action.[51] Ideologies demanding unconditional loyalty rest on a core principle: that national interests always trump individual interests.[52] Citizens live to serve the state, an idea antithetical to liberal democracies like the United States. Regimes justify these strong demands on their citizens in several ways. Communism, for instance, uses class as a rationale, fascism uses the cult of personality, nationalism uses the nation, and religious fundamentalism uses references to a higher spiritual power. In the modern era, states have preferred ideologies that combine nationalism to these other principles.[53] Nationalism defines individual interests in terms of the nation's interests by making appeals to a shared set of characteristics, such as a common culture, ethnicity, and language.[54] Nationalist appeals have historically motivated citizens and soldiers alike to make sacrifices for their country.[55] Any other kind of less demanding ideological appeals, such as the liberalism of most democracies, fail to increase regime control over the population. Liberal ideologies argue that the state exists to protect the liberties and the safety of individuals. In other words, the state serves individuals and by definition does not demand unconditional loyalty from citizens.

Even though governments promote certain ideologies, citizens do not always embrace them. To enforce its demands for unconditional loyalty, a regime must possess a monopoly on coercive power. This entails not only control over the organs of state power, like the police forces, but also the domination of civil society, free associations normally independent of the government. Under these circumstances, regimes either co-opt or quash any organization that could oppose them.[56] Coercive threats provide another method to ensure that the population meets its ideological demands.[57] The ability to dominate civil society and to use terror selectively against the population allowed Hitler to establish a high degree of control over Germany, which Mussolini could not do in Italy.[58] Resistance existed in both countries and both leaders constantly worried about opposition, but only Hitler's regime could dominate all the institutions of government as well as civil society. In contrast, Mussolini's regime faced resistance to its policies from the military, fellow politicians, and a monarch, especially as the war turned against Italy (see Table 2.3).

TABLE 2.3: Indications of Regime Control

High Degree of Regime Control	Low Degree of Regime Control
Ideological Control: Demands unconditional loyalty	Ideological Control: No demands for unconditional loyalty
Ideology claims national interests always trump individual interests	Ideology claims individual interests normally trump national interests
Ideology embraces combinations of nationalism, communism, fascism, and theocracy	Ideology embraces forms of liberalism
Ideology emphasizes hatred of other ideologies	Ideology emphasizes conditional national sacrifice
Political Control: Enforces ideology	Political Control: Cannot enforce ideology
Regime controls civil society	Civil society independent of regime
Civil society cannot check regime power to repress and terrorize	Civil society checks regime powers to repress or terrorize

Autonomy of Military Organizations. Countries can also create cohesive armed forces by granting their military organizations the autonomy to train for warfighting. Charged with the national defense, military organizations take responsibility for training a country's citizens to become soldiers. To accomplish their battlefield tasks, the armed forces instill their own principles to motivate their members. When newly inducted soldiers arrive, the military's ideas about the role of the individual will either contradict or reinforce what citizens have already learned from society.[59] Even though training aims primarily to acclimate soldiers to combat situations, these experiences also reinforce the need for members to act as a team and to trust their leaders.[60] They emphasize norms of loyalty and trust by training for warfighting.[61] Sociologists and social constructivist scholars have recognized common experiences as one way to forge strong bonds of loyalty among members. The common experience of formal training spreads these norms by creating a shared sense of a common fate, interdependence, and teamwork.[62] Training for warfighting also creates bonds of loyalty between personnel to their respective military institutions.[63] These loyalties help them retain their cohesion when high rates of personnel turnover, due to casualties, for example, sever small-group bonds. Soldiers fight for their buddies, but loyalties to the organization often facilitate these same "buddy bonds."[64] As David Segal and Meyer Kestenbaum observe, "[N]ot all social cohesion is limited to small groups.... This sense of imagined community is precisely what may seem to distinguish the armed forces from the rest of society and simultaneously bind members together."[65]

TABLE 2.4: Indications of Military Autonomy

High Degree of Military Autonomy	Low Degree of Military Autonomy
Autonomy to train and demand unconditional loyalty:	Lack autonomy to train and demand unconditional loyalty:
Fear of regime does not hinder training	Fear of regime hinders training
Training is a priority over indoctrination	Indoctrination more important than training
Domestic divisions do not interfere with training	Domestic divisions prevent training

Training also helps personnel fight with flexibility on the battlefield. The opportunity to train creates familiarity with tactics, techniques, and procedure. Equally important, it establishes a baseline level of knowledge personnel share. As Hew Strachan explains, "Surprise can destroy collective cohesion on the battlefield and training is the best antidote."[66] This expertise also facilitates what sociologists describe as "swift trust," or the ability of members of the same organization to accomplish tasks even though they lack a prior relationship. In times of war, for example, individuals from shattered units often find themselves thrown together in an ad-hoc manner to accomplish a mission.[67]

Two conditions can undermine organizational autonomy for training. First, when a regime believes that the armed forces pose a potential threat to its survival, it will disrupt the military's internal routines. Worried regimes will force the military to emphasize indoctrination over training, dictate the selection of officers, and exert control within units on the battlefield. These policies, in turn, create a climate of fear within the military, undermining the trust between personnel and their officers.[68] In such an environment the armed forces will find it difficult to instill their own norms of loyalty and trust within their ranks.[69]

The military's autonomy to train might also suffer from class, ethnic, or political cleavages within its ranks. Because soldiers are usually a reflection of society, conflict among different national groups could emerge within the armed forces.[70] Becoming embroiled in domestic political disputes risks creating rifts in the ranks of the military. Sometimes the military can insulate itself from domestic turmoil and rise above it to defend the nation. If it cannot maintain its autonomy to train, the same cleavages dividing the country will undermine the military and sap its cohesion (see Table 2.4).

Hypotheses on Military Cohesion: Four Types of Armed Forces

Together the degree of regime control and organizational autonomy of the armed forces interact to produce four types of national militaries capable of fighting with differing levels of cohesion. Specifically, each type of military varies in its staying power and battlefield performance. Each of these armed forces represents a hypotheses and set of causal logics connecting regime control and organizational autonomy to military cohesion. Table 2.5 summarizes these four hypotheses, and Table 2.6 outlines cohesion theory's causal logics.

Messianic Militaries: High Regime Control and High Organizational Autonomy. The combination of both high regime control and high organizational autonomy creates *messianic* militaries, which fight with a high degree of cohesion. They are messianic because the regime's ideology proposes a national mission that motivates the population as well as the armed forces. The promise of a better future is one element of the national mission.[71] This goal inspires citizens to work hard as a collective today for what they can achieve tomorrow. National missions frequently unify a country against a common set of enemies, both foreign and domestic.[72] Because it trusts the military, the regime gives it the autonomy to forge strong organizational loyalties. During war, messianic militaries exhibit a high degree of cohesion for two reasons.

First, they will fight with strong strategic power. The national armed forces will resist the pressures to collapse even when victory looks unlikely. Because of a high degree of regime control, soldiers will either internalize the regime's ideological demands for unconditional loyalty or act as though they have internalized them. Over time, a hard-core group of followers embrace the regime's demands for unconditional loyalty.

At the same time, less zealous soldiers will face pressure from hard-core followers to display allegiance to the regime's wartime goals.[73] The regime and its supporters will employ a combination of positive as well as negative incentives to motivate the country, especially the military. On the positive side of the ledger, the regime's uncontested ideology makes national sacrifice a virtue. Government propaganda, the primary and only sanctioned source of information, will constantly reinforce these beliefs. Should these appeals fail to motivate, the regime will threaten a variety of punishments to coerce citizens and soldiers.

Under these circumstances, members of society and the military that lack the devotion of hard-core regime supporters have no choice but to show loyalty, since exiting society or voicing opposition risks imprisonment and, potentially, death.[74] Compliance, then, with the regime's demands for unconditional loyalty, therefore, comes from a mix of true believers and those pressured to behave like true believers. These hard-core followers will show unfailing loyalty to the regime. They will compel the less motivated to continue the struggle, either through social pressure or through threats of punishment.[75] With a high degree of autonomy, military organizations reinforce this unconditional loyalty through training.

Second, messianic militaries fight with determination and flexibility on the battlefield. Coming from a society in which the regime enforces an ideology of unconditional loyalty, recruits enter the military believing that sacrificing for the greater good of the nation is a virtue. After all, individuals live to serve the nation. The military can, therefore, draw from a pool of recruits already exposed to norms of unconditional loyalty, with many of them ready to risk their lives in defense of the nation. The autonomy to train also creates military personnel with strong loyalty and trust. Not only do loyalties form among small units, but individuals also feel strong bonds to their military organization, which equips and trains them. As a result, soldiers will take greater risks when fighting for one another, showing great determination on the battlefield. Additionally, personnel have learned through training to trust one another. This trust helps units fight with flexibility, forming new units out of shattered ones, overcoming panic to conduct missions, and abandoning doctrine when it no longer works. States rarely create messianic militaries, either because they do not permit organizational autonomy for the armed forces or because a government cannot establish a high degree of control over the population. Historical examples include the armed forces of Nazi Germany, Imperial Japan, and North Vietnam.

Authoritarian Militaries: High Regime Control and Low Organizational Autonomy. States with a high degree of regime control over society, and armed forces with low organizational autonomy, field *authoritarian militaries.*[76] They fight with strong staying power but only moderate battlefield capacity. These militaries draw their cohesion primarily from the regime's ability to demand and enforce unconditional loyalty from the population. At the same time, strong regime control also undermines the autonomy of a country's military organizations, preventing them from forming strong internal bonds of their own. These governments worry about coup-proofing their militaries, rather than military effectiveness on the battlefield.[77] The state does not trust the military with the

TABLE 2.5: Hypotheses on Military Cohesion: Four Types of National Armed Forces

High Degree of Regime Control	Authoritarian Militaries Strong Staying Power: National military collapses only when an adversary possesses crushing material superiority Moderate Battlefield Performance: Most units fight with determination but lack flexibility	Messianic Militaries Strong Staying Power: National military collapses only when an adversary possesses crushing material superiority Strong Battlefield Performance: Most units fight with determination and flexibility
Low Degree of Regime Control	Apathetic Militaries Weak Staying Power: National military collapses quickly as probability of victory decreases Weak Battlefield Performance: Only the best units fight with determination and flexibility	Professional Militaries Moderate Staying Power: National military collapses slowly as probability of victory decreases Strong Battlefield Performance: Most units fight with determination and flexibility
	Low Degree of Organizational Autonomy	High Degree of Organizational Autonomy

autonomy to create loyalty and trust within its ranks. Without these military bonds, however, most units will fight with determination but lack the ability to fight with discipline (see Table 2.5).

Authoritarian militaries display two types of behavior during war. First, they fight with strong staying power. The military will avoid collapse under all strategic circumstances. With strong control, the regime can demand unconditional loyalty from the armed forces and deter any dissent that might arise from within it. A hard-core group of regime supporters will fight no matter the strategic conditions. Those same supporters will also use threats of punishment to inspire less motivated soldiers to keep fighting.

Second, authoritarian armed forces fight with determination but lack discipline on the battlefield. Fearful the military might turn against them, the regime exerts a high degree of control over the armed forces, preventing them from exercising the autonomy they need to train and to build bonds of trust among personnel. On the one hand, members of the armed forces fight with great determination on the battlefield. On the other, the regime creates a climate of fear within the armed forces that undermines the trust needed for flexible operations. Officers fear taking risks, units encounter trouble deviating from established doctrine and tactics, and units suffer from paralysis when command and control links collapse. The Soviet Union during World War II and Iran during its 1980–88 war with Iraq fielded an authoritarian military.[78]

Professional Militaries: Low Regime Control and High Organizational Autonomy. When regimes possess a low degree of control over society but permit a high degree of autonomy for the armed forces, they will produce *professional* militaries. I label them as professional because with autonomy the armed forces can concentrate on training for warfighting and, thereby, create their own internal bonds. Under these conditions they develop a corporate identity centered on their martial expertise and responsibility as the nation's defenders.[79] The armed forces serve as loyal organizations, with the singular purpose of preparing and fighting the country's wars.

With a focus on training, strong organizational bonds develop, enabling most units to fight with high levels of determination and flexibility on the battlefield. In combat, most units fight with determination and discipline. These regimes know they cannot exert a high degree of control over society and, therefore, give the armed forces the autonomy to train its personnel. Motivated by military loyalties, members of the armed forces share a sense of unity that inspires sacrifice as well as mutual trust. These traits help most units fight hard even under unfavorable operational circumstances.

Without strong regime control, however, professional militaries will slowly lose their ability to fight as the probability of victory decreases. As events on the battlefield lower the likelihood of overall victory, the armed forces begin to experience pressures to collapse.[80] Members of the armed forces start to ask if the cause for which they fight merits risking their lives. Reasonable, self-interested individuals should pose this question when engaged in war where victory seems uncertain or unlikely. This justification relates closely to what Charles Moskos called "latent ideology," or the belief that a soldier's social system is worth defending. Here I expand the concept of latent ideology to include whether the military's wartime goals looks attainable.[81]

In the midst of these growing doubts and questions about continuing the war, a hard-core group of regime supporters will not exist to coerce the armed forces to fight when defeat looks imminent, as they do in messianic and authoritarian militaries. Their reduction in staying power, however, happens slowly. The internal bonds of loyalty and trust inside the armed forces increase the endurance of military personnel. Representative professional militaries include the armed forces of liberal democracies like France during World War I and the United States, as well as armies of illiberal states like Wilhelmine Germany.

Apathetic Militaries: Low Regime Control and Low Organizational Autonomy. Finally, a low degree of regime control and a low degree of organizational autonomy interact to create *apathetic* militaries. These unmotivated armed forces fight with a low degree of cohesion. They will display two distinct behaviors during war. First, they will exhibit weak staying power. If successive defeats put the country at a strategic disadvantage, opposition to a war will quickly emerge inside the armed forces. Without a high degree of control, the regime will not possess the ability to suppress this dissent inside the armed forces. In short order, the military leadership will look for paths to end a conflict as units succumb to growing disorder. Second, without organizational autonomy, most units will not fight with determination and flexibility on the battlefield. Because military organizations cannot create a sense of unconditional loyalty among members, units will take fewer risks. Similarly, without a sense of trust among personnel, units will fight without flexibility, succumbing to panic, unable to deviate from doctrine when situations demand it. Examples of apathetic armed forces include the French Army of 1940, the Italian Army during World War II, and the armed forces of the Republic of South Vietnam.

How the Type of Armed Forces Can Change

Shifts in cohesion theory's explanatory variables—the degree of regime control, military autonomy, or both—can move a country's armed forces from one of the above categories to another. Sometimes these changes take place between wars. In Germany, for example, a professional military became a messianic military when Hitler's regime secured a high degree of regime control over society, while preserving the autonomy of the armed forces to conduct training. The armed forces of interwar France, in contrast, moved in the opposite direction. Their social cleavages eroded military autonomy, changing the professional French armed forces into an apathetic military.

Important Background Conditions

In addition to my argument, four background conditions can also influence military cohesion. In the following case studies, I try to control for these confounding factors. The first of these background conditions concerns armed forces fighting to protect their homeland. Armed forces defending their homeland will likely receive some boost in their cohesion. Individuals defending their own country will feel as if they fight to protect their way of life, and potentially, their survival. Moreover, as social psychologists note, individuals in these situations will also display a tendency to form in-group loyalties in the face of out-group threats. Hence, governments often appeal to national defense or nationalism to energize their populations to fight. Thus the stakes are higher and could compel even self-interested members of society to demonstrate more zeal in their wartime determination.

However, defending the homeland in and of itself does not represent a sufficient explanation of military cohesion; whatever short-term boost in cohesion that occurs by protecting the homeland will likely wear off if large pre-existing cleavages undermine regime control, organizational autonomy, or both. Put another way, defending the homeland will likely fail to motivate armed forces raised and sustained by divided societies and military organizations. In these instances, individuals might lack the enthusiasm to fight for a country they feel does not represent their interests. Simply put, homeland defense can increase military cohesion in the short term, but it alone will likely prove insufficient to sustain it.

Second, the balance of forces could explain why some militaries fight hard. The balance of forces on the battlefield, then, represents a potential confounding variable to my argument about the sources of military cohesion. Militaries with superior capabilities, all things being equal, will likely exhibit high levels of staying power, which seems a reasonable intuition. To that end, I control for material conditions through case selection. My cases try to focus on armed forces fighting defensive or losing campaigns that have equal if not inferior capabilities compared with their adversaries. Only the United States in the Vietnam War is an exception to this selection rule. Although the United States possessed superior firepower and manpower to use against the North Vietnamese, they could not match their endurance.

Third, a country's manpower policy can also play a role in the cohesion of the armed forces. Governments occasionally try to insulate the military from societal divisions that risk weakening its internal bonds.[82] These policies aim to foster a sense of loyalty inside military organizations that does not exist in the rest of the country. Professional militaries with long service times, for instance, represent one organizational design that can potentially separate the armed forces from debilitating cleavages in society. By virtue of their training, the longer individuals serve in the military, the lower the probability that societal divisions will affect them. This isolation from the rest of society, however,

TABLE 2.6: Causal Logics

Type of Military and Causal Logics		Military Cohesion
Messianic		
Hard-core regime followers prevent collapse of staying power	→	Strong Staying Power
Hard-core regime followers fight and pressure others to fight	→	Strong Battlefield Performance
Loyalty and trust inside military promote battlefield determination and flexibility	→	
Authoritarian		
Hard-core regime followers prevent collapse of staying power	→	Strong Staying Power
Hard-core regime followers fight and pressure others to fight	→	Moderate Battlefield Performance
Fear inside military prevents flexibility on the battlefield	→	
Professional		
Staying power collapses slowly because of loyalty and trust inside military	→	Moderate Staying Power
Loyalty and trust inside military promote battlefield determination and flexibility	→	Strong Battlefield Performance
Apathetic		
Staying power collapses quickly because of lack of loyalty and trust inside military	→	Weak Staying Power
Without loyalty and trust inside military, most units lack battlefield determination and flexibility	→	Weak Battlefield Performance

is not absolute. During war, when manpower needs become greater than in peacetime, military organizations will draw upon the rest of society to replenish their ranks. As a result, the societal cleavages the military tried to wall off will eventually infect the armed forces and sap their will to fight. To control for this separation from society, and to improve the chance of useful comparisons, I select conscript militaries for each of my case studies (see Table 2.6).

Lastly, the book's argument applies both to a country's regular military units as well as the irregular forces governments sometimes use to control their populations. These include organizations a regime might rely on to protect the state from foreign as well as domestic threats, such as Saddam Hussein's special security forces. Because their fate is tied to a government's survival, and because they might receive their own special training, these units will likely display more determination in battle. Their cohesion stems from a strong ideological support of the regime and strong organizational loyalties.

ALTERNATIVE ARGUMENTS

As a further test of my theory's explanatory power, I also examine three common arguments scholars provide for military cohesion: small-unit ties, nationalist ideologies, and democratic governments. At the end of each case study, I briefly compare cohesion theory to the three aforementioned existing alternative explanations for why militaries fight hard. My aim is not to accomplish an exhaustive test of cohesion theory against these alternatives, since they remain largely undeveloped. Instead, I show that my argument can shed more light on the selected historical cases by building on these existing arguments and folding many of their insights into cohesion theory.

Small-Group Theory

The first alternative, small-group theory, explains military cohesion by focusing on an organization's manpower and training policies. Specifically, this argument focuses on cohesion at the level of small units, such as platoons or squads. Military cohesion, proponents claim, depends on the creation of primary group ties. Face-to-face contacts and common experiences foster bonds of loyalty among small groups. These connections sustain soldiers in combat.[83] They motivate soldiers to fight with determination and staying power.[84]

The body of research supporting this argument lists several reasons beyond the creation of primary groups for small-unit cohesion. However, their core argument centers on the set of manpower and training policies required to promote the formation of primary groups. The central hypothesis, then, is that primary groups form out of shared experiences among individuals in units small enough for constant, face-to-face contact. In particular, primary groups stand a good chance of forming and sustaining military cohesion when small formations (platoons, squads) train together, fight together over time, as well as refit and accept replacements as a unit. Individual replacement policies, like those of the United States used in World War II and the Vietnam War, these scholars argue, undermine small-unit cohesion by preventing the formation of primary groups.[85] Individual replacements disrupt primary groups by forcing strangers to become comrades.[86]

Nationalism

A second theory points to the motivating power of nationalism as the source of military cohesion. Scholars across different theoretical traditions in international relations have noted the power of nationalism to motivate a country's armed forces. They point to the French Revolution as the birthplace of nationalism.[87] Michael Desch argues that nationalism motivated some of the most determined militaries of the modern era.[88] Similarly, Dan Reiter's study of Imperial Japan shows why nationalism can empower militaries to fight with staying power and determination on the battlefield.[89] According to Barry Posen, states recognize this potential and promote nationalism when facing security threats that require them to raise and to train mass armies. In response, their competitors must imitate by promulgating nationalist ideologies of their own.[90]

Although these scholars correctly identify nationalism as a potential source of cohesion, a question remains: When do these beliefs motivate militaries to fight hard? Put another way, these scholars have not specified the conditions under which nationalism successfully motivates militaries. Sometimes nationalist appeals rally a military, and sometimes they fall flat. Without these conditions, there is not an obvious way to conduct an empirical test of this argument.

Nevertheless, to provide a more developed and testable version of the nationalist argument, I look to areas of argument among these scholars. Proponents of the nationalism argument appear to agree that strong nationalism consists of three characteristics: a sense of national unity, a patriotic defense of the homeland, and a belief in self-sacrifice based on ethnic ties.[91] Other versions of the nationalism argument might exist, but these features seem the most commonly discussed sources of motivation. Therefore, in the book's case studies, when these three characteristics of nationalism are present within a country, we should expect their armed forces to fight with a high degree of cohesion.

Democracy

Lastly, some theories of international politics emphasize the importance of a country's domestic institutions to explain military cohesion. The latest innovation in this line of argument, linking a state's war-fighting capabilities to its regime type, contends that democracies can strongly motivate their soldiers in combat. Democracies are unmatched in promoting group solidarity among their citizens, and therefore, their soldiers. Their capacity to inspire their armies stems not simply from appeals to self-interest but also from the fact that members of democracies come to see their society as worth defending on the battlefield. As the argument goes, self-interested citizens embrace the political culture of individualism promoted by democratic governments. Appealing to a sense of self-interest, individualism is the belief that a government should intrude minimally into the lives of its citizens. The more democracies defend individualism, the greater their popularity among their constituents. Moreover, because they uphold an individualist political culture, citizens equate the interests of democratic governments with their own personal interests.[92]

Specifically, democracies fight with determination and flexibility on the battlefield but display only moderate staying power. Dan Reiter and Allan Stam offer three hypotheses to explain the cohesion of democratic militaries.[93] The first hypothesis is that democratic governments field armies with high morale because soldiers are fighting for a popular form of government.[94] The more ardently citizens endorse their government, according to this logic, the harder they will defend it. Democracies win the support of their citizens because they best promote their individual interests. As the authors put it, "[The] popularity of democratic governments means that citizens are more willing to fight in service of the state."[95] Individualism also cultivates a greater sense of egalitarianism in these societies. These egalitarian beliefs translate into norms of self-sacrifice prompting democratic soldiers to display a higher tendency to risk injury or their life to defend their country.

Second, democratic armies fight with determination because they trust their officers. In democracies, individuals advance in society because of merit and not from class connections or ethnic ties. Democratic militaries reflect this commitment to advancement through achievement in the way they select their officers. Still further, democratic soldiers can depend on their political leaders to use the military wisely. Reiter and Stam argue:

> Soldiers are more likely to accept the dangers of the battlefield and place their lives at risk if they are serving in a military overseen by a government grounded in democratic political institutions. They are more likely to perceive the war effort and the leadership itself as reflecting their own interests if the need for popular consent constrains the government and it can be removed from office if it fails to hold up its end of the social contract. . . . [S]oldiers are more confident that a democratically elected government will obey the laws and abide by its promises because failure to do so may result in its removal from power. In essence, the source of legitimacy lies in the rule of law rather than the cult of personality, as is the case in many autocracies.[96]

Finally, Reiter and Stam qualify their claims about democratic armies by noting that their morale declines over time. Although their militaries are, as they argue, highly motivated by virtue of their popular forms of government and because their soldiers trust the decision-making of their leaders, democracies lack a tolerance for long wars. Governing with consent, however, is a double-edged sword. Democratic leaders know their constituents will not tolerate long wars without results. Again, Reiter and Stam note, "The accumulation of casualties without victory erodes consent for war within both society and the military, causing all to question how elusive the expected victory truly is. Recognizing that few would wish to fight in a drawn-out, losing effort, we would expect, then, that the morale of democratic soldiers would decline as the war goes on."[97]

TESTING COHESION THEORY

The case studies in subsequent chapters serve two purposes. Primarily, they aim to show that cohesion theory is a plausible explanation for a military's will to fight. Each case study shows a correlation between my theory's explanatory variables, regime control and organizational autonomy, and my study's dependent variable, the degree of military cohesion displayed by national armed forces. Through process tracing, a feature unique to case studies, I also illustrate the causal logic of my four hypotheses. I select cases of militaries fighting wars where victory looks uncertain. In most of the cases, militaries defend their homeland. This selection criterion allows me to control for the inspiration soldiers find when protecting their native soil. The case studies also test my theory against the three competing arguments: small-group theory, nationalism, and democracy.

What kind of evidence would demonstrate the plausibility of cohesion theory? The cases support the theory if strong regime control produces national armed forces with strong staying power. In the theory's specific terms, messianic and authoritarian militaries will resist internal pressure to collapse, fighting even when defeat looks likely. The government's hard-core supporters, having embraced the regime's ideology, will carry on the fight and pressure the less zealous to continue the struggle no matter the strategic circumstances. They will rely on a combination of threats and ideological appeals to prevent the armed forces from breaking. In contrast, the staying power of professional armed forces will decay slowly when victory appears unlikely, while apathetic militaries quickly succumb to internal pressure to collapse when the odds of winning a war seem out of reach.

The cases will also support cohesion theory if a high degree of autonomy for a country's armed forces produces militaries whose units fight with determination and flexibility on the battlefield. Because they possess the freedom to train without domestic interference, most units of messianic and professional armed forces will attack and defend despite high losses. Still further, most units

TABLE 2.7: Cases and Predicted Military Cohesion

The Cases: Country and Year	Major Battles within the Cases	Cohesion Theory Prediction
Chapter 3: Messianic German Military, 1944–45	Retreat from France, West Wall Ardennes Offensive, Defense of the Rhine	Messianic Military: Strong Staying Power and Strong Battlefield Performance
Chapter 3: Professional German Military, 1917–18	Cambrai, Amiens	Professional Military: Moderate Staying Power and Strong Battlefield Performance
Chapter 4: Apathetic French Military, 1940	Dinant, Sedan, Montcornet, Laon, and Arras, Weygand Line	Apathetic Military: Weak Staying Power and Weak Battlefield Performance
Chapter 4: Professional French Military, 1914–18	First Marne, Second Marne	Professional Military: Moderate Staying Power and Strong Battlefield Performance
Chapter 5: Authoritarian Soviet Military, 1941	Bialystock-Minsk, Smolensk, Kiev, Moscow	Authoritarian Military: Strong Staying Power and Moderate Battlefield Performance
Chapter 6: Messianic North Vietnamese Military, 1965–73	Ia Drang, Operation Dewey Canyon	Messianic Military: Strong Staying Power and Strong Battlefield Performance
Chapter 7: Professional United States Military, 1968–72	Tet, Hue, Khe Sanh, Hamburger Hill	Professional Military: Moderate Staying Power and Strong Battlefield Performance

will display the flexibility to resist panic, adjust their plans and tactics, as well as fight when their command and control links are disrupted. According to cohesion theory, units from authoritarian militaries will lack flexibility, since a repressive regime undermines trust within the armed forces. At the same time, these same units will display determination in combat, as the regime's supporters in the military will fight hard and pressure others to do the same. Without organizational autonomy to train, only the elite units in apathetic armed forces will fight with determination and flexibility on the battlefield. Table 2.7 summarizes the following case studies and cohesion theory's predictions for each.

3 Germany, 1944-45

THE TENACITY OF THE GERMAN ARMY in World War II offers compelling evidence in support of *cohesion theory*. In its earliest victories, the Wehrmacht exhibited more cohesion than militaries opposing it. More important, in the face of impending defeat, the German armed forces still demonstrated a ruthless capacity to fight to the bitter end. This performance stands in stark contrast to the behavior of the French Army in 1940. When confronted with German breakthroughs, and when suffering mounting casualties, the French military capitulated relatively quickly. Some commentators point to this superior stamina when they describe the Wehrmacht as the best army of the twentieth century.[1]

Why did the Wehrmacht display determination and flexibility on the battlefield? What were the sources of its staying power? This chapter argues that the strong cohesion of Germany's *messianic* military stemmed from a high degree of regime control and organizational autonomy. Unified and galvanized by an ideology based on National Socialism, German soldiers endured a long, costly war even when victory was no longer possible. Strong bonds inside its military organizations only reinforced the regime's demands for unconditional loyalty, motivating most units to fight with determination and discipline on the battlefield. Some might argue that fear of the Soviet Union drove German units to fight when victory was no longer possible. While concerns about Soviet reprisals certainly motivated the Wehrmacht, cohesion theory also predicts the combination of strong regime control and organizational autonomy that allowed the German military to fight hard against the Soviets as well as the Western Allies. As the historian Omer Bartov suggests, this marriage between the regime's ideology and organizational bonds created a military capable of fighting resolutely against overwhelming odds:

> In September 1939 Germany launched what turned out to be a world war, and although its people marched to battle without much enthusiasm, and its resources were not totally mobilized, as of winter 1941 Hitler's Reich found itself up to its neck in a vast military confrontation, fielding millions of soldiers, straining both its physical and mental capacities to the limit. Ultimately, the mass of Germany's population became involved in one way or another in the war, and a growing proportion of its men, young, middle-aged, and old, workers, bourgeois, and aristocrats, Nazis and former socialists and communists, were recruited and sent to the front, turning miraculously into Europe's toughest and most determined troops, mostly fighting with extraordinary cohesion almost until the bitter end. For throughout the war, combat morale in the Wehrmacht generally remained high, mutinies were almost unknown, and an excellent system of manpower organization, draconian punishment, and extensive indoctrination combined to hold combat units tightly together, while a series of astonishing victories made it easier to withstand even greater defeats in the hope of fortune's wheel turning once more in Germany's favor.[2]

This chapter uses cohesion theory to explain two puzzles about Germany's military performance. First it describes why the Wehrmacht fought hard against the British and U.S. forces in the West as well as against the Red Army in the East. Historians attribute the tenacious German defense to the country's fear of a vengeful Soviet invader, which is certainly an accurate prediction. Nevertheless, this does not explain why the Wehrmacht could hold together and resist the democratic armies of the Western Allies. Second, cohesion theory explains why Germany collapsed in 1918 but not 1945. Simply put, the Nazi state exerted greater control over German society than the Wilhelmine regime, preventing the armed forces from dissolving when defeat appeared likely.

This chapter consists of five parts. The first three parts discuss the cohesion of the German Army in World War II, from 1944 to 1945. I examine the Wehrmacht's performance against the democratic Allies who attacked Germany from the West. I focus mostly on the Western Allies to control for arguments that Germans fought hard to forestall Soviet retributions for the Wehrmacht's war crimes on the Eastern Front.[3] Through a necessarily detailed discussion of the battles within these two campaign years, I demonstrate the high degree of cohesion displayed by Germany's messianic armed forces. As cohesion theory would predict, most German units fought with determination and flexibility on the battlefield, despite the increasing likelihood that they would lose the war.

The fourth part of the chapter briefly compares the cohesion of the Wehrmacht in World War II with the cohesion of German forces from 1917 to 1918. During World War I, the combination of organizational autonomy but a low degree of regime control produced a *professional military*. Although German forces could rely on strong organizational bonds within the armed forces for motivation, the German Army of World War I lacked the regime control that characterized Nazi Germany. Cohesion theory, therefore, predicts a higher degree of cohesion for Hitler's army than for the Kaiser's army. Wilhelmine Germany made the most of nationalist appeals, but it could not both demand and enforce unconditional loyalty from the military in the same way Hitler's regime could. Without a high degree of regime control, the armed forces would fight as long as victory looked possible. The failure of the Spring Offensives of 1918, however, broke the back of Germany's professional military. Failures on the battlefield fueled domestic opposition, further weakening the military's staying power. These differences between the two German armed forces offer additional evidence for the range of cohesion theory's explanatory power.

GERMANY'S MESSIANIC ARMED FORCES, 1944–45

A large part of the German Army's cohesion stemmed from a high degree of regime control over the country. Nazi Germany obtained a level of unity that had eluded it before World War I. By pledging to create an egalitarian society, and by exerting control over most of the country's institutions, Hitler managed to achieve a level of national harmony that his Wilhelmine predecessors had failed to accomplish. As a result, the German military received highly motivated recruits to fill its ranks, and the regime would keep domestic support under difficult strategic circumstances. Internally, the German armed forces retained a high degree of autonomy to foster strong organizational bonds through training. As a result, most units could fight with determination and flexibility on the battlefield.

The Nazi Regime Possessed a High Degree of Control over Germany

When Adolf Hitler took the reins of government in 1933, Germany stood in the midst of an economic crisis exacerbated by deep political divisions. Under the weight of the Depression, both the working and middle classes started to see the gains they had made after World War I slip away as Germany's economic condition worsened. Political parties on the Left, especially the communists, took advantage of the situation by advocating, often violently, for radical changes in government. To them, the political circumstances seemed ripe for the kind of revolutionary upheaval their Bolshevik brethren had undertaken in Russia decades earlier. Similarly, revolutionary elements on the Right, like the Nazi Party, also blamed the Weimar Republic for the deteriorating conditions in the country. They too sought to do away with the government they held responsible for Germany's postwar malaise.[4] Their workers' paradise, however, envisioned a country unified not along class lines but through appeals to German nationalism.[5]

Against this backdrop, Hitler gained a prominent place in German politics. Although the Weimar Republic claimed to represent the will of the majority, a strong minority of conservatives in the bureaucracy, courts, and the military held a disproportionate share of government power.[6] These were the true brokers of German political power. Faced with growing unrest from the Left, they sought out Hitler as an ally to shore up the government's position. For these political elites, bringing the Nazi Party into government represented the lesser of two evils.

Over the preceding years, Hitler's party had made steady progress in local and national elections. For the conservatives in government, the Nazis' increasing electoral prominence made them an ideal counterweight to the communists. Moreover, the republic's elderly president, von Hindenberg, and his chancellor, von Papen, saw in the Nazis their best chance to manipulate a political party that could balance the communists. Heeding the advice of his colleagues, Hindenberg brought Hitler into his cabinet to serve as chancellor on January 30, 1933. One of Papen's friends fatefully remarked on this occasion, "We have him framed in."[7]

In the next few years, however, Hitler boxed in his adversaries. Through a series of legislative maneuvers, he gradually began to accumulate more political influence than his opponents. His conservative allies believed, not incorrectly, that Hitler sought to pursue a nationalist agenda that would, to their delight, debilitate the country's democracy. In their optimism, they misread his intentions as he adroitly used his office to secure greater political power for himself and the Nazi Party.

Months into office, the new chancellor compelled a listless Reichstag, the German parliament, to pass the "Enabling Law," which allowed Hitler to rule by decree.[8] A year later, after Hindenberg's death, the German Army pledged its undying loyalty to Hitler. By 1938, Hitler brought the foreign ministry and the military, the centers of the old aristocracy's power, under his control.[9] The Wehrmacht decided to support Hitler in return for the promise that it would retain its status as the nation's armed force and that he would not replace the German military with a revolutionary army.[10] At the same time, the Nazi's encountered little or no resistance from other state institutions, including universities and industrial and professional organizations.[11] Even though the Nazi Party garnered support by serving as an umbrella for different interests, it was Hitler and his clique that reigned in these groups and dictated actual policy.[12]

The Regime's Ideology Demanded Unconditional Loyalty. Although controlling the machinery of government gave Hitler great influence and control over German politics, the key to his authority rested with the appeal of an ideology demanding unconditional loyalty to the German state and nation. In addition to his personal charisma, Hitler's values won him wide support among the German populace. Nazism, according to Karl Bracher, "was a conglomerate of ideas and precepts, of concepts, hopes, and emotions, welded together by a radical political movement in a time of crisis."[13] Hitler's promises of a brighter future for the German nation resonated with most of the population. This ideological program rested on three principles.

The first tenet of Nazi ideology held that individual Germans ought to view their own interests through the lens of the nation's interests. Among the many principles that composed the Nazi constellation of beliefs, the central conviction held that the preservation and the goals of the German nation came before whatever parochial interests each citizen might pursue. Because they bore the responsibility for guaranteeing the welfare of the nation, individuals served the state and not vice versa.[14] Karl Bracher perhaps summarized this best when he wrote:

> The simple schema into which flowed the conservative-authoritarian, antidemocratic, nationalist, irrationalist chain of ideas can be summed up as follows: Man can only live as a member of a nation, and therefore, the nation transcends group interests. It is strong only as a cohesive unit, and therefore, true "socialism" welds the classes together rather than dividing them; it upholds national idealism instead of Marxism and liberalism; duty, loyalty and disciplined followers instead of human rights; and, instead of a selfish pluralism, the monolithic leader state, which alone is able to overcome the weaknesses of centuries of German division and ensure optimal power.[15]

One of the rationales for the primacy of the state in German society stemmed from the Nazi worldview that international politics entailed a never-ending struggle for survival among racially defined nations. Propagating the social Darwinist views of an earlier generation, Hitler argued that human societies, like other species, lived in a competition where only the most fit would survive.[16] This concept resonated with many Germans because it reflected a widely growing trend of nationalist-racist thinking, rooted in popular thinking reaching back to the 1890s.[17]

To belong to the nation, citizens needed to prove their racial ties to the German people.[18] In many ways, Nazi views on race represented an intensification of already existing views on the subject.[19] Unlike the preceding version of nationalism that incorporated Germans of Jewish descent, Nazi views of the nation rested on a very narrow definition of what constituted the German race. Although they had experienced intermittent periods of anti-Semitism, German Jews could live and participate in Wilhelmine Germany, with many playing prominent roles in society. In contrast, Nazi ideology perpetuated highly stringent requirements for membership in the German national community. Tragically, these views would propel Germany to see the world in such stark terms that they would perpetuate great atrocities against many ethnic minorities in Europe, with much of their fury directed at those of Jewish descent.[20]

It was in this world of racially constituted nations fighting for survival that Hitler promised to lead the German people to victory over their enemies at home and abroad. According to Nazi ideology, this German national mission consisted of both positive and negative goals. In the first place, Hitler promised to create a nation that would move beyond the stagnation of the Weimar period to witness an era of social equality.[21] As Michael Burleigh explains, Hitler offered his people a national mission to replace the societal malaise afflicting Germany: "Nazism offered intense inclusivity in a society that had been scarred by deep divisions, dynamism where there was stagnation, and a sense of lofty purpose, almost a national mission, in a society where material interests seemed all-pervasive."[22] The Nazis pledged to lead a social revolution where national unity would replace class distinctions.[23] David Schoenbaum describes it as "a revolution of class and a revolution of status at the same time."[24] Instead of political divisions based on class distinctions, Hitler's ideology called for his people to think in terms of what was best for the German national community. Only when the interests of the nation as a whole trumped class interests, the Nazis argued, could Germany's middle-class and workers alike realize their full potential.[25]

Hitler's ideology, however, envisioned numerous enemies, foreign and domestic, opposing Germany's mission to create a national community devoid of social cleavages. To bolster national cohesion, Hitler demonized all of Germany's adversaries, creating a sense of urgency for the country to work together in order to achieve its revolutionary objectives. Among these opponents, a conspiracy between Jewish and Bolshevik groups posed the most insidious threat to national progress. The Nazis proclaimed that these Left-wing interests undermined German unity by appealing to class differences, agitating the worker against the rest of the nation.[26] This line of argument justified the Nazis' harsh crackdown, using the powers bestowed on the government by the newly passed Enabling Laws, against political opponents on the Left.[27]

Hitler directed much of Germany's energy toward what he saw as its foreign enemies. Here he could draw on a wellspring of resentment that many in the country, especially those in the military, still maintained about their bitter defeat in World War I. A good number of Germans felt that domestic subversion, the "stab in the back," had caused them to lose the war. They harbored even deeper anger about what they saw as the unjust terms of the Versailles peace, especially its war reparations and the restrictions on Germany's armed forces.[28] Nazi ideology called for the German people to reassert themselves in international politics, to finally claim their position as the predominant European power. Good to his word, Hitler reneged on the Versailles Treaty and sent Germany on the road to an ambitious rearmament campaign. Undeterred by the international reaction to his remilitarization drive, he observed, "A balance of power had been established without Germany's participation. This balance is being disturbed by Germany's claiming her vital rights and her reappearance in the circle of the great powers."[29]

In addition, Hitler forcefully argued that Germany's survival depended on the acquisition of *Lebensraum*, or "living space." With its ever-growing population, the country, as the Nazis argued, required greater resources to meet its needs. Moreover, in a world where fierce competition among nations was the rule and not the exception, Germany could not depend on other countries to supply it with the raw materials required to maintain a healthy level of economic growth.[30] To this end, Hitler reasoned, "For Germany . . . the only possibility of carrying out a sound territorial policy was to be found in the acquisition of new soil in Europe proper."[31] Eastern Europe and the Soviet Union served as the primary targets of Hitler's territorial ambitions. As he predicted, "We will resume where we left off six centuries ago. We are putting a stop to the old Germanic movement toward the south and west of Europe and turning our attention to the land in the east. We are putting an end to the colonial and trade policies of the prewar era and are going over to the soil policies of the future. But today if we speak of new land, we think above all of Russia and its . . . border states."[32] For Hitler, then, the critical battles Germany needed to win for its survival were not in the West, but in the East against what he saw as the looming Slavic menace.[33]

The Regime Possessed Physical Control over German Society. Why did German society embrace this ideology demanding unconditional loyalty? Although the regime at first used the threat of coercion to convince many citizens to toe the party line, by and large, many Germans enthusiastically accepted the Nazi agenda.[34] At least three factors contributed to the country's acceptance of Hitler's ideology, creating the conditions for the Nazi state to control the German population. Hitler garnered support by persuading the German nation to follow him on his quest to create a better future for the country.[35]

Nazi ideology appealed to the population, firstly, because it pledged to lift Germany out of the misery of the Weimar era. In particular, Nazi calls for social equality based on race appealed to a country where nationalism had always held an allure. These Nazi conceptions of equality based on race found a wide audience that crossed class lines.[36] Hitler's ideology, for instance, attained the backing of the conservative political establishment that had backed Hindenberg through his nationalist policies and assurance that they too would find social advancement in Nazi society.[37] When he came to power, a large segment of the population, especially Germany's Protestant subculture, already viewed the concept of a national community as an ideal goal.[38]

Tough economic conditions, moreover, made Germany ripe for any leader willing to promise some kind of way out of the Depression. As MacGregor Knox notes: "[N]ot merely the unemployed voted for Hitler. The fundamental component of the Nazi vote was Germany's solid citizens, to whom the Fuhrer offered leadership with 'the self-assurance of a sleep walker' and nationalist solutions. What above all distinguished 1932 Germany from 1922 Italy was the atmosphere of apocalyptic crisis that invested with plausibility and attractiveness the Nazi claim to be 'the opposite of everything that now exists.'"[39]

Promises of economic improvement also swayed the working class to Nazi views. In a country where they already displayed nationalist sympathies, Hitler's desire to create a nation in which race supplanted class as the crucial societal division pleased Germany's working classes.[40] Although some scholars contend that German workers ultimately resisted the Nazi message because the government could not improve their plight,[41] the balance of the evidence seems to prove otherwise. Their reasons for casting their lot with the Nazis were simple; as one Social Democrat from the Ruhr remarked at the time, "[T]hey had four, five, even six years of unemployment behind them—they would have hired on with Satan himself."[42] To reinforce the regime's concern with Germany's less fortunate, the Nazis sponsored several charity groups to aid the poor and, thereby, spread their ideological message.[43]

Hitler also pledged not only to allow common Germans to improve their economic plight but also to give them a society devoid of the instability that had rocked the Weimar Republic. They could overlook elements of the Nazi terror because the regime directed its coercive apparatus against the "enemies of the state." Social undesirables, and not the average citizen, they believed, would fall victim to the government's repression. At the end of the day, Germany seemed a better place under Hitler's strong leadership. As Robert Gellately explains, "On balance, most people seemed prepared to live with the idea of a surveillance society, to put aside the opportunity to develop the freedoms we usually identify with liberal democracies, in return for crime-free streets, a return to prosperity, and what they regarded as good government."[44]

In addition, Nazi ideology won the hearts and minds of the German people by dominating many of the country's institutions.[45] Through a strategy of divide and conquer, Hitler prevented Germany's various interest groups in the bureaucracy from subverting his power. He carefully played the role of mediator among the different facets of German society that competed for influence and governmental resources. In this way he exercised control over the large German bureaucracy, thus enabling him to spread the Nazi message.[46] The government also indoctrinated the country's youth by overseeing their curriculum in German schools and by recruiting them into organizations, such as the Hitler Youth and the National Labor Service.[47]

The Nazi Party also could rely on its own paramilitary security force, the Schutzstaffel, or the SS, to protect the regime from internal enemies. Fanatically loyal to Hitler, the SS oversaw the regime's concentration and extermination camps. The SS also served as a counterweight to any effort by the German Army to seize the reigns of government power from the Nazi Party. In addition, fighting units of the Schutzstaffel, the Waffen SS, fought with great ruthlessness and tenacity during the war.[48]

During the war, the regime kept public opposition to the conflict to a minimum. While many Germans displayed little enthusiasm for the war before Hitler's invasions of Poland and France, they eventually rallied to his cause. In fact, the horrible consequences of the war, including the enormous costs in manpower as well as the terrible bombing campaigns launched against Germany by the Allies, seemed for some Germans to vindicate Hitler's view of international politics as a brutal struggle for survival among nations.[49] Among the general population, the regime encountered few opponents while at war. On the contrary, the country expressed great outrage when they heard about the failed assassination attempt on Hitler in July 1944.[50]

Causal Logic: A High Degree of Regime Control Bolstered German Military Cohesion. This high degree of Nazi control over German society enhanced the cohesion of the Wehrmacht. Because the Nazi-dominated German society supplied the military with its manpower, new recruits entered the armed forces imbued with National Socialist tenets. With little organized opposition to its ideological views, the regime created a hard-core group of regime supporters within the military willing both to fight under all circumstances and to pressure others to do the same.[51] This relationship infused the Wehrmacht with strong staying power, preventing the armed forces from disintegrating when victory looked impossible. Similarly, this hard-core group improved the German military's battlefield performance, inspiring soldiers to fight with determination. In particular, two aspects of the regime's ideology improved staying power and battlefield performance.[52]

First, Nazism inspired the Wehrmacht's soldiers to sacrifice themselves to advance Hitler's vision. *Landsers*, or German infantrymen, believed that they fought to remake their society and to ensure their country's survival against the host of enemies that plotted to destroy it, especially those from the East. This view, as Stephen Fritz describes, sustained the average soldier on the front, "The staying power of the average German soldier, his sense of seriousness and purpose, what often went beyond sacrifice, courage, and resolution to fanaticism, depended in large measure on the conviction that National Socialist Germany had redeemed the failures of World War I and had restored, both individually and collectively, a uniquely German sense of identity."[53] The Nazi ability to demonize Germany's enemies by depicting them as barbaric also drove its soldiers to fight to the bitter end. Such images created a powerful sense that all Germans needed to stand united against an inhuman adversary.[54]

Second, the German Army retained its faith in its leadership, especially in

Hitler's ability to find some way of leading the nation to victory in the face of impending defeat. When the war went badly for Germany, both its citizens and its armed forces retained their trust in Hitler's ability to guide them through difficult times.[55] Even in the most desperate conditions, German soldiers expressed optimism. In the midst of the Falaise Pocket, one *Landser* wrote: "It doesn't look very good . . . but nonetheless there is no reason to paint too black a picture . . . . There are so many good and elite divisions in our near-encirclement that we must get through somehow." Under similarly dire circumstances another German proclaimed: "Shoulder to shoulder we fulfill our duty, like the old comrades, and are firmly resolved to fight and to triumph, so that our models, the fallen comrades, will not be sacrificed in vain. Their death is my duty . . . better to fight honorably and die than to steal life."[56]

When ideology failed to motivate, the regime relied on coercion to maintain the Wehrmacht's cohesion. The same tools of terror and murder that enabled the Nazi regime to control German society also became useful for keeping the armed forces fighting. The regime could depend on its hard-core supporters, usually SS units, to enforce draconian punishment for a lack of obedience. Officers and enlisted men of the Wehrmacht understood they were equally vulnerable to the regime's punishments. Threats of execution for dereliction of duty applied not only to individual soldiers but also to their officers.

The Wehrmacht Possessed a High Degree of Military Autonomy

Another source of the Wehrmacht's cohesion derived from its high degree of autonomy to train its personnel. With the independence to train, the armed forces could instill in its soldiers bonds of unconditional loyalty and trust. With its emphasis on unity, the virtues of self-sacrifice, and the need for trust within the ranks, the German military reinforced what many of its soldiers already believed when they entered the military. This emphasis on training had always separated the German military from the armed forces of other countries.[57] Thus, after training in the armed forces, the German *Landser* was driven by both deeply ingrained ideological convictions and loyalty to his fellow soldiers.

The German military promoted three types of norms through its training that contributed to the cohesion of its units. One of the ideals taught recruits that they should view each other as members of a team. Soldiers learned that they shared the common fate of acting together to survive combat. A sense of unity, their instructors explained, would now replace whatever class distinctions might have separated them in civilian society. As one German soldier wrote in a letter home, "We are all the same in our service for our people, no one is asked his origins or class, whether he is rich or poor . . . . Snobbery, class consciousness, envy, and idleness are left out on the street. This is the way from 'I' to 'We.'"[58] Many recruits had already understood the importance of viewing themselves as part of a team from their experiences in the Hitler Youth as well as the National Labor Service. As Stephen Fritz explains, "Military training began early in life with the Hitler Youth, Labor Service, and finally the Wehrmacht. The Hitler Youth instructed Germans that they 'should serve a community,' and 'live a life of comradeship, be harder and ready to fight, carry the will within to greater deeds.'"[59]

Officers commonly used the language of family to reinforce this sense of togetherness. During training, instructors would serve as tough disciplinarians, but it was common for these same drill sergeants to fraternize with their young charges. One soldier recalled in a letter: "There above stood my sergeant, with his arms crossed, shouting: 'Come on, get going, don't plead exhaustion as an excuse!' Many ran as if they were already drunk, but there was no mercy. [But] in the evenings, 'big-mouthed' Schmidt comes into the barracks and sits informally with us, and laughs and jokes and sings with us just like a good comrade."[60]

By stressing teamwork, the German Army managed to promote good relations between its officers and enlisted men. According to Martin van Creveld:

> The bond between officers and men was emphasized more strongly still by the use of a comprehensive term, "soldiers," to describe them both, and by the regulations that required them to salute not merely their officers but each other too. Having spent much of their training period in the company of enlisted men, officers were freely permitted to fraternize with them off duty and even encouraged to do so by the tenets of National Socialism. Possibly as a result of all this, in interviews with prisoners of war, "nearly all NCOs and officers of the company grade level were regarded by the German soldier throughout the Western campaign as brave, efficient, and considerate."[61]

This attempt to homogenize differences among the rank and file was in many ways unique to the German Army. Stephen Fritz notes that this attribute strengthened the cohesion of the Wehrmacht:

> Still, some aspects of the *Landser* experience differentiated them from other soldiers such as the American GI. Comradeship, for example, though certainly not unique to the *Landser*, nonetheless represented a more persistent and pervasive theme in his everyday military life. Wehrmacht leaders, in fact, raised the concept of camaraderie almost to the level of strategic doctrine, seeing in tight-knit, cohesive groups of com-

> rades the same spiritual means by which to overcome, at least partially, the material superiority of the enemy. As an organization the German army went to great lengths to instill and nurture a sense of family; recruitment, training, and replacement policies, as well as relations between officers and men, aimed at creating and maintaining in the *Landser* the sense of belonging to a purposeful and powerful community of men who were enduring the same hardships and sharing the same fate.[62]

The need for trust among officers as well as enlisted men represented a second ideal instilled in German soldiers. Part of this effort entailed creating a spirit of comradeship based on treating new recruits with respect during training. The result, as Stephen Fritz suggests, enabled enlisted men to hold their leaders in high esteem: "On the basis of admittedly impressionistic evidence, it would seem there was less petty, personal harassment of troops in the German army than in the Anglo-American armies. The Wehrmacht made a concerted effort to promote a strong spirit of comradeship between junior officers and their men."[63] Promotions based on merit also furthered this sense of trust in the military. Officers felt that if they worked hard, their superior would reward them with advancement.

This sense of trust also translated into battlefield practices. The Wehrmacht separated itself from other militaries of the time by emphasizing mission-oriented tactics, or *Auftragstaktik*. They developed a command philosophy based on risk taking and action. German officers learned to exercise individual leadership and initiative to pursue combat objectives. An atmosphere where senior officers entrusted their lieutenants with wide latitude to accomplish battlefield objectives made this kind of leadership style possible.[64]

A constant focus on training for warfighting helped forge strong bonds within the ranks. Unlike the militaries of some countries, the Wehrmacht did not see its mission as serving as the national gendarme. Instead, the military was united in its efforts to ready itself to fight the country's wars. This concentration on preparation led the armed forces to adopt a harsh and realistic training environment, conditions that further separated it from similar militaries of the period.[65] Because they saw constant preparation as an important organizational goal, the Wehrmacht constructed a unit replacement system that allowed soldiers to train as a group before going to the front. Many observers have noted that training practices such as these allowed the German military to cobble together new units from ones fragmented in battle because men were accustomed to working in groups and found it easier to re-form units.[66]

The Wehrmacht possessed a high degree of autonomy to instill these norms.

Relations within the military suffered from none of the divisions that might have hampered the armed forces from getting its organizational message across to its members. By drawing noncommissioned officers (NCOs) from similar backgrounds, the armed forces facilitated good relations between their NCOs and enlisted men.[67] Moreover, the organization maintained the morale of its officers by providing an environment where they could experience both career advancement and a higher standing in German society.[68] Nazi ideology, argues Stephen Fritz, also reinforced the harmony in the Wehrmacht, "Concern, friendship, sincerity, idealism: clearly this was a complex and dynamic relationship, something that could not easily be created through the establishment of an NSFO. Yet once it developed, it could result in a formidable bond."[69]

With little to fear from the rank and file of its armed forces, the regime permitted the military to maintain its own organizational culture. This stands in marked contrast to the situation in the Soviet Union at the time. Concerns over the loyalty of the military drove Stalin to undertake a bloody purge of his officer corps. Officers also shared power with political commissars. Only in the midst of its war with Germany did the Soviet leadership remove these political officers from its combat units.[70] The German case differed in several respects from the Soviet experience. At the time of Hitler's ascension to power, most Germans already held the armed forces in high regard.[71] Over time, the Nazi regime managed to secure the allegiance of the military. Through a large conscription effort, the Nazi regime won wide support in the ranks of the Wehrmacht.[72] The expansion of the officer corps allowed many loyal Nazis to take a prominent role in the armed forces.[73] Although some officers conspired to remove Hitler from office once the war began to go badly for Germany, a majority in the military still supported the Fuhrer.[74] Their efforts to assassinate Hitler, ironically, actually bolstered loyalty to the regime among German soldiers.[75]

Unlike other countries with a high degree of regime control, the Nazis permitted the German Army to indoctrinate itself.[76] General Blomberg, the head of the army at the time, and himself a devoted Nazi, decreed in 1935 that the Wehrmacht would play a role in educating society: "Service in the Armed Forces is therefore the last and highest step in the general education of a young German, from parental home, through school, Hitler Youth, and the Labor Service. The educational goal of the Wehrmacht is not only the basically trained soldier and the master of a weapon, but also the man who is aware of his nationality and his general duties towards the state."[77] German officers, and not Nazi Party officials, took the responsibility for giving soldiers political instruction.[78] Similarly, the government allotted funds for the Wehrmacht to publish its own pro-Nazi political newspapers.[79] Such actions only made sense in a society where most recruits entered the military with considerable exposure to Nazi ideology.[80]

Causal Logic: High Degree of Organizational Autonomy Bolstered Military Cohesion. The autonomy to train reinforced Nazi ideology and, thereby, contributed to the cohesion of the German Army. Specifically, organizational autonomy helped the Wehrmacht fight with strong battlefield performance. In combat, Germany's fighting men believed that they shared a common fate, and this motivated them to continue the struggle, even under dire circumstances. A sense of loyalty bolstered the determination of most units. Similarly, bonds of trust enabled most units to fight with flexibility.

Many see the Wehrmacht's cohesion deriving from strong primary group ties, connections that developed exclusively from German training practices. However, this reasoning neglects a simple fact: many German units lost their primary groups through attrition and remained cohesive. Units like the Gross Deutschland Infantry Regiment, for example, which lost many of its members, fought with determination and discipline to the very end of the war.[81]

For Omer Bartov, the tenacity of the German Army stemmed not only from training but also from what individual soldiers took from Nazi ideology:

> In fact, we may derive some insight into the relationship between the people and the regime by suggesting that while *real* "primary groups" do not fully explain combat motivation due to their unfortunate tendency to disintegrate just when they are most needed, the idea of attachment of *ideal* "primary group" composed of a certain category of human beings clearly does have a powerful integrating potential. This kind of "primary group," however, is in some respects the precise opposite of the one presented by the original theory, for it is very much the product not merely of social ties but also of ideological internalization, whereby humanity is divided into opposite groups of "us" and "them."[82]

THE GERMAN WESTERN CAMPAIGN IN 1944–45

The last few years of World War II provide a good place to test the cohesion of Germany's messianic military. Like other cases in this study, Germany fought on the defense for high stakes. Moreover, the focus on the Western Front pitted Germany against the democratic Allies and not its more feared foe, the Soviet Union. As the evidence demonstrates, the Allies nevertheless needed crushing material superiority to defeat the Wehrmacht.

Recovering from the Defeat in France: Evidence of Staying Power, Battlefield Determination, and Flexibility

On June 6, 1944, D-Day, the Allies crossed the English Channel and landed a large force in northern France. After pushing the Germans out of Africa, and after making significant gains in Italy, the Allies stood poised to approach the western borders of Germany. With overwhelming superiority in manpower, American and British forces consolidated their foothold in Normandy and broke out to the rest of France in late July.[83] The German Army confronted a difficult strategic situation with the weight of its war effort in the East. Against the Soviet Union, Germany had committed 228 divisions, leaving 58 divisions in the West, and only 15 at Normandy.[84]

To stop the flood of Allied armies pouring southward, the Germans mounted their first significant counterattack on July 25. In the end, this ambitious assault against American forces at Avranches made the situation worse for the Germans. The Allied armies not only repulsed the attack but also then proceeded to trap the overextended 5th Panzer Army and Panzer Group Eberbach in a pocket near the town of Falaise. This resulting encirclement cost Germany sixty thousand soldiers, as well as great masses of guns, tanks, and equipment.[85] Although it was a devastating defeat, the Wehrmacht managed to pull fifty thousand of its troops from the jaws of this trap, which later provided the framework to rebuild the Germany Army in the West.[86] The Allied breakout eventually gained momentum and swept most of the German forces out of France by the end of August.[87]

On September 4, 1944, Hitler reinstated Field Marshal Gerd von Rundstedt as commander of German forces in the West. For Rundstedt, this was the second time he had taken charge of Germany's Western defenses; his last tour of duty had ended two months earlier. As he surveyed his armed forces' headlong retreat eastward, the situation in the beginning of September appeared dire. Montgomery's forces had pushed their way into Holland and stood ready to turn the northern flank of the West Wall, or what the Allies called the Siegfried Line. In Belgium, the American First Army, under General Hodges, quickly moved toward the German frontier, while on the right General Patton's Third Army steamed across Lorraine on its way to Metz and the German Saar Basin. Still further south, the 6th Army Group advanced up the Rhone River to the southern part of the French border with Germany.

Rundstedt spent the month of September turning a disorganized retreat to the Siegfried Line into an organized defense of the German frontier and its remaining territory in the Low Countries. To contain the Allied tidal wave approaching Germany, the wily field marshal needed to make good use of limited resources. He planned to engage in an elastic defense in depth, establishing consecutive lines that could give way and provide an opportunity to counterattack forces before they achieved a serious breakthrough.

Undertaking such a defense appeared difficult when considering the material superiority of the Allied forces. The Allies rode to the Siegfried Line with almost 50 full-strength divisions, 2,000 tanks, and 14,000 aircraft. In stark contrast, the Westheer, or the German Army in the West, could rely on only 30 divisions (almost half of these listed as partially fit), close to 100 tanks, and 570 aircraft. The picture looked even worse given that German divisions contained almost 3,000 fewer soldiers than their Allied counterparts.[88]

German forces arrayed themselves into four army groups in the West (see Map 3.1). Defending the Netherlands to the town of Roermond was Army Group H (Student) with the 15th Army (Zangen) and the newly constructed 1st Parachute Army (Schelmm). On its left, Army Group B (Model) held the German frontier from the Netherlands to Luxembourg. Model commanded three armies from north to south: the 6th Panzer Army (Dietrich), the 5th Panzer Army (Manteuffel), and the 7th Army (Brandenberger). Army Group G, commanded by Balck,[89] guarded the Saar with the 1st Army (Knobelsdorff), opposite the 3rd U.S. Army. Guarding the border with France and Switzerland, Army Group Oberrhein (Himmler) deployed the 19th Army to block the advance of the Allied 6th Army Group.

The Allied commander on the Western Front was General Dwight Eisenhower. Along the Siegfried Line and the western border of Germany, Allied forces consisted of three army groups running from north to south: the 21st British Army Group, the 12th U.S. Army Group, and the 6th Army Group. The British 21st Army Group, under the command of Field Marshal Montgomery, held a front stretching from the western Netherlands all the way to the German frontier near Aachen. Running from north to south, Montgomery's 21st Army Group directed two armies: in the north, Canadian First Army (Crerar); and in the south, the British 2nd Army (Dempsey). A few months later, Eisenhower would also place the 9th U.S. Army (Simpson) under Montgomery's command. The 12th U.S. Army Group, commanded by General Omar Bradley, was on Montgomery's right flank and operated south of Aachen to the rest of the German border. Bradley controlled the U.S. 1st Army (Hodges), and in the

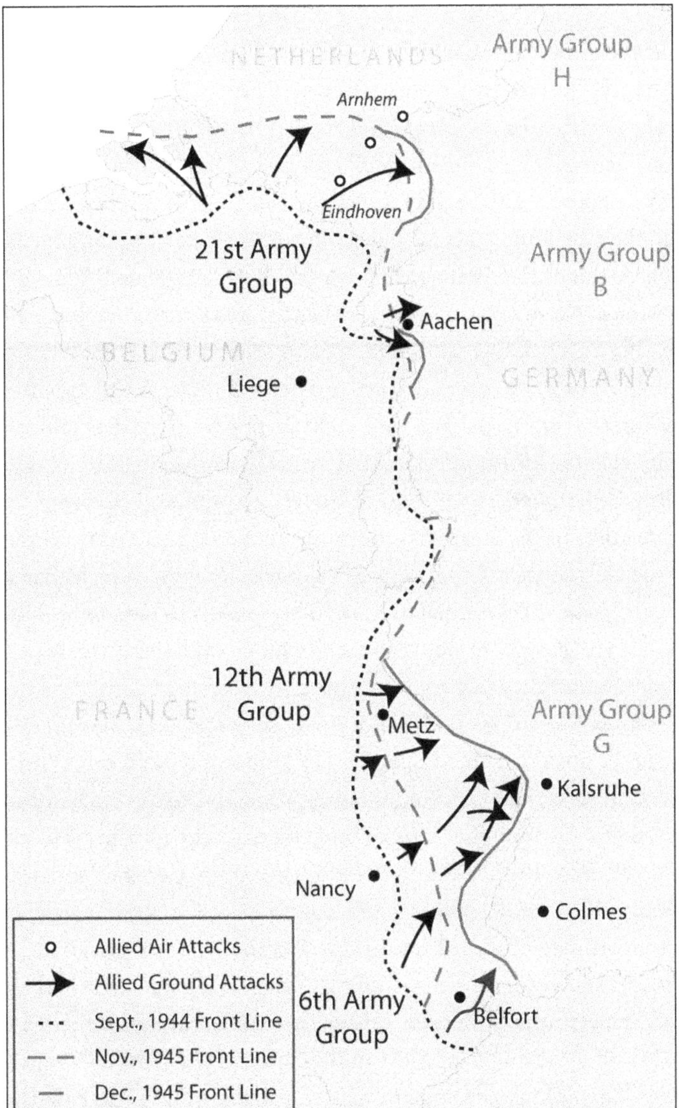

MAP 3.1. Allied Approach to Western Germany, Fall 1944.

south he fought a constant battle to curb the enthusiastic forward movement of General Patton's U.S. 3rd Army. The 6th Army Group (Devers) contained the U.S. 7th Army (Patch) and the 1st French Army (de Lattre).

While the Allies debated their strategy for invading Germany, the Wehrmacht began to recover from its defeat in France. Making use of their underground factories in the Harz Mountains, created by the technocrat Albert Speer, the Germans were able to replace most of the SP field guns lost in the retreat from France. Although tank production could not replace the losses of that year, the new mobile guns would improve the defenses along the Siegfried Line. The Germans also managed to raise the men and materiel for thirty-two new divisions, twelve of them motorized and partially armored panzer or panzer grenadiers. Unbeknownst to Rundstedt, Hitler and his OKW Staff planned to devote most of these new forces for an offensive through the Ardennes. Thus the new forces coming on line were held in reserve by Hitler and remained largely unavailable for Rundstedt to use in his defensive operations.

In early September, Hitler placed Model in command of Army Group B, and the defenses of the Siegfried Line opposite the 12th U.S. Army Group. Even though the German military had staged a miraculous recovery, Model still had limited capabilities and needed to husband resources to defend his part of the West Wall.[90] In Model's mind, keeping the Allies off balance represented one way for the Germans to buy recovery time.

For example, to stiffen the resolve of those forces now defending the city of Aachen, he promised them relief if they could only hold out a little longer against the newly arrived U.S. 30th Infantry Division and the 3rd Armored Division, already threatening the city. Cobbling together the remaining forces in the region, largely from the 9th and 116th Panzer divisions, General Model launched a counterattack across the Pell Marshes just north of Aachen in early October.

The unfortunate object of this attack was the U.S. 7th Armored Division, which Model's forces sent reeling back six miles in a panic. Before their attack halted, German forces prompted a disorderly retreat among American forces, at the same time causing confusion among the rear elements of Montgomery's 2nd Army. This initial foray against the Allies was minor compared with the battles of attrition for which Model prepared his limited forces in the months to come. Model needed to delay the Americans from breaking through the West Wall to make time for the German Army to find the right time to launch an offensive of its own.[91]

As the worsening weather thwarted Allied close air support, and as logistical problems mounted, the American drive eastward began to lose steam. The initial assaults by the 1st Army against the West Wall cost the Americans ten thousand casualties. After slugging it out with the Germans there, Eisenhower paused to reconsider his options about the best ways for the 12th Army Group to punch through the Siegfried Line.

Throughout the fall of 1944, the Allies would attempt to penetrate the German defenses along the West Wall. Instead of a quick victory, however, they were met with stiff resistance from the outnumbered and outgunned German defenders. On several occasions, Allied forces would mount an attack only to find themselves on the end of limited but painful German counterattacks. Although the autumn campaign failed to produce a significant breakthrough for Eisenhower's forces, American units did manage to take the city of Aachen by the end of the year. As winter approached, the Allies believed that the German Army stood on the verge of collapse. While the Allied coalition prepared to make another push into Germany, the still highly motivated Wehrmacht launched an offensive of its own. Ironically, this ambitious offensive through the Ardennes, while daring, only worsened Germany's strategic situation, by depleting its already low reserves in manpower and equipment.

German Defense of the West Wall: Evidence of German Determination and Flexibility on the Battlefield

After the German retreat from France, Allied planners believed that they would defeat the Third Reich in short order. From the north, in Holland, to the south, along the borders of Lorraine, the Allies met cheering crowds of people happy to finally see an end to German occupation. Armed with optimistic estimates, the Allies believed that they would quickly pierce the Siegfried Line and continue their rapid pursuit of German forces. However:

> Little did they know, those confident planners housed in chateaux and elegant hotels, that it would cost the U.S. Army 140,000 casualties (79,000 battle and 51,000 non-battle)—more than the whole total of the Vietnam War—to fight the Battle of the West Wall over the coming three months. And even then they would not succeed. For in that coming December, a whole U.S. Army, 80,000 strong, would be reeling back from the Wall in defeat, shocked by the impact of the great German counter-offensive. Soon it would seem to the planners that the Great Wall of Germany, running mile after mile along the hills of that remote frontier, was impregnable.[92]

The American attack on the West Wall began on September 12 in the southern sector of 1st U.S. Army when the 28th Infantry Division crossed the Sur River in strength. The German defenders in the area were largely Volksgrenadiers, militia forces composed of men normally unfit for duty in the armed forces. For two days, the 28th Division took heavy casualties as it tried to breach the line of pillboxes. By September 17 the division had not penetrated the wall and called off the attack after losing fifteen hundred men, one-tenth of its strength before the assault.[93]

On the left flank of the 28th, the 4th U.S. Infantry Division made a drive for the town of Prum. The 2nd SS Panzer Division, the "Das Reich" Division that had participated in countless battles on the Eastern Front, stopped them for several days until the Americans withdrew. Under the weight of German counterattacks, elements of its lead battalion, the 22nd, shattered, suffering 55 percent casualties. The 4th would not capture its objective for another four months.[94]

To the north, in 1st U.S. Army's sector near the town of Aachen, American forces made better progress. General Collins, the commander of 1st Army's 7th Corps, ordered the 1st Infantry Division ("The Big Red One") and the 3rd Armored Division ("The Spearhead") to breach Aachen and push forward to the Rhine, some forty-miles west of the ancient German town. The West Wall along Aachen, what German propaganda minister Joseph Goebbels described as the "Holy City," contained two lines of fortifications about five miles apart. General Schwerin, in charge of the defense of Aachen, had few forces at his disposal. He sent the remnants of his 116th "Greyhound" Panzer Division to the north of Aachen, to prepare for a counterattack. He left behind a motley collection of Hitler Youth as well as "stomach and eye battalions," men suffering from stomach and eye ailments, to defend the city. Schwerin had little faith in the amount of resistance these "Christmas Tree Soldiers" could mount against the approaching American onslaught. They would prove him wrong.

On September 13, the 3rd Armored Division tried to breach the city from the south but failed to pass the first line of defenses. Events did not proceed smoothly for the Americans as they advanced toward Aachen with a force of twenty Shermans and supporting infantry. Armed with the deadly *Panzerfaust*, a one-shot antitank weapon developed on the Eastern Front, the German forces hid along the road to Aachen awaiting the approach of the American tanks. Springing their ambush, the German forces managed to disable first the lead tank and then those behind it. In a few minutes, four damaged Shermans blocked the road advancing in a single column. As the German trap unfolded, a brigade of motorized infantry attacked the Americans' right flank. Armed with their Ferdinand SP Guns, their armor-piercing rounds easily disabled six more tanks in the column of Shermans. The first effort at taking Aachen ended with an American retreat. The "Christmas Tree Soldiers" that General Schwerin had left in Aachen would hold together under the American assault for three days. At the same time, 145,000 of the original 165,000 inhabitants began to evacuate the city.[95]

American forces would make their way close to Aachen as a result of the efforts of the 1st Infantry Division. With persistence, the "Big Red One" managed to slip past the first line of fortifications. By the evening of September 14, they deployed around three different sides of the city from the north. Once more, the 3rd Armored Division attempted to break through to the city, but they still could not match the progress of their sister division as the tenacious defenders pushed them back yet another time. The next night, September 15, the 3rd managed to push past the defenders and sweep through both lines of the West Wall's fortifications. Following its original instructions, the 3rd Armored Division attempted to drive toward the Roer River, opening an avenue of approach to the Rhine.[96]

Concerned by the sudden American breakthrough, General Erich Brandenberger, the commander of the 7th German Army opposite Hodges's 1st U.S. Army, ordered his forces to push the American forces back at all costs. Closing the breach early on September 16, the severely depleted 9th Panzer Division counterattacked. Once more, the Germans sprung a trap on the advancing American Shermans. On the road to the Roer River, through abandoned villages, the 9th Panzer attacked the American tanks from two sides. After losing 40 percent of their tanks, the 3rd Armored Division withdrew. Although their casualties had been high, and even though the German counter had pushed them back, U.S. forces managed to break through the West Wall, threatening to take Aachen.

Once more, the Germans tried to keep the Americans off balance and responded quickly to the threat. With the knowledge that the balance of forces was against them, Brandenberger sent the newly arrived 12th Infantry Division into the fray. Late on the evening of September 16, elements of this division, built around experienced NCOs from other units and new recruits, attacked the dug-in positions of the 3rd Infantry Division. After an initial artillery barrage, the "Wild Buffalos," the nickname given to the 12th Infantry from its days on the Eastern Front, threw a force of 15,000 men against the Americans.

> In its fury it was almost Wagnerian: a great elemental frenzied counterpoint . . . hardly had the barrage died away, leaving the [Americans] ashen-faced and shocked, than

> the men of the 27th Fusilier Regiment of the 12th Infantry Division began to charge the American positions in disciplined waves, bayonets fixed. They might have been practicing on some peacetime training ground . . . . Far to the rear, in some cases as far as five miles back, the division artillery took up the challenge . . . . Great gaps appeared in the waves of running *Fusilliere.* Still they came on, goaded by some strange freaks of fancy as men are at such moments, or carried away by a wild unreasoning blood-lust . . . . Perhaps a dozen of them reached the American line . . . they came and came again.

All three of the division's regiments displayed a similar ferocity in their attacks. The counterattack lasted one day, decimating the "Wild Buffalos" and failing to push the 3rd Armored Division away from Aachen.[97]

Final Defense of Aachen: German Determination and Flexibility on the Battlefield

While Montgomery's effort to seize Antwerp, known as Market Garden,[98] fizzled in the north, the focus of Allied operations turned southward as 12th Army Group set out once again to capture Aachen. In early October the 1st and 30th Infantry Divisions consolidated their hold around the city. On October 8 American forces conducted a massive aerial and artillery bombardment of Aachen. The next day, the American commander demanded that German forces in the city surrender or face destruction. The 246th Volksgrenadier Regiment, about five thousand formerly wounded or ill soldiers, represented the bulk of the German defense in Aachen. Their commander, Lieutenant Colonel Maximilian Leyherr, rejected the ultimatum three days later.

Upon receiving the German reply, the American forces immediately began another bombardment of Aachen. From the air, the Allies dropped 62 tons of bombs on the beleaguered German defenders. At the same time, they threw another 169 tons of artillery rounds against the city. The next day, after this colossal display of firepower, the 1st Infantry Division began a slow advance against the German positions. Inside the city, American and German forces engaged in horrific house-to-house fighting.[99]

Against overwhelming odds, the 246th Regiment launched a counterattack, only to withdraw in the face of aerial attacks from American P-47s. Undaunted, the Germans mounted another attack on October 16, but U.S. forces rebuffed them. The two assaults cost the 246th Regiment two-thirds of its soldiers, forcing the Germans to withdraw and regroup. As the defenders retreated to consolidate their forces, American units encircled the remainder of the city. Facing

a terrible situation, the German commander in Aachen pleaded with Model to give him permission to leave the city. Model instead instructed him to "Fight to the last man! If necessary, have yourself buried in the ruins."[100] For two days, elements of the shattered 9th and 116th Panzer divisions tried to break the Allied encirclement. On October 21, with dwindling supplies, the German defenders capitulated. The Allies took twelve thousand prisoners, most of these the remaining citizens of Aachen. In the nearly one month of fighting, the 1st U.S. Army lost nearly ten thousand casualties in taking the ruined city.[101]

Offensive in the Ardennes: Evidence of German Staying Power, Battlefield Determination, and Flexibility

As weather conditions worsened in mid-December, grounding British and American aircraft in the theater, Hitler sprung his surprise offensive against the Allies. Since August, Hitler had directed his generals to plan for an ambitious attack that he hoped would cripple the Allied forces in the West and give him the breathing room he needed to throw additional forces against the Soviet steamroller on the Eastern Front. To provide the units necessary for the offensive, Germany's planners began to tap into all available manpower reserves. Hitler's staff not only trimmed existing forces but also conscripted youths, old men, and the unfit, most of whom had seen little to no military action, to form *Volksgrenadier* divisions. In an effort to compensate for their inexperience, the German military appointed veteran commanders and noncommissioned officers to lead these units of the "people's infantry." At the same time, the Germans cobbled together panzer divisions from the cadres of units that had survived their battles in France (see Map 3.2).[102]

The drive mirrored the German attack on France in that this operation also called for armored spearheads to cross the Meuse River quickly. This time, however, the German forces could not depend on air support and the same steady stream of supplies. For these reasons, von Rundstedt and Model voiced grave doubts about the offensive, recommending a more limited attack to dislodge the Allies from Aachen. They recognized that German forces lacked the same capabilities they had possessed in May 1940.

Still further, they pointed to the difficult terrain east of the Meuse that threatened to delay the German advance. Hitler found their attitude overly pessimistic and ignored their advice. Disagreeing with his commanders overseeing the defense in the West, Hitler gambled that surprise and speed would compensate for the material weakness of his forces.[103] Moreover, the troops, as Rundstedt later

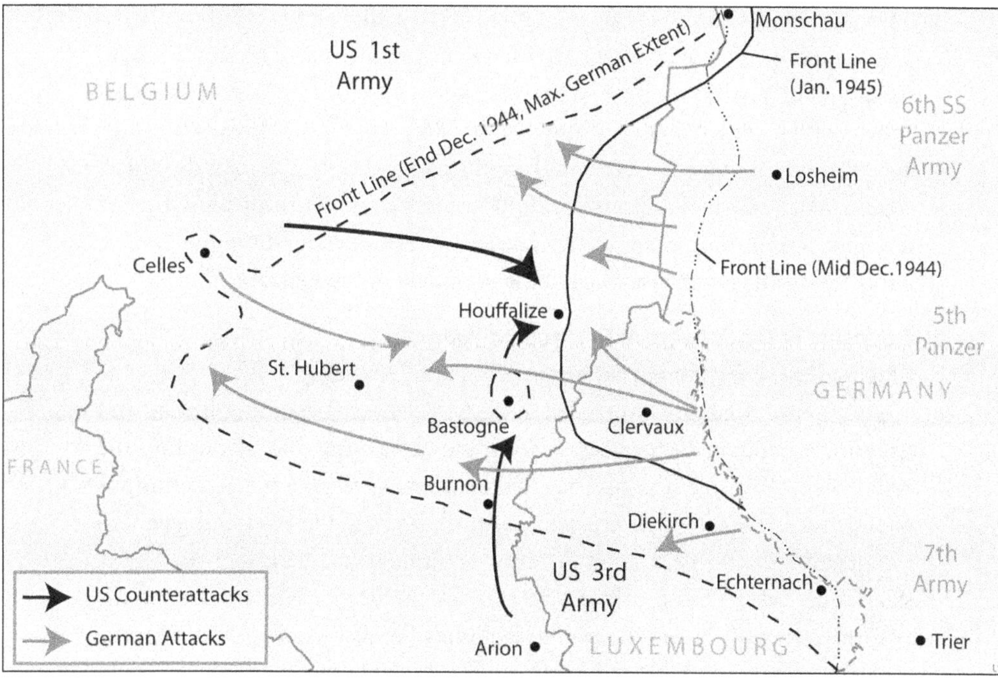

MAP 3.2. German Counteroffensive in the West, 1944–45.

noted, maintained a high degree of faith in Hitler: "The morale of the troops taking part was astonishingly high at the start of the offensive. They really believed victory was possible—unlike higher commanders who knew the facts."[104]

Along a sixty-mile front in the Ardennes, the Germans arrayed thirteen infantry and seven panzer divisions. The assault force consisted of nearly 1,000 tanks and 2,000 guns, assets equaling those deployed on the Eastern Front.[105] Some 450 additional tanks in five more divisions composed the bulk of the reserves allocated for the operation. In two columns arching northward, these forces sought to sweep through southern Belgium and Luxembourg en route to capturing the critical port of Antwerp. Sepp Dietrich's 6th SS Panzer Army represented the *Schwerpunkt*, or the main thrust, of the attack. According to Hitler's plans, the 6th would cross the Meuse west of the city of Huy. To the south, Manteuffel's 5th Panzer Army was to follow in a parallel course. German forces would conduct other attacks to protect the flanks of this main effort. On Dietrich's right, Zagen's 15th Army would strike at Aachen to the south and then push its way through the hole in Allied forces opened by the 6th SS Panzer Army. Brandenberger's 7th Army would drive further south to protect Manteuffel's left flank.

Allied commanders had not expected an attack through the Ardennes. Dubbed the "ghost sector," this area had remained quiet enough through the fall that American forces used the area as a staging ground for new forces and units recovering from previous engagements. Bradley saw an assault in the area as only a remote possibility. Holding a stronger opinion, Montgomery believed the Germans incapable of staging any major offensives.[106] As a result, two fresh (the 99th and 106th) and two resting (the 4th and 28th) American divisions guarded the area.

The attack commenced on December 13, 1944, with artillery barrages in the Ardennes. General Hodges, whose 1st Army controlled the sector where the German forces planned to advance, thought the activity amounted to nothing more than a spoiling attack to disrupt Allied operations already underway to the north and south of the Ardennes. Thus, when their infantry attacked at 0530 hours, the Germans achieved complete surprise. Although word reached the Allied Headquarters slowly, Eisenhower quickly ordered the 7th Armored Division from Simpson's 9th Army and the 10th Armored Division from Patton's 3rd Army to reinforce the area.[107] The extreme north and south of 1st Army's area of operations mounted a vigorous defense, holding off the German advance. Hanging on by a thread in the north, Hodges's 99th Division managed to hold off Dietrich's attack. The 1st, 2nd, and eventually, the 9th arrived to hold the 6th SS Panzer east of Malmedy.[108] Further south, the 4th Armored Division contained the forward elements of Brandenberger's *Volksgrenadiers* from the 7th Army.

However, the center of 1st Army's sector fared much worse. Manteuffel's 5th Panzer Army made quick work of the American 28th and 106th divisions and drove toward the cities of St. Vith and Bastogne, a critical juncture of the roads leading to the Meuse. While two regiments of the 106th surrendered quickly to the German forces, the remaining elements of the division put up a stout defense at St. Vith. The newly arrived 7th Armored Division also bolstered the town's defenders, holding out against the German advance until December 22 before withdrawing across the River Salm. At Bastogne, the 101st Airborne and portions of the 10th Armored Division arrived as Manteuffel's forces surrounded the city. When asked to surrender, the American commander refused with his now famous reply of "Nuts!" The siege at Bastogne would tie down nine critically needed German divisions.[109]

As events quickly unfolded, Eisenhower ordered Patton and Montgomery to prepare themselves for a counterattack to relieve the mounting pressure on

1st Army. Hodges regrouped his remaining forces at the Meuse as ULTRA intelligence now alerted the Allies to the German plan of advance.[110] To the north, Dietrich's thrust westward had stalled badly. On December 22, to facilitate a breakthrough in his sector, Hitler committed most of his remaining reserves to the 6th Panzer Army. Since the greatest German gains occurred in the south with the 5th Panzer Army, this was a critical error.[111]

Weather conditions improved enough on December 22 for the Allies once again to take advantage of their control of the skies. Allied air forces flew nearly thirteen hundred sorties the next day. On Christmas Eve, two thousand Allied aircraft struck thirty-one separate targets. These air strikes badly damaged the already weakened German supply network. More important, they hampered the mobility of the panzer units, which also had to cope with the poor roads and the stiffening Allied resistance.

Smelling defeat in the air, Rundstedt requested a withdrawal, but ever confident in his operational genius, Hitler refused. By December 24, Manteuffel's drive to the Meuse began to stall near the town of Foy-Notre Dame, as he desperately required reinforcements. To the rear, the siege at Bastogne remained a thorn in 5th Panzer's westward thrust to the Meuse. Unfortunately for Manteuffel, most of those reserves had already gone to aid Dietrich's forces. In a final bid to achieve air superiority and alleviate the pressure on their ground units, the Luftwaffe threw everything it had against enemy airfields on New Year's Day. They destroyed 156 Allied aircraft but lost 300 of their own planes in the process. German air forces never recovered from this attack.

Unable to move any farther, Manteuffel found himself in the middle of a pincer movement executed by Hodges from the north and Patton from the south. On January 3, American forces attacked toward the town of Houffalize in 5th Panzer's rear. The same poor road conditions and weather that had hampered the Germans delayed the Americans from cutting of Manteuffel's forces until January 16. Now accustomed to the terrible conditions, the German forces took advantage of the slow American move toward Houffalize to execute a speedy withdrawal.[112]

The Regime Keeps the Armed Forces Fighting: Evidence of Staying Power

Although most of their forces escaped the Allied counterattacks in the Ardennes, the desperate German offensive had cost the Wehrmacht 100,000 soldiers out of the 500,000 it had initially sent to cross the Meuse. They also wasted nearly 800 of their tanks and 1,000 planes committed to the operation.[113] The Allies suffered similar losses (81,000 casualties), but they could draw upon a bountiful supply of resources, while the Germans could not. Since their defeat in France, Germany had lost close to three quarters of a million men.[114] To make matters worse, the Allied bomber offensive had "nearly shut down the entire German transportation system. The Germans could no longer move to combat units the meager weapons and ammunition Speer's remaining factories were grinding out under the terrible conditions of winter 1944–45. In effect, the offensive in the Ardennes used up Germany's strategic reserve not only in the west but throughout the Reich as well."[115]

While the battlefield conditions for the German Army worsened, Hitler also began to lose confidence in the loyalty and determination of his military, especially its officer corps. Still reeling from the almost successful attempt on his life carried out by a conspiracy of high-ranking officers of the Wehrmacht, Hitler and his political allies started to exercise greater control of the armed forces. The Fuhrer worried not only that the military might refuse to carry out his orders faithfully, but also that they might try again to seize the reins of government.[116] Referring to the German officer corps, and the General Staff in particular, Hitler raged in 1945: "It is intolerable to me that a group of intellectuals should presume to press their views on their superiors. But such is the General Staff system, and that system, I intend to smash."[117]

To both assert his control over and to motivate the armed forces in their increasingly hopeless defense of the German homeland, Hitler relied on several measures. First and foremost, he exerted greater political pressure on the rank and file of the military. As Karl Bracher explains:

> More and more, the Army was held in distrust and subjugated to the ideas and the encroachment of the SS, and thus, Himmler became the most powerful figure in the Army as well. This development was responsible for the "National Socialist guidance officers" in the Army—informers and propagandists modeled on the commissars of the Red Army. Ideological-political reliability and fanaticism became the highest ideals, transcending all considerations of military expediency; costly, sacrificial defensive battles and hopeless offensives led deeper and deeper into chaotic self-destruction; this, for Hitler, was the only possible alternative to "final victory."[118]

Second, with events rapidly turning against the German Army, Hitler began to impose ruthless punishments on his soldiers. His flying courts-martial of SS officers shot men who did not fight or failed to blow up bridges. These traveling courts-martial enforced his proclamation of January 1945, which stated: "In the future, I shall impose draconian punishments on any attempt at concealment, whether deliberate or arising from carelessness or oversight."[119] Under this order, and ones similar to it, summary justice, kinship arrests, and executions became more common.[120] In the presence of these roving executioners, soldiers had greater incentive to remain with their units and to keep fighting. By some estimates, fifteen thousand soldiers were executed for desertion.[121]

Third, Goebbels's Propaganda Ministry flooded the front with leaflets both warning and imploring soldiers to continue the fight. One of these warned:

> All Germany's agricultural wealth will be placed at Moscow's disposal and planned famine will be used as a means of repression. German labor will be used as war reparation and thousands of slave laborers will be torn away from their families and deported to Siberia. German women will be carried away by human beasts, raped and assassinated. German children will be taken from their parents, deported and brought up as Bolsheviks. The German people, as such, will be obliterated and the survivors will have nothing to live for. The need and sufferings of the moment are nothing compared to the extermination planned for us by our enemies. All Germany rises against this fate and will fight as one great National Socialist army.[122]

Some German soldiers bought this line of reasoning and continued to fight. In the words of one historian, many *Landsers* carried on the struggle more out of their ideological convictions than out of fear of punishment:

> Even in the West, against enemies who threatened their prisoners with Canada and Arizona rather than Siberia, the Germans fought with dogged inspiration. It was no accident that the unit that gave the Allies the greatest difficulty in the Normandy campaign was the 12th SS *Panzerdivision* "Hitler Jugend," formed from young enthusiasts under the leadership of Eastern Front veterans. So powerful was the force of belief that wounded "Hitler Jugend" prisoners refused medical care, preferring to "die for the Fuhrer" until quelled by the threat of transfusions with Jewish blood.[123]

Some might suggest that these brutal disciplinary practices might have undermined the motivation of the German Army. Such a view, however, would ignore the numerous diary entries of those soldiers writing at the time who expressed their ideological commitment to the Nazi cause. As Bartov observes, "As the war became ever more painful and disillusioning, these documents reflect a powerful conviction of the necessity to go on fighting for a good cause against a demonic enemy; hence the perceived lack of alternative is in no way merely the function of a ruthless penal system."[124] While the armed forces definitely pursued more draconian policies, resulting in a significantly higher rate of executions than in World War I,[125] these harsh disciplinary measures had more to do with Nazi ideology than desperate measures to keep the armed forces fighting. This severe military discipline reflected the kind of punishment deemed necessary by Nazi society when dealing with its numerous domestic and international adversaries.[126] Under these circumstances, crimes by soldiers became political because they threatened the very survival of the Nazi state.[127] For many soldiers, the desperate nature of the German cause justified this severe system of military justice.[128]

Overview: The German Western Campaign in 1945

Germany confronted a terrible strategic situation in February 1945. Not only had the Soviet Union seized Warsaw in January, but its armies also stood poised to push deep into the German homeland. To counter this looming threat, Hitler shifted sixteen divisions and large quantities of artillery from his already depleted forces in the West to the Eastern Front. Still further, the German military also sent most of the new equipment coming out of its crumbling armaments industry to the East. Out of the 1,781 newly manufactured armored vehicles available in February, only 67 went to the beleaguered defenders protecting the western approaches to the Rhine.

The Allied armies in the West quickly capitalized on the overstretched German defenses. In the beginning of February, Eisenhower directed his forces to attack the western side of the Rhine in multiple places. Because they had little faith in their ability to defend the Siegfried Line, Hitler's generals begged permission to withdraw their forces to the eastern side of the river. The German dictator, stubborn to the end and refusing to bend to military realities, demanded that the Wehrmacht defend every foot of the homeland. Although the Allies encountered stiff German resistance, especially along the northern approaches to the Rhine, they succeeded in making significant gains. By March 3, fifteen German divisions west of the Rhine, whose leaders Hitler had denied permission to retreat, stood on the verge of encirclement.

As the Allies tried to cross the Rhine at Remagen, the Germans tried desperately to destroy the bridge. The frantic Germans fired several V-2 rockets, sent frogmen to set explosives, and even tried to use their largest artillery piece, the 130-ton "Karl Moser," to send 4,400-pound shells onto the Ludendorff Bridge. Each of these desperate measures to bring the bridge down failed, permitting Bradley to slip five divisions across the Rhine. On March 17 the Ludendorff Bridge finally fell in a deafening groan of crashing metal. Fortunately for the

Allies, they had constructed several pontoon bridges by this time and were well on their way to establishing a strong twenty-five-mile-wide and ten-mile-deep salient across the Rhine. In response to the Allied victory at Remagen, Hitler replaced von Rundstedt as commander of German forces in the West for the final time with Field Marshal Albert Kesserling, who when he arrived at the front informed his staff, "Gentlemen, I am in the new V-3!"[129]

Farther south, the Allied 6th Army Group conducted Operation UNDERTONE, an advance to the Rhine in the Saar region. Allied forces overran the West Wall on March 20, mauling the German 7th and 1st armies. When news of the events reached Hitler he immediately ordered a counterattack, but his officers informed him that the "cupboard was bare."[130] The Allies incurred 17,220 casualties while inflicting an estimated 113,000 on the opposing German forces they had encircled.[131] As they had done at the Falaise Pocket in August 1944, however, the resourceful Germans still managed to slip a few units out of the pincer and across the Rhine.

Allied ground forces now stood poised to cross en masse over the Rhine into the German heartland. Eisenhower's impressive armies along the German border consisted of eighty-five divisions containing two million soldiers.[132] Germany, in contrast, faced a dreadful situation:

> The German forces that defended the Reich were a shell of their former selves. Although . . . factories still produced equipment in quantity, German manpower reserves were dwindling. The Wehrmacht could neither reconstitute the mobile reserves wasted in the ill-fated Ardennes counter-offensive nor adequately sustain existing forces with replacements and fuel. What reserves there were went east to fight the advancing Red Army . . . . Nevertheless, German soldiers continued to fight, held together by ideology and ruthless discipline, and driven by the fact that they were defending their own soil.[133]

In total, the Germans had lost 325,000 men in their defense west of the Rhine.[134] With fifty-five weak divisions in the West, the end was near for the Third Reich.[135]

On March 21, Allied air forces bombed the approaches to the Rhine. Simpson's 9th U.S. Army stormed through Wesel on March 27, executing a southeastward thrust behind Model's Army Group B. Hodge's 1st Army reached Marburg and started to turn north behind Model. The two American armies met at Lippstadt on April 1, to completely ensnare the German defenders. True to form, Hitler declared the entire Ruhr pocket a fortress and forbade Model from attempting any breakout. The damaged industrial center of Germany could sustain the German forces for only fifteen days. Afterward, most of the 317,000 German forces surrendered. Fearing the Americans would turn him over to the Soviets, Model wandered into the countryside and shot himself.[136]

As the British and Canadian forces made their way to the Elbe against continuing pockets of German resistance, the American armies pushed along a north and northeasterly route across central Germany, bypassing the entire southern area of the Rhine and the Czechoslovak border. Eisenhower now faced a choice between seizing Berlin and driving toward the Bavarian Alps to prevent the Germans from establishing a national redoubt for a last stand. The Supreme Allied Commander worried that such a possibility might tie down American forces needed in the Pacific.[137] Thus Allied forces proceeded to the Elbe on a broad front with special attention to preventing the Germans from erecting a defensive enclave in the south.

With the destruction of Army Group B, Kesserling had a 125-mile gap in his center and, thus, could mount only a weak defense against the massive Allied drive eastward. Although resistance was sporadic, the Germans still had the ability to inflict casualties on their opponents. In early April, at Paderborn, a major tank training center for the Wehrmacht, Hodges's 3rd Armored Division encountered a force of students and instructors that put up a stiff fight. The American unit lost several of its tanks in a fierce engagement before bringing superior firepower to bear.[138]

The scene was repeated on April 11, as American forces met a unit of young paratroopers of the 3rd Parachute Division near the suburbs of Cologne. Before the Germans were subdued they shot up thirty of the 13th Armored Division's Sherman tanks with their 88mm guns.[139] Although the war appeared all but lost, many German soldiers still fought with determination. As MacGregor Knox notes, "Nor did the end come easily in 1945; German forces were so reluctant to give in—even to the Americans—that they killed 9,373 U.S. ground troops and aircrew in April 1945, only slightly below the monthly average for the entire U.S. advance from Normandy to the Elbe."[140] By late April the Allies had reached the Elbe, all but ending German opposition. With Soviet armies laying siege to Berlin, Hitler committed suicide on April 30. Their leader gone, German forces finally laid down their arms on May 5, 1945.

Evaluating the German Army, 1944–45

In the face of the onslaught from the combined democratic armies of Britain, France, and the United States, a messianic German military fought with a high degree of cohesion (see Tables 3.1 and 3.2). Although they lacked any appreciable close air support, they could not match the Western Allies in terms of material resources, and their manpower reserves suffered from devastating defeats in France and in the Soviet Union, the Wehrmacht demonstrated strong staying power. They accomplished this feat while at the same time sending the bulk of their forces to engage in a losing struggle on the Eastern Front. Table 3.3 summarizes some of the major battles of this period on the Western Front. It also provides figures for two battles on the Eastern Front. On most occasions, German forces per capita inflicted more casualties than the Allies, and they did so at a large rate. Only at the end did German superiority in casualty infliction falter.

Some might argue that Germany's determination stemmed from the fear of a Russian victory. This counterargument attempts to explain the Wehrmacht's cohesion by looking at its rather bleak strategic situation. Because it had conducted a murderous campaign in the Soviet Union, Germany could not find a political settlement in the war against an Eastern invader bent on revenge. Germans, therefore, had no choice but to fight to the death. A similar version of this argument might claim that the Allied demand for unconditional surrender forced Germany to fight to the bitter end.

By itself, fear of Russia does not explain the Wehrmacht's cohesion. At best, this is a partial account. It seems that if Germany worried solely about losing to the Soviet Union, they could have made the Western Front campaign much

TABLE 3.1: Assessing Staying Power:
Messianic German Armed Forces, 1944–45

Evidence
Most top military leaders obey regime, even when they disagree with Hitler's strategy
Regime uses coercion and hard-core followers to keep military fighting
German military mounts large-scale counterattacks against Allies
Throughout period military inflicts greater ratio of casualties on opponents
German armed forces fight until overwhelmed by material superiority of Allies

PREDICTION: Strong Staying Power.
Military Collapses Only When Adversary Possesses Crushing Material Superiority.

TABLE 3.2: Assessing Battlefield Performance: Messianic German Armed Forces, 1944–45

Battles	Evidence
Retreat from France	Most units retreat in good order after initial panic Most units capable of defending and attacking New units cobbled together from shattered units
West Wall	Weaker German defenders defend and attack Units surrender only after vigorous defense
Ardennes Offensive	German units willing to suffer large casualties in attack German units capable of fighting with materially superior opponent
Defense of the Rhine	Several units fight even when cut off from command and control New units cobbled together from shattered units

PREDICTION: Strong Battlefield Performance—Most Units Fight with Determination and Flexibility.

easier for Allied forces. According to this argument, Germany's leaders and soldiers should have thought early in 1944 that it was better to lose to the Western Allies than to a Soviet Union eager to settle the score. The historical reality, however, is that they demonstrated an almost similar level of determination in the West as they did against the Soviets. Germany's leaders and its soldiers were motivated by an ideology that made compromise impossible. True, Germany sent more forces to the Eastern Front, but Soviet forces were larger. In the end, the German leadership chose to defend both fronts because it could rely on the staying power of its armed forces. Unlike the Kaiser's armed forces, Hitler could force his military to fight until the Allies destroyed it (see Tables 3.3 and 3.4.).

A PROFESSIONAL GERMAN MILITARY DURING THE LAST TWO YEARS OF WORLD WAR I

This section compares the performance of the German Army from 1917 to 1918 with the Wehrmacht of World War II. I argue that the combination of a high degree of organizational autonomy but low degree of regime control produced a *professional military* in Wilhelmine Germany. The country, led by Kaiser Wilhelm II, lacked a regime with control over the population. Germany's government could not demand and enforce unconditional loyalty from its citizens in the same way that Hitler's regime did decades later. The cohesion of the Reichsheer reflected strong internal bonds creating a sense of loyalty and trust among officers and enlisted men. The German military still struggled with the same class tension found in larger society, but managed to focus on training in the years prior to the war. Emphasizing training created units capable of fight-

TABLE 3.3: German and Allied Battle Casualties, September 1944 to May 1945

Battles	German Forces	Allied Forces	Germany Casualties	Allied Casualties	German Per Cap Loss Infliction	Allied Per Cap Loss Infliction[a]	German Loss-Infliction Ratio	Allied Loss-Infliction Ratio
Market Garden[b] Sept. 17–26	15,000	72, 115	10,000	17,200	1.15	0.14	8.29	0.12
Aachen[c] Oct. 8–21	12,000	100,000	10,600	5,000	0.42	0.11	3.78	0.26
Hürtgen Forest[c] Sept. 14–Dec. 1	80,000	120,000	28,000	31,000	0.39	0.23	1.67	0.60
Siegfried Line[d] Sept.–Dec.	350,000	500,000	95,000	140,000	0.40	0.19	2.11	0.48
Ardennes[e] Dec. 16–Jan. 28	330,160	489,219	74,459	63,901	0.19	0.15	1.24	0.80
Vistula-Oder[f] Jan. 12–Feb. 2	400,000	2,203, 600	300,000	194,191	0.49	0.14	3.5	0.27
Rhineland[g] Feb. 8–Mar. 24	600,000	4,000,000	310,000	28,310	0.05	0.08	0.66	1.55
Fall of Berlin[h] Apr. 16–May 2	1,117,000	2,062,100	937,378	361,267	0.32	0.45	0.71	1.42

[a]"Per Cap Loss Infliction" measures how many casualties, controlling for force size, one side can impose on an opponent. Attacker Per Cap Loss Infliction, then, is defender casualties/attacker force size. "Loss-Infliction Ratios" compare the per capita magnitude of casualties one side inflicted on the other. Attacker's Loss Infliction Ratio, then, is attacker per capita loss infliction/defender per capita loss infliction. Accurate data for force sizes remain unavailable for some battles. As a result, I estimate German and Allied forces by type of unit historians have identified as common for the period. See John Ellis, *World War II: A Statistical Survey* (New York: Facts on File, 1995). On German forces on the Western Front, see Charles B. MacDonald, *The Siegfried Line Campaign* (Washington, DC: Army Center for Military History, 1990), 14–19; and Niklas Zetterling, *Normandy 1944: German Military Organization, Combat Power and Organizational Effectiveness* (Winnipeg: J. J. Fedorowicz Publishing, 2000).

[b]For Market Garden, see Michael Clodfelter, *Warfare and Small Armed Conflicts: A Statistical Reference, Vol. II: 1900–1991* (Jefferson, NC: McFarland and Company, 1992), 527–28.

[c]For Aachen and the Hürtgen Forest, see Robert Sterling Rush, *Hell in the Hürtgen Forest: The Ordeal and Triumph of an American Infantry Regiment* (Lawrence: University of Kansas Press, 2001); and Clodfelter, *Warfare and Small Armed Conflicts*, 528–32.

[d]Charles Whiting, *Siegfried: The Nazis' Last Stand* (New York: Jove Books, 1983); and Clodfelter, *Warfare and Small Armed Conflicts*, 528.

[e]Trevor Dupuy, David Bongard, and Richard Anderson, *Hitler's Last Gamble: The Battle of the Bulge, December 1944–January 1945* (New York: Harper Collins, 1994), 464–71.

[f]David Glantz and Jonathan House, *When Titans Clash: How the Red Army Stopped Hitler* (Lawrence: University of Kansas Press, 1995), 298–300 and 246, fn. 23; Earl F. Ziemke, *Stalingrad to Berlin: German Defeat in the East* (New York: Dorset Press, 1986), 417.

[g]Clodfelter, *Warfare and Small Armed Conflicts*, 530–31.

[h]Max Hasting, *Armageddon: The Battle for Germany, 1944–1945* (New York: Knopf, 2004), 467–76; and Clodfelter, *Warfare and Small Armed Conflicts*, 515.

TABLE 3.4: U.S. Monthly Battle Deaths, June 1944–April 1945

Month	U.S.	Month	U.S.
June	17,784	December	19,548
July	19,711	January	16,631
August	13,091	February	16,467
September	14,197	March	20,325
October	14,539	April	18,751
November	16,876	Total	187,920

SOURCE: Michael Clodfelter, *Warfare and Small Armed Conflicts: A Statistical Reference, Vol. II: 1900–1991* (Jefferson, NC: McFarland and Company, 1992), 384.

ing with determination and discipline on the battlefield. Nevertheless, without a high degree of regime control, the Kaiser's armed forces could not continue the struggle when victory no longer appeared likely.

In the first three years of the war, the Reichsheer displayed a high degree of cohesion, especially in containing the multiple Allied efforts at a breakthrough in 1917. All of this changed, however, one year later. After the failure of Ludendorff's Spring Offensives in 1918, the German armed forces steadily lost staying power. Old domestic disputes set aside in 1914 began to resurface, creating disorder within the armed forces. Throughout 1918, especially during the summer, the German military suffered some critical defeats. With a successful military resolution to war increasingly unlikely given battlefield conditions, Germany's political leaders felt compelled to sue for peace in November 1918. Later the military would claim that the public "stabbed the military in the back." The truth was that Germany's armed forces failed to win the war in the spring of 1918 and then could not defeat the Allied coalition, especially with the arrival of American forces in Europe. The Kaiser could not force the armed forces to continue, as Hitler did in 1945.

A Professional Germany Military

The Second German Reich emerged from a series of wars from 1864 to 1871, when Prussia organized several German principalities, outside of Austria, into a single powerful state. In these successive conflicts, Prussia, under both the adept political leadership of its chancellor, Otto von Bismarck, and the military skill of General Helmuth von Moltke the Elder and his talented General Staff, defeated first the tiny Kingdom of Denmark, then the Austro-Hungarian Empire, and lastly France, which many at the time regarded as the most powerful country on the continent.[150] This last victory gave Prussia—once a minor power—the momentum it needed to secure its control over Germany. Fresh from his victory over the French and surrounded by the splendor of Versailles's Hall of Mirrors, King Wilhelm I of Prussia proclaimed himself the Emperor of Germany in 1871.

In seven short years, the emergence of Germany had altered the European balance of power. A disparate collection of small states in central Europe had become a dynamic empire with a growing population, a formidable military, and a vibrant economy. For the remainder of the nineteenth century, Bismarck worked adroitly to keep the peace among Europe's great powers. The German chancellor, however, had to work much harder at keeping the peace at home.[151]

With a burgeoning economy, and an expanding middle class demanding more political rights, the German government grappled with how best to accommodate the country's various interest groups. In addition to a Kulturkampf that pitted Protestants in northern Germany against Catholics from the south, Bismarck also had to contend with an increasingly strong Left-wing political opposition in the form of the Social Democratic Party. The changing landscape of the German electorate put Bismarck and subsequent chancellors in the position of maintaining a delicate balance of power among various economic classes. Germany's political leaders ultimately relied on a domestic coalition of the old aristocracy, heavy industry, and the military to counter the rising political clout of the working and middle classes.

Low Degree of Regime Control. Prior to World War I, then, Germany suffered from two broad societal cleavages that contributed to the regime's low degree of regime control over the population. Some of these divisions reflected regional variations in terms of culture and religion.[152] By far the most important split in German society, however, took place along class lines.[153] As the German economy rapidly modernized, the country's demographics also changed quickly, with the population increasing by 60 percent from 1871 to 1910. In a few decades, Germany went from a predominantly rural economy to an industrial power with the capacity to rival the British and American economies. Half of the population, for example, relied on farming for employment in 1875, but by 1910 only one-third of Germans worked in rural areas.[154]

This dynamic economy gave birth to an expanding working class, as well as a sizable middle class, whose labor fueled the country's industrial surge. As these groups increased in size, they demanded a larger share of German political power. To this end, Germany's workers relied on labor unions and the Social

Democratic Party (SDP) as their primary political outlet. In the Reichstag, the German parliament, the SDP's growing presence threatened to unseat the traditional ruling class, who were a predominantly rural-based aristocracy with strong connections to the military.

At the time, Germany's Social Democratic Party was the largest and the best organized socialist party in the world. With an ever-widening base of supporters, they advocated a truly revolutionary political agenda. The SDP, for example, called for the overthrow of the country's rigid social and political institutions; they demanded reforms to the economy to soften the hardships of unbridled capitalism on the German worker; and they sought wide-ranging democratic changes to the semiauthoritarian government. Because the SDP continually garnered enough votes to establish itself as a significant presence in the Reichstag, the government and its conservative allies needed a strategy to deal with these political upstarts. First, the conservatives sought to quell the Social Democratic Party by implementing a series of reforms, including improvements in education and unemployment insurance. Later they took even stronger measures, such as temporarily outlawing the party from 1878 to 1890, and even going as far as plotting a counter-revolutionary coup with the military should the domestic situation worsen. Despite these efforts the Social Democratic Party could claim 1 million members, and 2 million of Germany's workers belonged to labor unions.[155]

Eventually, conservative politicians and their partners in the German government turned to nationalism as a means both to divert attention away from domestic issues and to steal support from the SDP. An alliance of conservative interest groups promoted an increasingly nationalist foreign policy that they hoped would co-opt the middle-class supporters of the more Left-wing political parties. By arguing that Germany needed its own "place under the sun," deserving recognition as a great power, these conservative elites sought to direct the energies of the country away from domestic concerns.[156] Rather than focus on internal problems, the German government, aided by its allies in heavy industry and the old aristocracy, advanced a bold foreign policy that would sustain public support for the established political order. The entire nation's attention, as the thinking went, would focus on external enemies rather than the petty squabbles among different interest groups.[157]

These groups did not anticipate, however, that the Left might use nationalist arguments to its advantage as well. Instead of ending these ongoing class struggles, the other less conservative parties appropriated the nationalist ideology advanced by the government and used it to win more supporters. These political parties would call for additional military spending and would criticize the government when it deviated from the hard-line foreign policy it originally had advocated. In particular, "navy leagues," or groups of individuals in favor of a German maritime presence that would rival Britain's, sprouted around the country demanding a navy larger than the one supported by the government.[158]

As a result, the nationalism promoted by the government did very little to unify the country, but rather became another tool with which various political parties could vie for supporters. Put differently, because groups used nationalism to compete with one another, it lacked the same kind of all-encompassing nature that Nazism would later use to unify the country. Instead of a centralized government, promoting an ideological message the public generally accepted, nationalism before World War I served as yet another forum where various politics parties could carry on their disagreements. Unlike Hitler's ideology, German nationalism in this era attempted to paper over the differences among the country's interest groups. Because it did not claim to resolve Germany's class cleavages, it was not a catchall ideology that motivated the public to see its individual interests in term of the group's interests. German nationalism, thus, could not count on the same widespread institutional support that characterized Nazism of the 1930s.[159]

These domestic differences, however, did not prevent Germany from going to war with a sense of unity. On the eve of the World War I, the country's various interest groups agreed to set aside their domestic disputes to deal with the grave international crisis they now confronted. In the summer of 1914, the leaders of the Social Democratic Party called for societal unity in the face of French and Russian aggression. As one leading socialist wrote: "We do not want our wives and children sacrificed to the bestialities of the Cossacks."[160] Historians have referred to this decision to enter the war unified as the "Spirit of 1914." While Germany's left-right factions did not believe the war would resolve their disagreements, they did believe it was necessary that they work together to defend the Fatherland.[161] A temporary respite in the struggle among the classes ensued. For these reasons, Germany entered the war united, but without a regime capable of exerting a high degree of control over society.

High Degree of Military Autonomy. The same disputes that tore apart German society also affected the military. In terms of the class conflict that dominated the politics of the era, the German officer corps considered itself and the armed forces a conservative bulwark against radical domestic change. Many on the Left shared this view and considered the German Army yet one more institution that the old guard refused to open to other classes.[162] Military necessity prior to World War I, however, changed this view as the officer corps opened its ranks to admit experts from all classes to serve in the more technical branches such as artillery and logistics. This burst of practical egalitarianism, for instance, brought the future General Ludendorff into the officer corps.[163]

In spite of these tensions, the German armed forces obtained a high degree of organizational autonomy for training.[164] Even though it was regarded as a conservative institution, the German Army remained popular, especially in the years just before World War I. Both the Left and the Right could agree that the entire society benefited from a strong military capable of protecting the country's interests. Because none of Germany's political parties wanted its constituents to see it as unpatriotic, they were mostly unified in their support for the military.[165] Moreover, the German Army largely stood apart from the rest of society, almost as if it existed as a state within the state.[166]

With a solid basis of support at home, the conditions existed for the Reichsheer to create strong internal bonds through training. As continental war seemed increasingly likely, the government did its best to prepare the armed forces, supplying it with more men and materiel.[167] Still further, the training of the German Army promoted norms of unconditional loyalty and trust. In the last decade before World War I, officers increasingly received promotions based on merit. The old bias that favored officers from rural areas also began to disappear.[168] Officers believed that the changing nature of warfare made it necessary to promote trust among officers and their men. At the same time, training manuals emphasized the desirability of treating subordinates with respect.[169] More important, the army put all of its recruits through a training regimen whose rigor and realism rivaled that of any military in Europe.[170]

An Overview of the 1917 Campaign: Evidence of Staying Power

With the failure of the Schlieffen Plan at the Battle of the Marne in 1914, the German Army quickly found itself in strategic stalemate against British and French forces on the Western Front.[171] The possibility of quick victory against France, as envisioned by the German General Staff under Helmuth von Moltke the Younger, faded into the background when the 1st and 2nd German armies had to withdraw from the approaches to Paris. In the subsequent last months of 1914, the German and Allied armies raced northward from the Marne, in what was later called the "race to the sea," as both sides tried to find an area where they could encircle the enemy.[172] The reality of a prolonged war set in after the unsuccessful German attempt to break through the Allied lines in Flanders at the First Battle of Ypres.

Throughout 1915, the German Army concentrated its offensives on the wide plains of the Eastern Front. In the West, they endured a succession of British and French offensives. One year later the Germans, now under the military leadership of Falkenhayn, turned westward again and mounted their own large-scale assault on the French fortress at Verdun. Falkenhayn attempted to bleed the French white by forcing them to commit their dwindling supplies of manpower in a costly defensive battle at Verdun. The Germans correctly estimated that France viewed Verdun as a symbolic piece of territory and would sacrifice large numbers of soldiers to defend it.[173] What they did not anticipate was losing 330,000 of their own men to inflict 378,777 French casualties.[174]

After the bloodletting of 1916, in the following year the German Army returned to the defense against the Western Allies. This enabled them to devote most of their forces to knocking Russia out of the war. The British and the French seized upon this opportunity to attempt another set of offensives. None of these attacks, however, punched the hole in the German lines that the Allies hoped would bring their enemy to its knees. The year 1917 was, nevertheless, difficult for the Germans. In the spring, the British assault at Arras accounted for 84,000 German casualties.[175] The German Army also successfully defended its positions along the Aisne, in the midst of the failed Nivelle Offensives, at the cost of 163,000 killed and wounded.[176] Several months later, the German defenders effectively countered a large British offensive attack at Ypres, which included a bitter retreat from the town of Passchendaele. These last actions inflicted 260,000 casualties on Germany's already weary armed forces.[177]

The Battle of Cambrai: Evidence of Determination and Flexibility on the Battlefield

The Battle of Cambrai took place on the heels of the mired British offensive at Ypres. It is a good battle to discuss in some detail because it demonstrates how the German Army still had enough resolve to overcome surprise by counterattacking, an undertaking that would become more difficult one year later. By November 1917, the Allies had struggled to achieve the decisive victory they thought was possible when the year began. For the British and the French, 1917 seemed on the verge of ending in disappointment. After the British Army's painful assault in the muddy fields of Flanders, Field Marshal Haig sought an area where he could mount a spoiling attack elsewhere along the German lines to regain the initiative and to restore national morale.[178]

Haig resorted to an earlier plan that his staff had developed in the summer of 1917 that called for a tank incursion between the Canal du Nord and the St. Quentin Canal. The original scheme, developed largely by Colonel John C. Fuller of the British Tank Corps, envisioned a limited raid in a relatively quiet sector that would demonstrate the power of armored forces. Searching for the elusive breakthrough of the German lines, Haig had bolder aims in mind. He adjusted the plan's objectives and prepared his forces for an armored thrust toward the Bourlon Woods near the city of Cambrai, a critical German logistics hub along the Hindenberg Line.

The new British strategy called for six assault divisions of 476 tanks to make up the first wave of the assault, with several additional Allied divisions held in reserve as an exploitation force. In mid-November, under the cover of darkness, the British moved their tanks into position. Two days before the attack, on November 18, German troops captured a prisoner who revealed the timing of the planned assault. German military leaders, however, discounted the report because the sector near Cambrai had seen little action and was relatively quiet.

Thus, when the British launched their assault on the morning of November 20, they took the German defenders by surprise. By striking at an undermanned part of the German front line, the British had achieved tactical superiority in terms of numbers, which allowed them to overwhelm the first German units they encountered. Haig's forces also changed their artillery tactics. In the past, long artillery bombardments had preceded the Allied attacks, but on this occasion rolling bombardments accompanied the British advance. This produced clouds of smoke that not only made the Germans think that a gas attack had occurred but that also blinded them in their trenches. They were stunned when the British tanks and the follow-on infantry quickly streamed into their rear positions. Making use of these advantages, the British captured forty-four hundred German prisoners by early morning.[179] Haig and the Allies believed they stood on the verge of the victory they had tried all year to achieve.

By the afternoon spirits ran high among the Allies, since the six British divisions had created a five-mile-long breach in the German defenses and pushed five miles deep to the tiny village of Flesquieres. It was there, however, that they encountered a section of German resistance which, unlike that of the previous engagements of the day, refused to collapse. Unable to overpower the

Germans at Flesquieres, the British found that this small pocket had slowed the momentum of their attack. Outnumbered, the Germans withdrew, much to the surprise of their opponents, on the evening of November 20. Their departure allowed the British to occupy the village the next day. Late on November 21, the British reached the outskirts of the Bourlon Woods, one of their primary objectives.

British forces realized they needed to exploit their breakthroughs before the Germans had an opportunity to react. Time was of the essence, as Ludendorff and the German General Staff had finally realized the extent and the nature of the Allied attack. Luckily, they had recently reinforced the Cambrai sector with units from the Eastern Front. Impressed by the British gains, they sent these newly arrived forces to quickly patch up the hole torn into their lines. These reinforcements restored the balance of forces. For the next seven days German and British forces settled into the familiar stalemate pattern of warfare, and fierce fighting ensued. Since their tank forces had exhausted themselves during the initial breakthrough, the British offensive had lost its momentum. In heavy fighting, the area near Bourlon Woods would switch back and forth between both sides.

On November 29, the Germans mounted a massive attack of their own. The British lines had become overextended, with a large northeastward salient providing an inviting target for a German assault. Relying on the infiltration tactics that would later become the signature of the Spring Offensives of 1918, the Germans attacked the exposed British flanks from the north and the south. Only the timely arrival of British forces prevented the Germans from recapturing all of the Allied gains. By the time the fighting stopped in December, both sides had suffered nearly forty thousand casualties.

The Campaign in 1918: A Professional German Army Loses Staying Power

At the beginning of 1918, the German military had started to buckle under the pressures of a prolonged war. The quick victory promised by the German General Staff in the summer of 1914 had not materialized. Nor had the French Army collapsed under the weight of Falkenhayn's offensives in 1916. Germany also began to suffer from its heavy losses in manpower, which reduced its edge. Along the Western Front, for example, the Germans maintained a slowly eroding edge in rifle strength of 1.56 million to 1.39 million men (see Figure 3.1).

In the meantime, the German economy, and therefore the public, began to

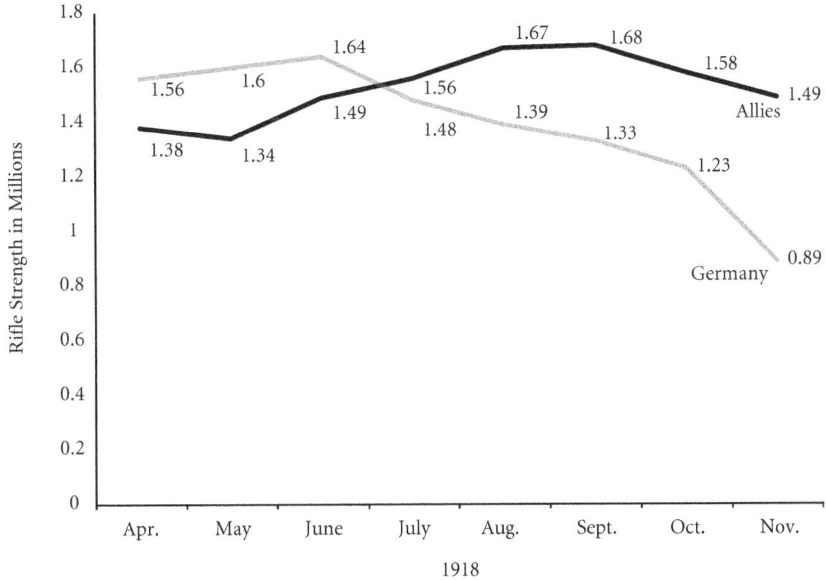

FIGURE 3.1. Allied and German Rifle Strength, 1918. Source: Arthur L. Frothingham, *Handbook of War: Facts and Peace Problems* (New York: Committee on Organized Education National Security League, 1919), ch. 9.

suffer from the shortages imposed on them by the British blockades. Starting in the winter of 1917, which was unusually cold, food and fuel shortages touched the lives of most Germans. The per capita caloric consumption of the German public plummeted from the summer of 1914 to 1918.[181] More important, wages started to stagnate and the gap separating the working and middle classes narrowed. All of these changes threatened to reopen the class tensions that German society had set aside to conduct the war as a unified nation.[182] The declining conditions at home translated into higher rates of crime as well as an increase in the number of strikes.[183] To make matters worse, the Socialist Democratic Party began to voice its opposition to the war by calling for more moderate war aims.[184]

With the defeat of Russia in 1917, however, it appeared that Germany might win the war after all. The Peace of Brest-Litovsk, concluded with Russia's new Bolshevik government, permitted the German military to devote more of its forces to the Western Front. A climate of optimism returned to the home front, as it seemed that Germany's hopes for victory had reemerged with the peace in the East.[185] The country's military leaders realized that they faced a rapidly closing window of opportunity to deliver a knockout to the British and French before the already arriving American forces established themselves in large numbers.[186] Germany could simply not match the ever-widening gap in Allied men and material. To this end, Ludendorff and his generals planned for an ambitious set of offensives in the West during the spring of 1918. They understood that these attacks represented the last role of the die.[187] Germany faced defeat unless their offensive succeeded before the American presence reached its full potential.

Unfortunately for Germany, Ludendorff's offensives in the spring of 1918 did not achieve the strategic breakthrough that he had planned. Although Germany's new infantry tactics had produced significant gains early, they could not exploit them. Already in May, he had the gnawing feeling that his gamble had failed.[188] In an attempt to secure the decisive victory they desperately needed, Ludendorff would mount several attacks in the West during the late spring and early summer of 1918. By July the German forces had overextended themselves and held exposed positions far from the relative safety offered by the well-fortified Hindenberg Line.

The Allies, now holding an increasing advantage in manpower, went on the offensive starting with the Second Battle of the Marne in mid-July 1918. Weeks later, at the Battle of Amiens, the German Army experienced a defeat that saw its units dissolve under the pressure of combat for the first time. Although the Germans would continue to resist until November, the decay in the army's staying power would increase. In the fall, Allied forces cleared the German salient at St. Mihiel in the south and penetrated the Hindenberg Line in the north with breakthroughs at St. Quentin as well as Cambrai. By October, starting with the Meuse-Argonne offensive, the Allies began a steady advance eastward and the Germans conducted a fairly well organized retreat back toward their own border.

The failure of the spring offensives had an adverse effect on the German armed forces. At home, the Social Democratic Party became the voice of a growing public sentiment against the war. In the summer and fall of 1918, the number of strikes increased and the momentum for domestic political change grew.[189] By mid-October, Germany's military leaders worried about the need to keep the army intact in order to prevent disorder at home.[190]

Although the military could mount an effective, and at times, spirited defense, the staying power of the armed forces began to deteriorate rapidly.[191] On the one hand, the German Army suffered 760,000 casualties after the spring offensives. On the other, this same military also saw 750,000 to 1,000,000 of its soldiers refuse combat in one form or another after the spring offensives.[192] This steady erosion in stamina reflected the fading probability of a German victory. For the first time in the war, the Allies began to inflict more casualties per month on the Germans than their opponents inflicted on them. There were also other indicators that the staying power of the German military had begun to decline. After each battle, the total number of German prisoners increased. The front-line troops knew that at home the public had begun to turn against the war. For this reason, they greeted the replacements with jeers of "Strike breakers!" and "You're prolonging the war!"[193] By the end of October 1918, Prince Rupprecht declared the army "unreliable."

The Battle of Amiens: Evidence of Further Decline in Staying Power

The Battle of Amiens was a harbinger of events to come for the German Army for the remainder of the late summer and fall of 1918. Throughout the year, German forces would still offer stiff resistance, but the number of deserters and prisoners would steadily increase. At the time, many of the German military and political leaders considered Amiens the turning point of the war. For the first time, German units simply surrendered under pressure. Both the strain of the war and the unsettling fact that victory appeared unlikely, with the failure of the Ludendorff Offensives, had taken a toll on the military's will to fight.[194]

Two days after the French eliminated the German salient near the Marne River, an action better known as the Second Battle of the Marne, the Allies mounted their second major offensive of the summer. The Allied commander-in-chief, the French Marshal Foch, wanted to push the Germans away from Amiens, where their artillery could threaten the critical railway links to Paris. General Rawlinson's 4th British Army, consisting of seventeen divisions and four hundred tanks, provided the key units for the assault. From Amiens, they attacked the twenty understrength and tired divisions of the German 2nd and 18th armies, which occupied positions north and south of each other along the Somme River.

On the morning of August 8, the British conducted a rolling artillery barrage that provided cover for their advancing armor and infantry. Unlike the previous Allied attacks from 1914 to 1917, at places such as Aisne, the Somme, or Ypres, the initial German defenders in 1918 offered little resistance. The British forces stormed the German trenches and, incredibly, advanced ten miles. By the afternoon, they had accepted the surrender of 16,000 enemy soldiers. In total the Germans had suffered 27,000 casualties on August 8. These unprecedented events shook the core of the German Army. When he received word that several German units had dissolved under the pressure of the British assault, Quartermaster Ludendorff described the episode as the "Black Day of the German Army."[195]

After the impressive gains of the first day, the British progress slowed steadily. A few days later, to the south of Rawlinson's forces, the 3rd French Army struck the German forces whose flanks had become exposed by the significant British gains. They quickly liberated the city of Montdider from the Germans. After a brief pause, the 3rd British Army, to the north of Rawlinson, joined in the attack on August 21. Concerned about the pressure being exerted on his already exhausted forces, Ludendorff ordered a general retreat along a thirty-mile front, which proceeded in a more or less orderly fashion. On September 3, as the Allied advance continued, the Germans withdrew to behind the Hindenberg Line. The defense near Amiens had cost Germany 75,000 casualties, with more than 30,000 of these losses the result of surrenders. In contrast, the British and French lost 22,000 and 24,000 men, respectively.[196]

The Decision to Capitulate: The Loss of Staying Power

After the Battle of Amiens, it was only a matter of months before the Germans decided to sue for peace.[197] As the Kaiser later wrote, the failed offensives in the spring marked the beginning of the end for Germany. Unable to cope with the deteriorating military situation, Quartermaster General Ludendorff resigned his post on October 26. Two days later, the German Army refused to put down a naval mutiny in Kiel. The opposition to war increased with massive

TABLE 3.5: Assessing Staying Power: Professional German Armed Forces, 1917–18

Evidence
Slow decay in staying power as war grinds to a stalemate: decline in discipline
Defeat of Russia bolsters staying power
Failure of Spring Offensives of 1918 lead to decay of staying power: organized surrenders in fall

PREDICTION: Moderate Staying Power, Military Collapses Slowly as the Probability of Victory Decreases.

TABLE 3.6: Assessing Battlefield Performance: Professional German Armed Forces, 1917–18

Battles	Evidence
Cambrai	German defenders at Flesquieres offered stiff resistance
	German units recover from surprise to defend and counterattack
Amiens	Mixed evidence for prediction
	First large surrenders by German units in 1918
	After defeat, German units conduct orderly withdrawal

PREDICTION: Strong Battlefield Performance, Most Units Fight with Determination and Flexibility.

protests that started in Munich and later spread to the rest of the country. In November a civilian government replaced the Kaiser and sued for peace. The war ended in a German defeat on November 11, 1918 (see Tables 3.5 and 3.6).

CONCLUSION: COMPARING GERMANY ACROSS THE WORLD WARS

The cohesion of German armed forces varied greatly between the two world wars. Cohesion theory can account for these changes by pointing to an increase in regime control from World War I to World War II. Germany went to war with a professional military in 1914, as the public attempted to set aside class differences and its armed forces built strong internal bonds, training its soldiers to value unconditional loyalty and trust. It lacked, however, regime control over the German society, pushing and enforcing an ideology demanding unconditional loyalty. German nationalism in 1914 was a popular vision, but both the Left and the Right appropriated it for their own political gain. Nor did the Wilhelmine regime possess the coercive apparatus to impose its will on German society. Nationalism, then, did not provide the depth of commitment for war among the German public that it did under Hitler. As a result, the staying power of the armed forces deteriorated in 1918 when it became clear that failure of the Ludendorff Offensives meant Germany would not win the war. Even though the German Army still fought hard in 1918, it could not win the battles that would lead to a German defeat of the Allies and achievement of lofty German war aims.

Nazi Germany's messianic army, in contrast, displayed a high degree of cohesion, maintaining determination and discipline on the battlefield while resisting all pressure to quit when victory appeared impossible. Hitler's regime

TABLE 3.7: German Casualties, World War I

	Casualties	Initial Forces	Casualties/Initial Forces
1914	952,440	1,800,000	53%
1915	898,535	2,126,000	42%
1916	1,312,488	2,911,700	45%
1917	883,979	2,920,000	30%
1918	1,498,138	2,911,700	51%
	Total Casualties	Total Mobilized	Casualties/Total Mobilized
	7,348,724	13,314,756	55%

SOURCES: Kevin D. Stubbs, *Race to the Front: The Materiel Foundations of Coalition Strategy in the Great War* (Westport, CT: Praeger, 2002); John Ellis and Michael Cox, *The World War I Databook* (London: Aurum Press, 2001); Randal Gray and Christopher Argyle, *Chronicle of the First World War, Vol. I: 1914–1916* (New York: Facts on File, 1990); Michael Clodfelter, *Warfare and Small Armed Conflicts: A Statistical Reference, Vol. II: 1900–1991* (Jefferson, NC: McFarland and Company, 1992).

TABLE 3.8: German Casualties, World War II

	Casualties	Initial Forces	Casualties/Initial Forces
1940	190,451	2,760,000	7%
1941*	833,050	3,315,000	25%
1942*	1,080,950	2,600,000	42%
1943*	1,601,445	3,430,000	47%
1944*	1,562,400	2,460,000	64%
1944–45*	4,966,107	4,480,000	111%
	Total Casualties	Total Mobilized	Casualties/Total Mobilized
	13,448,000	17,900,000	75%

SOURCES: See Table 3.7.
*Eastern Front Only.

not only promoted an ideology demanding unconditional loyalty but also held the power to enforce these demands in the armed forces. As the materially superior Allied armies advanced to the borders of Germany, the Wehrmacht not only conducted a resolute defense, but also mounted a daring and significant counterattack against overwhelming odds through the Ardennes in December 1944. To stiffen the resolve of the armed forces, Hitler and his ardent supporters began to apply a series of coercive measures and flooded the population with propaganda demonizing the Allied forces. All of these measures bolstered the military's staying power, a key dimension of military cohesion. Unlike Wilhelmine Germany's military, Hitler and his Nazi regime could rely on a hard-core group of supporters in the armed forces to continue the struggle and to encourage those around them. It took the crushing material superiority of the Allies to defeat German resistance on both fronts. (Tables 3.7 and 3.8 compare German casualties in the two wars.)

Evaluating Alternative Explanations

These two cases provide mixed evidence for the nationalism argument. According to this explanation, German armed forces in both wars should have fought with determination and flexibility on the battlefield, while displaying strong staying power. However, only Nazi Germany showed the predicted strong staying power, fighting until destroyed. A key problem with the nationalism argument is that it ignores the difference in a regime's ability to enforce its ideology. In this instance, only Hitler's Germany could compel adherence to the regime's ideological demands.

This discussion also provides weak evidence for primary-group theory. Both countries practiced unit training and rotation policies, the key mechanism to create small-group ties. Nonetheless, both armed forces suffered high casualty rates. The need for large numbers of replacements to quickly rebuild units or create new ones shattered primary groups.[198]

Finally, the hypothesis that democracies fight harder than nondemocracies cannot explain the differing levels of cohesion displayed by the German armed forces of these two eras. In neither period did Germany possess the necessary political institutions that would qualify it as a democracy.[199] Still further, the performance of the Wehrmacht casts doubt on the claim that individualism inspires soldiers to fight harder than the more antiliberal norms of National Socialism. Nor can the democracy argument account for variation in an army's will to fight that can occur between two wars. As German society became less democratic, its armed forces became more cohesive. Put differently, cohesion theory can explain the variation in military cohesion that takes place *within* the German case. Regime type is not the most important variable determining German cohesion across these two wars. Instead, this chapter shows how changes in the degree of regime control and military autonomy caused the German military to fight with different degrees of cohesion in 1914 and 1945.

4 France, 1940

ON JUNE 21 1940, France's Third Republic surrendered to Nazi Germany. In less than two months, Germany had accomplished what it had failed to do over the four bloody years of World War I. Today, scholars continue to debate the reasons for France's rapid collapse in the summer of 1940.[1] The earliest explanations claimed that the moral decay of its society made France's defeat the last step in its decline as a great power.[2]

Recent research, however, attributes the outcome to a long list of factors, many of them unrelated to French domestic politics or a decadent society. These candidate causes include an outdated military doctrine,[3] poor coordination with allies,[4] problems inside the military organization,[5] intelligence failures,[6] and German strategy.[7]

This chapter uses cohesion theory to explain why France lost quickly and decisively in 1940. According to the theory, domestic political disputes created an *apathetic military* by undermining the French Army's autonomy and ability to train. The military could not insulate itself from the political divides and the deep feeling of war weariness that characterized the last years of the Third Republic. As a result, the French military lacked the cohesion to recover from the surprise blows of Germany's clever strategy. The armored thrust through the Ardennes put enormous pressure on a fragile French Army. Although some units fought hard, most could not muster the determination and flexibility on the battlefield to recover from the initial defeats. Reserve units, in particular, gave a lackluster performance during crucial breakthrough battles. In more cohesive armed forces, even reserve units display tenacity in combat, as the previous chapter demonstrates. These series of early defeats in May 1940 also sparked a cascade of panic throughout the French armed forces that quickly grew into a sense of despair about the military's strategic situations. Already disillusioned with their civilian masters, and worried about an insurrection, the French military forced an end to the conflict.

In addition, cohesion theory can account for crucial differences in the French and German militaries. If French armed forces had fought a much less cohesive opponent, the outcome of the Battle of France might have been different. Instead they faced the *messianic* armed forces of Nazi Germany, which showed more determination and took greater risks during critical battles. The two armed forces also differed in staying power. Politicians of the Third Republic could never exert a high degree of regime control over society, as Hitler and Stalin did in their countries. Without this control, the French government could not count on a group of hard-core supporters to compel the armed forces to fight as the country's strategic situation worsened. Such a capability allowed the Nazi regime to prolong the war even when victory looked impossible.

Cohesion theory can also explain differences between the French performance in World War I and World War II. Before both wars, the Third Republic experienced great domestic turmoil. Nonetheless, they endeavored to create a first-class military organization with strong internal bonds. On the eve of World War I, these efforts created a French Army that proved capable of recovering from the early defeats of 1914 and enduring a war of attrition with Germany.[8] Even though France's prewar "national revival" might not have resolved all of the country's political and military problems, it did bolster the cohesion of its armed forces. On the eve of World War II, however, similar efforts failed. The intensity of the disputes and a profound sense of war weariness undermined the autonomy of the military. A second national revival posed too great a hurdle for France's political and military leaders to overcome.

The remainder of this chapter consists of four sections. In the first section, I explain why a low degree of military autonomy produced an apathetic French Army in 1940. I discuss why political cleavages and fatigue from World War I prevented the military from forming the strong internal bonds it needed to fight with cohesion in World War II. The second section details the fragile cohesion of the French military as they attempted to counter the German invasion from May to June 1940. Section three briefly compares the cohesion of the French military in World War II with its cohesion in World War I. Again, a "national revival" gave the French Army a respite from domestic disputes to focus on training and building loyalties within the armed forces. While not overstating how much military effectiveness this revival produced, it at the very least gave the armed forces a respite from domestic political disputes and some of the resources it needed to defend France from Germany. This period saw the creation of a *professional* French military, which was armed with the cohesion to endure four horrific years of conflict. The comparison between the two periods shows the explanatory range of cohesion theory. In the final section, I compare cohesion theory with three alternative explanations: nationalism, primary-group theory, and democratic militaries.

AN APATHETIC FRENCH MILITARY IN 1940

French armed forces went to war in 1940 with fragile military cohesion. Domestic turmoil and war weariness disrupted the efforts by the military to create internal bonds of loyalty. The French public and its politicians were mired in familiar debates over how to manage the economy, corruption in government, and the future of the Republic as a democracy. Against this backdrop was also a debate over the role of the military in French society.[9] Politicians on the Left worried that the military would intervene in politics. To counter this fear, they sought to impose a "nation-in-arms" model on the armed forces as a way to prevent an insular military from becoming threatening to the regime. Some on the Right welcomed an end to the Third Republic. A frustrated French Army found itself caught in the middle of this seven-decade debate. It had weathered previous domestic storms to prepare for war, but in the years before 1940, the strains proved too difficult to overcome.

Low Degree of Regime Control and Military Autonomy

The relationship between the political Left and Right in France was not always turbulent. Before World War I, French politicians and the constituents they represented managed to set aside their differences to form the Union Sacrée. In defense of the sacred homeland, and in an effort to recapture the lost provinces of Alsace-Lorraine, France went to war unified in 1914.[10] Unlike in past eras, this sense of unity had enough staying power to endure the first decade after the Versailles settlement. A period of economic growth also contributed to this peaceful political climate.[11]

A variety of parties defined the contours of postwar French politics. From Left to Right, these parties included: communists, socialists, radicals (centrists), moderates, and conservatives. These labels, however, represent only a broad sketch of the different official political organizations vying for power in France at that time. Beyond the formal party system, French civil society was rich with a variety of interest groups, such as trade unions and veterans groups. These organizations would often assist the various political parties to win supporters, or in their most sinister form, intimidate opposing politicians with violence. In addition, it was not unusual for political leaders to change their party affiliation. The highly influential, and eventually notorious, Pierre Laval, for instance, began his career as a Marxist, and later became a protégé of the very conservative Marshal Pétain. Similarly, the last premier of the Third Republic, Paul Reynaud, adhered to a fairly centrist political line but dubbed himself an "independent."

Even during the prosperous and tranquil era of the 1920s, however, the traditional political disputes between the Left and Right were beginning to take shape. Political leaders on the Left sought improved working conditions as well as wages for workers. The more conservative politicians fought to guard the interests of the wealthy and middle class from the economic policies of France's more egalitarian ideologues. These debates appear fairly innocuous and similar to the arguments experienced in most contemporary democracies. Unfortunately, a history of military intervention in politics made these fault lines in French society fairly charged.

Fresh in the minds of the Left were the numerous moments the French Army had used force either to defend or to install reactionary governments. With the bloody crackdown against the Paris Commune of 1871 serving as a recent backdrop, politicians on the Left believed, and rightly so, that the military was very sympathetic to the more conservative elements in French society.[12] For them, the best way to curb this tendency was to create a conscript army whose professional or standing component was as small as possible. The Left desired a small standing army that could grow in size on the eve of war.[13]

Most military leaders and their allies on the Right agreed in principle with the need for a "National Army" rather than a "Professional Army" with conscripts serving multiple years in the military but argued for a sizable standing army in order to give the country more leverage in foreign policy. Fearing a rapid attack by Germany into the country's vulnerable industrial area in the northeast, venerable French generals such as Foch, Weygand, and Pétain saw the need for a larger standing army to serve as a covering force more than politicians on the Left believed necessary. This dispute immediately started to play out at the conclusion of World War I.[14] Although a conservative coalition controlled the government, the armed forces nonetheless underwent a series of contractions. In a concentrated period from 1921 to 1928, the legislature started to slash the terms of service from three years in 1921, to eighteen months in 1923, and finally to one year in 1928.

When the economic shocks of the Great Depression finally reached France in the early 1930s, political events took a turn for the worse.[15] As the economic conditions worsened, a coalition of conservative politicians controlled both houses of the French parliament. The government of Andre Tardieu took a very deliberate approach to the economic downturn, refusing to liberalize the country's currency and keeping wages low. To make matters worse, the Tardieu government also exacerbated tensions with workers by cracking down harshly on strikes. The voters being unhappy with the ruling coalition's languid efforts to stabilize the economy, the elections of 1932 placed a Socialist and Radical majority in Parliament. The electorate's swerve to the Left, however, did not serve as a harbinger of political stability. Instead, this Left-leaning majority was short-lived and marked the beginning of a period in which France went through five successive governments from December 1932 to February 1934. While the economy continued to worsen, France's political leaders seemed incapable of arriving at any consensus on how to deal with this worsening problem.

The government's continued paralysis only fostered greater antirepublicanism on the Right. In January 1934, bitterness simmered in Paris as various right-wing groups, whose ranks included a large number of war veterans, began to complain stridently about the failure of politicians to address the economy. Out of this expanding wellspring of discontent, a collection of antiparliamentary organizations began to emerge. These groups could trace their origins to the anti-Semitic and old antirepublicanism leagues of the last part of the nineteenth century that fanned the flames of the Dreyfus Affair and yearned for a return of the monarchy.

What separated this new generation of reactionaries from their predecessors were their increasingly violent tactics and a growing admiration of fascism. As one chauvinistic writer proclaimed: "How can we get rid of this weak and rotten regime? Who is the leader who will emerge in France, as he emerged in Italy and Germany?" Another author saw a similar path for the country, speculating: "The best minds envisage the experiment of an authoritarian government on the model of those of Italy and Germany." One newspaper proclaimed more directly: "There is only one solution—and circumstances will soon impose it—and that is a government of authority, supported by an irresistible popular moral force .... This is the task to which we dedicate ourselves. And not one among us will stop until it has been accomplished." These groups also had a plan for change and did little to hide their intentions. The head of the Taxpayers League proclaimed in late January 1934, with his words blazoned across the front page of a mainstream Parisian newspaper: "We are going to carry out a march on the Chamber of Deputies, and if necessary we will use whips and sticks to sweep out this chamber of incapables."[16]

In addition to disputes about how to fix the economy, French politicians engaged in a series of scandals that further undermined confidence in the government. Each of these suspicious events seemed to validate charges from both sides of the political spectrum about the corruption of the government. Eventually, these opponents of the government took to the street to express their outrage. On February 6, 1934, a colossal protest against the new government of Eduoard Daladier turned into a massive battle with Paris police on the Place de la Concorde. In a strange twist of events, the communists joined the protest, seeing an opportunity to pressure the regime. The government, unsure whether it could trust the military to come to their aid, hung its hopes on the police. An ensuing struggle led to the deaths of fourteen protesters and one policeman. In the end, the Daladier government gave way to the pressure and collapsed. Although the Republic was saved, another weak coalition took the reigns.

Fearing the growing power of the Right, the electorate took another swing in the opposite direction by returning a Left-leaning government to power. Leon Blum's Popular Front, a collection of socialists and communists, took control of the French parliament in 1936, after a dramatic increase in Communist Party and trade union membership. In much the same way that the Right was sparked by its triumph in politics, the victory of the Left-wing coalition emboldened its supporters to demand even more change. Even though the Blum government was sympathetic toward their aims, angry workers began a wave of sit-in strikes to urge the politicians to quicken the pace of reform. To accommodate their angry constituents, Blum and his ministers pushed through legislation to shorten the workday and improve wages.

Although these measures were long in coming, they effectively strengthened the impression among the middle class and wealthy that the Popular Front was ruining the economy at their expense. Thus, on the eve of World War II, the split between the political Left and Right widened. The Left believed that reactionary forces were on the verge of toppling the government, with the military either aiding their cause, or at the very least, standing aside. Capitalizing on this tension, Communist Party propaganda questioned whether it made sense for

French workers to take up arms in defense of a government that did not defend their interests. A similar view was expressed by the middle class and wealthy on the Right. Many of them considered Hitler an improvement over Blum and his Popular Front Allies.[17]

Growing War Weariness. As the political divide in France expanded, a general apathy toward war, always slightly beneath the surface of French politics, emerged to steadily erode the nation's morale. A French society already demoralized by political rifts also wanted to avoid another bloody repeat of the last war.[18] On the verge of another conflict with Germany, Frenchmen could easily recall the high price they paid for victory in World War I. France suffered the highest casualty rate of all the combatants, with more than 1.3 million dead, depriving the nation of 27 percent of all men aged eighteen to twenty-seven.[19] The country simply had no stomach for another war.[20]

Foreshadowing the antipathy toward fighting that would afflict France a generation later, a soldier wrote in his journal during the Battle of Verdun in 1916: "They will not be able to make us do it again another day; that would be to misconstrue the price of our effort."[21] Even the commander-in-chief of the French military, General Gamelin, recognized his country's aversion toward suffering the pain of war. A few months before the outbreak of hostilities, he warned the American journalist Walter Lippman that he believed that this time France could not suffer more than 100,000 casualties.[22] When France took up arms in 1939, even though it was politically divided, there was widespread agreement on one point: the nation wanted desperately to avoid a conflict similar to the bloodletting of 1914–18.

Causal Logic: Tensions Undermined the Autonomy of the Armed Forces to Train and to Establish Internal Bonds of Loyalty

While political dissension tore at the unity of French society, many of the same disputes were also tearing at the unity of its armed forces, preventing it from fighting with the same staying power that it had displayed in World War I. These controversies emanated from both outside and within the military. An increasingly contentious relationship between French civilian and military leaders represented one obstacle to fostering a sense of loyalty and trust among members of the armed forces. As a result, politicians on the Left moved successfully to change French manpower policy in ways that curtailed the military's ability to maintain a large, well-trained standing army that might intervene in French politics. Their victory would affect how the military prepared for war.

Problems within the French Army were even more serious. Rampant careerism, animosities between senior and junior officers, as well as class conflicts among the ranks went a long way toward undermining the organization's ability to create a sense of solidarity as it braced for the coming conflict.

Politicians from the Left, especially those of the Popular Front, did not trust the armed forces to stay out of the politics of the Third Republic. These politicians were not necessarily against the military; they simply feared a large professional army.[23] Their long memories recalled the numerous times the French Army had asserted itself into the political arena against them. From the revolutions of 1848, to the Paris Commune of 1871, the French Left remained wary of creating a large professional army that would serve as the Gendarme of the Right.[24] Also fresh in their minds was a more recent episode that had occurred in the Rhineland. After the Versailles Agreement ended World War I, a French general, against the advice of his superiors, had attempted to create a separatist movement that would forever split the region near the Rhine from Germany. It was with these events in mind that communist and socialist politicians harbored deep suspicions about the intentions of the armed forces.[25]

If the Left feared the armed forces, the military hierarchy viewed the civilian officials of the Republic with contempt.[26] In the view of most senior French officers, the politicians were ruining the French military. The armed force became increasingly estranged from the Third Republic. Within the armed forces there was a feeling of "public indifference, for an authoritarian, conservative state that would ensure the Army's material and moral welfare, the decline of its power and the intellectual impoverishment of the Army cadres."[27] General Maxime Weygand "was convinced [as he later wrote in a volume of memoirs concerning this period entitled *Mirages et réalité*] that the struggle had 'ruined the present and compromised the future.'"[28] Not only had they reduced the time new recruits served to one year, but they had also curtailed the army's professional component. As a result, the armed forces lacked enough professionals to inculcate its recruits with military values.[29] While the budget for armaments remained steady, and even grew to meet the growing German threat, the manpower policy prevented the military from training its forces in the way it saw best for fighting, whether on the defense or the offense.

Since the members of the French military establishment already knew that civilian leaders viewed them with suspicion, many chose different paths for dealing with politicians.[30] While some of these older generals, such as Maxime Weygand, opted for retirement, others, like the French commander-in-chief, Gamelin, decided that it was best to cooperate with France's elected heads of state. Although the political sympathies of many officers rested closely with conservative politicians,[31] Gamelin concluded it was best not to rock the boat of French politics by openly questioning civilian prohibitions over a large professional component for the army.[32] To this end, he often countered the attempts of other military officials, such as Charles De Gaulle, who argued that a sizable professional force was necessary in order to create independently operating armored units. Still others, mostly a cadre of junior officers, joined underground right-wing organizations that prepared for the day when either the monarchy would return or a more authoritarian leadership would emerge to save France from the moral decay.[33] The presence of these groups did not go unnoticed by the leaders of the Third Republic. In 1938, the Daladier government called on Gamelin to report if indeed these groups had established a large presence in the military.[34]

As a whole, the French officer corps decided that it had to play with the cards the politicians had dealt it. In the years preceding World War II, the stack of cards they received translated into an army composed mainly of short-term reservists. The French military concluded, as a result, that the best way to fight with such a force was to adopt a defensive doctrine emphasizing methodical set-piece battles based on the concentration of firepower from artillery.[35] With a larger-standing, better-trained force, the military might have contemplated going on the offensive. Indeed, in the early 1920s, a period of considerably amicable relations between civilian and military leaders, the French Army had pursued an offensive-minded military doctrine.

However, when constraints on manpower policy were matched up with the army's views about the lethality of artillery, seared into their memories from battles like Verdun, the officer corps quickly concluded that a mostly reservist force offered limited potential. Under these conditions, especially with a shortage of experienced officers, military planners decided that their best bet was to conduct tightly controlled battles that made use of French artillery. To this end, senior French officers anticipated the need to exercise great control over the conduct of any war. An army of reservists, in their view, lacked enough training to use its own initiative in combat.[36] Instead, battlefield operations would require detailed scripts because the armed forces trained its soldiers to react slowly to events on the battlefield.[37]

Considerable debate exists about how French manpower policy affected the way the army fought in 1940. Some suggest—like the claim made here—that France's defensive doctrine grew out of domestic political disputes. Others disagree, by pointing out that the decline in French power relative to that of Germany persuaded it to choose the defensive as the best way to fight. It seems that both arguments point in the same direction. However, France's deficit in relative power cannot account for why the style of defense was methodical. There are many ways to conduct a defensive campaign, such as fighting with fortifications, fighting in depth, and fighting with a mobile defense. In the French case, they emphasized the need for the military hierarchy to exercise great control over the fighting. This key facet of French military thinking grew out of the conviction of the officer corps that recruits with little training required a very rigid style of command.[38]

Trouble also existed within the ranks of the military. In particular, tensions arose in the French officer corps over the rate of promotions. In an effort to avoid confrontation with civilians, senior officers created an atmosphere in the armed forces that discouraged innovation.[39] Under the stress of the coming war with Germany, and as civil-military relations deteriorated, they did not want to see rapid change in the armed forces. Like any organization facing external pressures, the military began to settle into a static routine, falling back on established practices in order to cope with its circumstances. French generals, then, promoted officers not on the basis of merit but based on how well they adhered to conventional wisdom.[40]

Junior officers, consequently, began to feel that their superiors actively kept them from advancing in order to retain their hold over the organization.[41] Disillusioned with both the regime and their military leaders, they "expressed their traditional aversion for the weakness of the parliamentary regime, their fear of a mysterious—if not actually mythical—revolutionary subversion and their criticism of a high command bogged down in an outworn doctrine and overly subservient to the civil authorities."[42] To cope with this situation, many junior officers abandoned the military as a career. As a profession, serving as an officer in the military was neither very lucrative nor very popular in the eyes of French society. It became well known, for instance, that many officers would moonlight in the evenings as taxi drivers to compensate for the terrible pay.[43] The junior officers that remained in the military pursued advancement, then, by toeing the line established by their superiors. This propensity, as the historian Marc Bloch notes, inculcated the French officer corps with a widespread spirit of careerism.[44] Thus, in an organization with few slots for advancement, the surest path to promotion called for an unquestioning adherence to authority. In contrast, some officers saw membership in a right-wing paramilitary group as the best outlet for their frustrations. For them, the government deserved the blame for their plight.[45]

In addition, the class divisions of French society also created friction within the armed forces, preventing the French Army from operating as a united fighting force. As Eugen Weber described it:

> Taken as a whole, the officer corps left something to be desired. And no wonder: Pay was poor, promotion slow, prestige dim. Senior officers who may have performed heroically on the Marne or Somme had aged since then . . . . Too many were worn out because they stayed on as long as possible in order to avoid the ridiculously low retirement pay. Whatever the reasons, the effect was dim, and dimmer still because of latent hostility between general and subordinate officers, between staff officers and those who had come up in the trenches, between different—sometimes very different—generations.[46]

These tensions also reflected class divides. Enlisted men tended to come from a lower socioeconomic status than commissioned officers. French conscripts during the era came from a society that at best was ambivalent about the military, and at worst, hostile to it.[47] They came from either rural areas or from the numerous factories that dotted France's urban centers. For this latter group of recruits, serving under officers from middle-class or wealthy backgrounds was particularly galling. Outside the military, these workers were engaged in a bitter struggle against property owners to improve wages and to limit the number of hours they toiled in French factories.[48] Communist propaganda fanned the flames of this resentment, calling workers to avoid shedding blood for a capitalist cause.[49] As a result, French officers found it difficult to gain the respect of their men and not be seen as tools of the wealthy.

From the officers' perspective, the communists, and not Hitler, posed the greatest danger to the country.[50] Thus much of the French military saw the struggle between Left and Right at home as more meaningful than the conflict between France and Germany. Observers noted these class differences:

> In contrast to the more democratic Wehrmacht, French officers seemed to look after themselves much better than their men, with whom they had deplorably little contact. Although the wearing of helmet, gasmask and belt was obligatory at the front, officers were often to be seen strolling around in service caps, jackets un-buttoned, hands in pockets and cigarettes dangling from their lips. An American observer was shocked by the sleekly well-groomed officers with their "well-polished nails and brilliantined hair, compared with the soldiers in their grubby barracks and dreary canteens in which they could spend their fifty centimes a day."[51]

Class cleavages, civil-military disputes, and general war weariness took a toll on the cohesion of the armed forces. Colonel A. Goutard summarized the general atmosphere of the French military on the eve of World War II when he explained: "No one dared give an order for fear of criticism or of making a mistake. Military exercises were considered a joke, and work an imposition. Thus, instead of becoming more hardened, our army tended to lose its fighting qualities."[52] Alistair Horne agreed:

> In many units discipline left much to be desired; men did not salute their officers, and the officers often failed to exact such "external marks of respect." This seemed to be particularly the fault of reservist officers: "The events of 1936 had perhaps caused them to lose in civil life the habit of being obeyed," suggested Marcel Lerecouvreux of the 2nd Cavalry Division, and—whether right or wrong—his was a view widely shared among the French officer corps.[53]

A diarist of the time reported: "I observe every day social divisions and class resentments which complicate and sully the conflict."[54]

Still further, the French officer corps, suffering from careerism, could not cultivate military leaders who valued innovative thinking.[55] French military planners had already created a doctrine calling for a methodical style of battle. The marriage between this sense of careerism and the belief that the best approach to fighting battles required deliberate actions all but ensured that the French military would exhibit very little ability to adapt on the battlefield.

Before the start of hostilities, then, the French armed forces seemed unready for war. Military personnel seemed "lukewarm and reluctant to fight."[56] One historian notes: "An army's morale is one third creature comforts, two thirds self-confidence born of hard training and obstacles overcome. The former were in short supply; the latter was seldom in evidence. French and foreign observers were struck by the caution of the fighting troops."[57]

THE BATTLE OF FRANCE, MAY–JUNE 1940

In 1939, France found itself once again at war with Germany. Only this time, French armed forces entered the conflict with fragile cohesion. Below, I describe why an apathetic French military lacked battlefield determination and staying power to counter the German invasion. According to cohesion theory, most French units should have lacked determination and flexibility on the battlefield. Further, the French military should have also quickly lost its staying power after initial setbacks.

The analysis focuses on the important German breakthrough in the Ardennes, along the Meuse River. Most of Germany's panzer spearheads eventually forced a crossing of the Meuse on May 12 and 13, 1940. As they worked to establish bridgeheads across the river, they encountered a mixed level of resistance. A few French units fought doggedly to halt the German crossings, but most of the defenders collapsed under the strain of the assault. Aware of the gravity of the strategic situation, the French High Command reacted to the crossings by attempting seven series of half-hearted counterattacks from May 13 to 21. Each of these efforts to push the Germans back over the Meuse, however, suffered from French units that simply demonstrated little determination to fight and little flexibility.

As the German spearheads exploited their multiple breakthroughs and raced toward the English Channel, the bulk of the French forces regrouped along the Weygand Line. They formed a defensive position that stretched across the Somme and Aisne rivers. Already members of the French government contemplated negotiating a separate peace with Germany. The final assault on France began on June 7. For four days, the French Army maintained its positions. On June 11, however, the German forces penetrated the Weygand Line and French resistance quickly collapsed. After a change in government, France's military and political leaders concluded hostilities with an armistice with Germany on June 22, 1940.

Planning for War

World War II began with the German invasion of Poland in September 1939.[58] Although they had done little to prevent Hitler from seizing Austria in 1938 and Czechoslovakia in 1939, the Western Allies finally decided to take a stand against German expansionism with the invasion of Poland. They declared war against Germany, but did little in terms of military intervention to assist the ill-fated Poles. With great reluctance, the French Army mounted a limited offensive that amounted to seizing a little more than three miles of German farmland. The Allies contented themselves with both building up a large force on the western frontiers of Germany and contemplating action in other theaters, such as Scandinavia or the Middle East, where they could gain some leverage over their adversary. A "phony war" ensued as the two sides faced off in the West, preparing and waiting for a fight that would not come for another eight months. After the defeat of Poland, Germany's dictator, Adolf Hitler, ordered his military to prepare plans for the invasion of France. Germany's Plan Yellow, conceived by General Erich von Manstein and promoted by General Heinz Guderian, called for German forces to mount a two-pronged assault against the Allied forces in the West. The main thrust of the attack would come through the Ardennes Forest, relying on seven panzer divisions from Army Group A to cross the treacherously hilly terrain between the French 9th and 2nd armies. Meanwhile in the north, Army Group B, with three panzer divisions, would stage an elaborate feint, sending forces into Holland and Belgium, via the Gembloux Gap. Like a matador's cloak, the plan depended on British and French forces rushing to meet the forces of Army Group B, while the main drive from the south would split the Allied forces in two.[59] If their strategy worked, the German Army would cut the Allies in half, with the northern forces trapped in a pincer between Army groups A and B. An infantry heavy Army Group C would serve as a covering force sitting along the Maginot Line.

Almost as if they had decided to play the role cast for them in German strategy, the French war plan called for a rapid advance deep into Belgium to meet Dutch forces and begin a largely defensive campaign. The Allies hoped to create a continuous and static obstacle for German forces stretching from the English Channel to Switzerland. With this in mind, Gamelin placed the cream of the French Army on his left. Once the German attack began, the 7th Army, along with the British Expeditionary Force (BEF), would wheel north and west to the Dyle River with the intention of linking up with the remainder of Belgian forces. From this point, Gamelin later decided, the Allies should push farther northward in an effort to aid the Dutch near the town of Breda. To the right, the 1st French Army would advance into the Gembloux Gap, what the Allies viewed as the primary avenue for the German advance.[60]

From their perspective, the Maginot Line would prevent the Germans from mounting a quick offensive into France. Closely in line with this thinking, the French regarded the Ardennes Forest to the left of the Maginot Line as an impenetrable natural barrier. These impediments, in their view, left Belgium, the Gembloux Gap in particular, as the most likely route to France. The French General Staff thought that Germany would attempt a variation of the old Schlieffen Plan as the basis for its strategy. Had this reflected the actual German strategy, the Dyle Plan would have made some sense, because it would leave French forces in a position to conduct the bulk of their operations off their home soil. The northern area of France contained a large percentage of the country's industrial resources. As such, fighting a defensive campaign in Holland and Belgium would give the French time to build up their resources for a long war of attrition they confidently anticipated winning.[61]

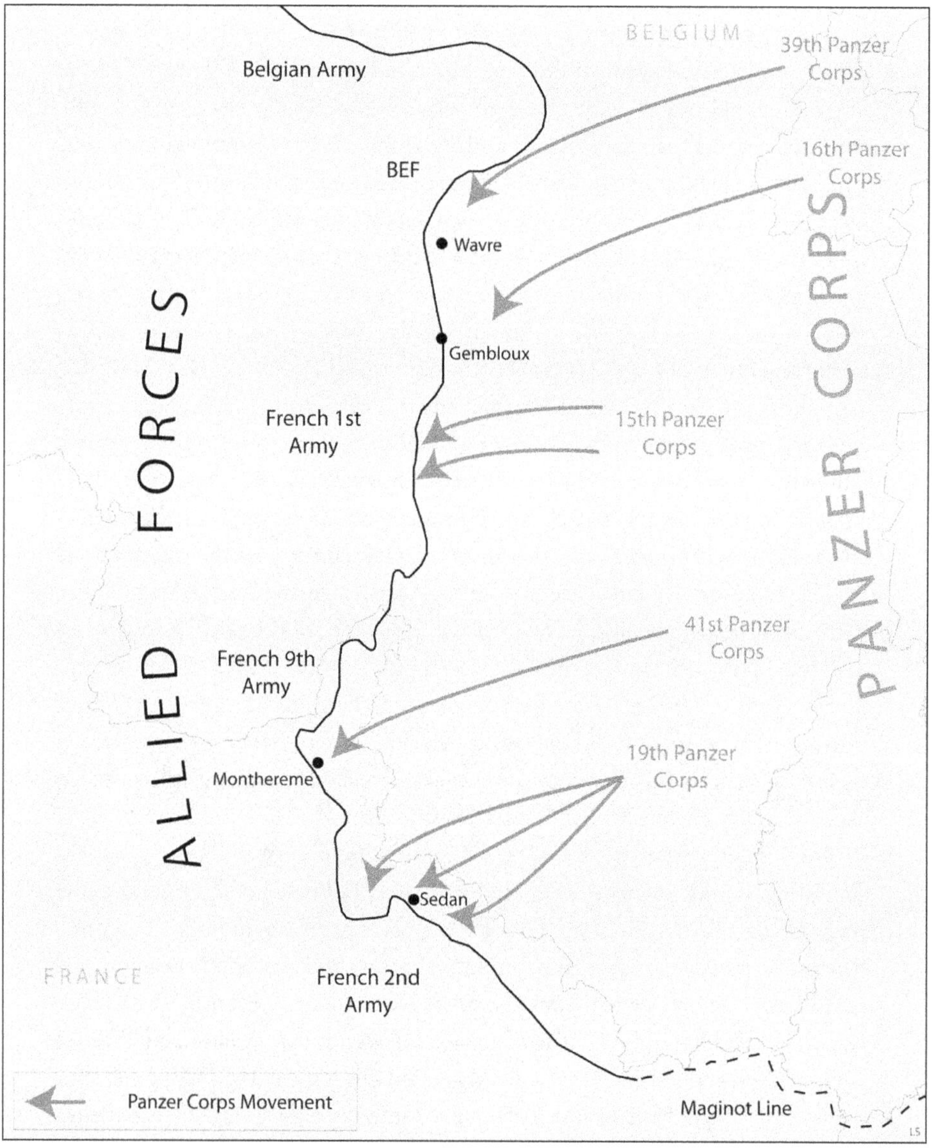

Map 4.1. Western Front, 1940.

Before the conflict, the balance of forces between Germany and the Western Allies was roughly equal. The Allies sent 3 million men to the theater, compared with the 2.5 million soldiers along the western German frontier, putting Germany at slight disadvantage in terms of manpower. Germany fielded 136 divisions, and the British along with their French allies could make use of 133 divisions. The Allied coalition included six armored divisions, dispersed throughout their formations. In contrast, the Germans sent ten armored divisions to the front, with seven of them devoted to Army Group A's thrust through the Ardennes. With respect to armored forces, this translated into an edge for the Allies, who possessed 3,000 tanks compared with 2,700 German tanks. In fact, by 1939, France had the best tanks in Europe, far superior to anything used by either Britain or Germany.[62]

Germany also faced a deficit in artillery pieces. Whereas the Allies could rely on 12,200 various guns, Germany could only field 7,710. Although the Germans suffered from fewer artillery pieces, they did hold an advantage in antitank guns, with 12,800 to 8,050 British and French guns. Parity more or less existed in terms of fighter aircraft: Germany maintained a small lead with 972, and the Allies, not counting Belgian and Dutch planes, could rely on 891. Germany, however, did hold a decisive lead in terms of bombers with 1,217 in the West, compared with 377 British and French bombers.[63]

The Start of Hostilities in May 1940

The Germans launched Plan Yellow on May 10, 1940, with simultaneous attacks on Holland, Belgium, Luxembourg, and France. After resisting for four days, and after the brutal bombing of Rotterdam by the Luftwaffe, the Dutch capitulated on May 15. The rapid defeat of Holland dashed Gamelin's hopes that his 7th Army would link up with Dutch forces near Breda.[64] German forces encountered Allied troops in Gembloux Gap on May 12. For two days, mechanized elements of Blanchard's 1st French Army fought Hoepner's 16th Panzer Corps to a standstill.[65] Except for the quick capitulation of the Dutch, Allied plans appeared to unfold as they had predicted. They now stood in place to counter where they thought the main German advance would take place (see Map 4.1).

Farther to the south, however, the main thrust of the German attack slowly made its way through the Ardennes Forest. The undulating terrain, with its treacherous roads, presented a formidable but passable obstacle for the large concentration of German armored forces. Three panzer corps with seven armored divisions, under the direction of Rundstedt's Army Group A, constituted the bulk of this effort. From north to south, these units included: the 15th Panzer Corps commanded by General Hoth (with the 5th and 7th Panzer divisions), General Reinhardt's 41st Panzer Corps (with the 6th and the 8th Panzer divisions), and the largest of these three, General Guderian's 19th Panzer Corps (with the 1st, 2nd, and 10th Panzer divisions). The last of these two corps were under the direction of General von Kleist. The German headquarters created Panzer Group von Kleist in order to exercise tighter control over Guderian, who had already established a reputation as a highly aggressive commander. They believed Kleist's tendency toward being extraordinarily cautious would balance Guderian's inclination toward risk-taking.[66]

According to Plan Yellow, each of these panzer corps would cross the Meuse at three points to establish bridgeheads for the remainder of Army Group A. After crossing, the armored forces would push westward into France between Arras and Amiens toward the coast. They aimed to pass through the French 9th Army, commanded by General Corap, and then turn northward to encircle the French 1st and 7th armies and the British Expeditionary Forces from the rear. As the plan dictated, German infantry would protect the flanks of the armored forces as they wheeled to the north.

The entire plan faced several obstacles, most notably the potential for French forces to mount a counterattack against the exposed flanks of the three panzer corps as they raced westward. German planners worried that their infantry might fall behind the quickly advancing panzers and thus fail to position themselves to thwart such an attack. In addition, the three panzer corps faced a tough test in navigating their way through the Ardennes, especially the 19th, which proceeded through Luxembourg one division at a time. Along the roads, several tanks and trucks would break down, blocking traffic until the Germans could remove them. Finally, the plan's success depended on the ability to convince the Allies that the main effort of their attack would come farther north. If the French screening forces along the Ardennes could delay the German advance and warn the French High Command about the attack, then the panzer spearhead would lose the element of surprise and the Allies might find a way to attack the exposed flanks of the armored forces.

The French placed General Corap's 9th Army and General Huntzinger's 2nd Army to guard the sector facing the Ardennes. Unlike the forces in the north, devoted to an advance toward the Dyle River, these two armies contained several divisions manned by reservists. Although they had eight months to prepare for an attack, the divisions in these two armies spent most of their time building pillboxes on the Meuse or conducting menial tasks, rather than undergoing additional training. Rather than practicing how to fire their weapons, divisional commanders used their forces for construction and gave them liberal amounts of leave. As a result, these already undertrained forces wasted valuable time that they needed to become familiar with their essential tasks. In fact, many of the soldiers in this sector had to rush to the front from furlough when the day of the German attack finally arrived.

Defense of Dinant: Weak French Determination and Flexibility

Hoth's 15th Panzer Corps skirted the northern edge of the Ardennes and planned to cross the Meuse between Dinant and Houx.[67] The 5th and 7th Panzer divisions, the latter commanded by General Rommel, proceeded on a route that took them through the towns of St. Vith, Charbrehez, Marche, and, finally, Dinant. Although the 7th Panzer Division took one of the easiest courses to the Meuse, it still encountered several problems with congestion along the roads as it made its way westward on May 10. After resolving their traffic problems, the 7th Panzer encountered another set of obstacles in the form of Belgian fortifications and a regiment of Belgium's Chasseur Ardennais. Rommel's forces quickly overwhelmed the Belgian units and continued their way westward.[68]

On the evening of May 11, German infantry units reached the key river and made their way through the narrow gorges in front of Dinant. As they had done at the Our River, the French forces did not put up much of a fight and withdrew once again to the western side of the Meuse. Finding all of the bridges at the river destroyed, Rommel's units prepared to cross the Meuse. As they assembled for a crossing, the French forces on the heights used their artillery to shell the infantry of the 7th Panzer Division. In the early hours of May 12, Rommel's troops discovered a weir near the town of Houx. Using this unguarded route, he sent some of his forces across to a small island in the middle of the Meuse.

Under the cover of smoke from nearby houses lit ablaze by German forces, Rommel's riflemen began crossing the Meuse at Houx early on May 13. At first, these lightly armed soldiers encountered stiff resistance, with artillery fire raining down on them. Then, by midmorning, Rommel's panzers began to arrive on the scene. Using fire from the tank guns to cover their advance, the German infantry began to cross the river.[69] Once on the other side, they encountered a few French tanks. With no antitank guns, the infantry relied on small arms fire to repel these defenders. Even though they outgunned the German attackers, French forces withdrew in a panic.[70] With the successful crossing at Houx, French units all along the Meuse between Dinant and Houx began to fall back, most of them in disarray.[71] It took the commanders of the local French forces three hours to inform their headquarters about the German crossing.[72]

Failed Counterattack at Dinant. By the evening of May 13, General Corap intended to throw Rommel back across the Meuse as soon as possible. Generals Martin and Duffet, the French officers on the scene, could not decide which of their units to use against the German infantry. Although confusion existed about what form the attack would take, General Corap's staff assured him, "The incident at Houx is in hand; General Martin is one hand with General Duffet."[73] Their indecision, however, impeded them from acting quickly, giving Rommel more time to consolidate his bridgehead.[74] Finally, they decided to use the 39th Regiment of the 5th Motorized Division (Regular), the 18th Infantry Division (A Reserve), a squad of armor, as well as three groups of artillery for the attack (Regular).[75] At 2000 hours, the infantry claimed it was not ready, postponing the counterassault until the next day.[76]

Unlike the French, Rommel wasted little time once his forces crossed the Meuse to push his bridgehead westward. On the morning of May 14, his forces moved three miles west from Houx and Dinant to the outskirts of Onhaye. By this time, the first elements of the 5th Panzer Division had begun to catch up with the 7th Panzer Division, also crossing the Meuse at Houx.[77] On the morning of May 14, Rommel worried about the prospects of a French counterattack because only his infantry had made it across the river in force. His fears were partially warranted because General Corap had ordered the 1st French Armored Division (Regular) the night before to move toward Houx. However, like the other potential French counterattack, the 1st Armored took too much time to make the twenty-five mile trip to the south. The entire unit did not begin moving until 1400 on May 14. Already late to mount their assault, one of the division's battalions lost its way.[78] More than a day after Corap ordered them to attack, the 1st Armored Division finally reached its designated position, only to discover a consolidated German bridgehead at the Meuse (see Map 4.2).

As this French armored force lumbered southward, they met Rommel's 7th Panzer Division dashing westward from Houx. Rather than engage that 1st Armored Division, which had halted to refuel, Rommel turned his forces toward the southwest, leaving the 5th Panzer Division to outflank the French armor later in the day.[79] At Onhaye, they met elements of the 4th North African Divi-

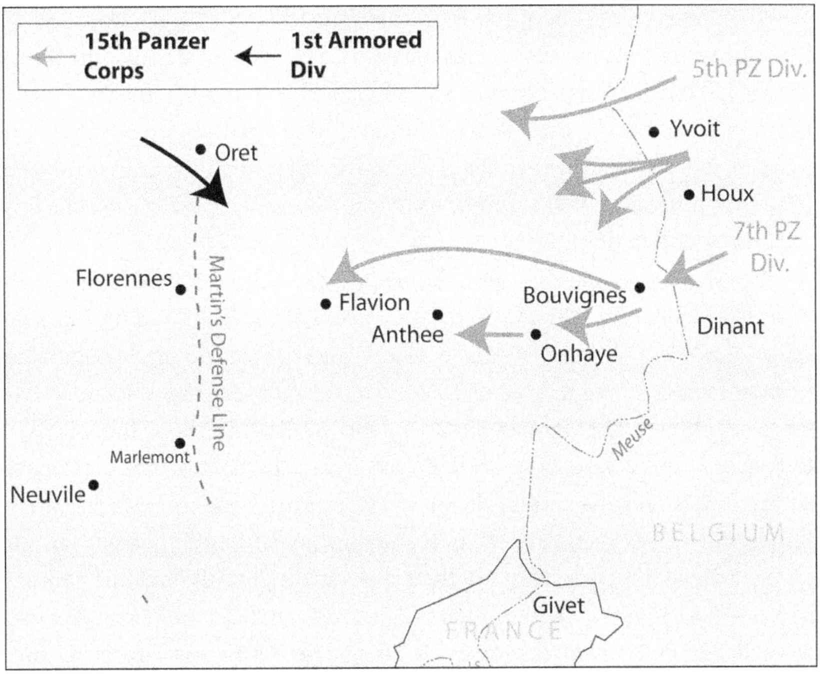

MAP 4.2. Meuse Crossing Near Sedan.

sion (Regular), a unit Corap held in reserve. They had not anticipated a German attack, since General Corap had already ordered the French 18th and 5th Motorized divisions to plug the gap at the Meuse. Confused by the German attack they offered disorganized resistance.[80] After a few hours of fighting they withdrew, even though they outnumbered and outgunned the German forces. Their decision to retreat exposed the flank of the 22nd French Division, causing it, in turn, also to pull back. In a matter of hours, the entire 9th Army would begin to retreat. On the afternoon of May 14, Corap's army stood on the verge of collapse.[81]

Defense of Sedan: Weak Determination and Flexibility

The strongest German thrust through the Ardennes came from the 19th Panzer Corps commanded by General Guderian, which approached the Meuse on May 13. Guderian selected Kirchner's 1st Panzer Division to conduct the initial crossing of the Meuse just north of Sedan at Gaulier. To the north, 2nd Panzer planned to drive to Donchery, while the 10th Division sought to make its way across at Wadelincourt. Because it bore the brunt of the attack, Guderian devoted the corps's artillery and the Gross Deutschland Infantry Regiment to Kirchner's forces. Although the plan appeared simple on paper, all of the pieces did not fall into place as Guderian had anticipated. For instance, the 2nd Panzer Division did not arrive until later in the day, and not all the corps artillery was ready for the attack. However, the German Air Force did manage to conduct a continuous bombing attack from 0700 to 1500 hours that softened considerably the French dug-in defenders.[82]

After the air assault, the German infantry crossed the Meuse, near Gaulier, at 1500 hours. Initially the French defenders of the 55th Infantry Division fought fiercely against the advancing Germans. Artillery and machine gun fire from pillboxes on the heights of La Marfee kept the German soldiers at bay for several hours. The German *Landers* made excellent targets as they exposed themselves during the attack. However, the French did not use their artillery to its maximum potential. Strangely, General Grandsard, the commander of the French 10th Corps overseeing the 55th Division, hamstrung the French defenders by ordering his gunners to use their shells sparingly. Although the gunners had access to plenty of ammunition for their guns, Grandsard concluded that he needed to husband his resources for a long fight.[83] This was a missed chance to obstruct the German advance, as Alistair Horne explains:

> "What targets! What an opportunity," wrote a French military critic, General Menu, for the artillery to rain down those "hammer blows," to practice those "sweeping concentrations" which constitute, in the manipulation of firepower, the high point of the five hundred specific pages of the "General Instruction on Artillery Fire!"[84]

Eventually, through the combined efforts of both the Gross Deutschland Regiment and infantry from the 1st Panzer Division, the Germans swept within reach of the French fortifications. By midnight, they had used explosives to destroy many of the pillboxes. Those pillboxes not destroyed by the infantry fell victim to direct fire from the division's tanks on the opposite banks of the Meuse. As the German infantry captured the high ground above Sedan, the French defenders began to pull back. Many of them chose to surrender rather than fight. Still other units retreated in a panic.[85] The defenses crumbled even though the German infantry remained highly vulnerable to counterattack without their supporting tanks. By midnight on May 14, however, 1st Panzer had constructed a pontoon bridge over the river, permitting the division's badly needed armored forces to finally cross (see Map 4.3).[86]

General Veiel's 2nd Division, manning the 19th Panzer Corps's northern sec-

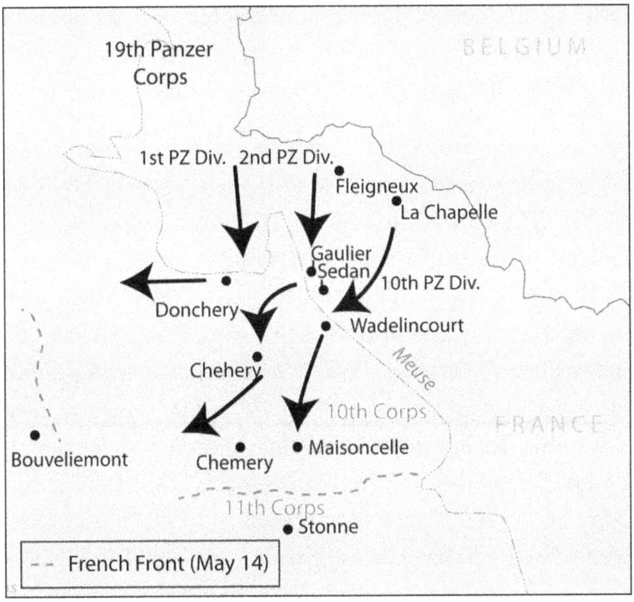

MAP 4.3. Meuse Crossing Near Dinant.

tor, attempted to reach the western side of the Meuse from Donchery on the afternoon of May 13. Using supporting fire from their tanks, the 2nd Panzer's infantry crossed the river by 2000 hours. Again, the French defenders offered very little resistance.[87] In contrast, the 10th Panzer Division found tougher going in its effort to traverse the Meuse south of Wadelincourt. Attacking late in the afternoon, it neglected to put the Luftwaffe's bombardment of French positions to good use. Those French units guarding the Meuse used the delay in the attack to dig in further. As a result, General Schaal's unit suffered the greatest casualties of the three panzer divisions.

Nevertheless, once the German infantry reached the high ground above Sedan, they managed to brush aside the defenders. By dusk, the entire 19th Panzer Corps had crossed the Meuse. Frustration mounted for the French commander-in-chief, as Colonel Goutard would later note: "At about 7 pm these troops withdrew to the south bank and blew up the bridge. The Sedan loop had been abandoned! 'And yet,' Gamelin wrote, 'they were supposed to defend it at all costs.'"[88] Alistair Horne compares the French performance at the Meuse in 1940 with the fighting at the First World War, "In attempting to isolate the reasons for the breaking of the Sedan gunners, one comes face to face once again with the twenty-four corrosive years separating the poilus of Verdun from the men of Sedan; here is the terrible harvest of those years of mutual mistrust, disunity, despair at the losses of '14–'18, *je-m' en-foutisme* and defeatism in France."[89]

First Failed Counterattack at Sedan. In the face of this deteriorating resistance, General Huntzinger rushed a battalion of infantry to quickly throw the Germans back. He rightly feared that his 55th Division (B Reserve) stood on the verge of collapse. Because Guderian's units made rapid progress in establishing a bridgehead, this first counterattack failed. With the Germans on the west bank, panic began to spread across the entire front. Once members of the artillery began to abandon their pillboxes and their guns, the French infantry broke. Faking orders to withdraw, several officers ordered their men to retreat, which they did with little discipline. Officers at all levels fell under the spell of the escalating panic.[90] Even though the German forces at the time lacked a sizable armored force on the French side of the Meuse, staff officers began to exaggerate reports of a huge tank force pressing their units back.[91]

Therefore, while Guderian's corps worked hard to consolidate its undeveloped bridgeheads, the 55th Division retreated westward. Huntzinger, frustrated by events, ordered the division to regroup for a counterattack, only to find his officers expressing great reluctance to follow his instructions. By late evening on May 13, panic had frozen the 55th as a fighting force. In describing this unit's collapse, Robert Doughty notes the failure of the French military to prepare its soldiers for war: "Burdened by the task of digging trenches and building bunkers, they neglected to train their soldiers adequately and to inculcate in them a will to fight."[92]

Adding insult to this grave injury, a few retreating units of Lafontaine's division found enough time to loot local towns as they withdrew.[93] With events deteriorating on its left flank, the 71st Division (B Reserve), unaffected by the engagements at Sedan, decided to retreat westward. At the same time, 2nd Panzer outflanked, and then annihilated, the brave 3rd Spahi Brigade (Regular) at La Horgne. The valiant effort of the Spahis was wasted by the overall lackluster French defense.[94]

Second Failed Counterattack at Sedan. General Huntzinger, commander of the 2nd Army and responsible for Sedan, thus faced a terrible situation on the morning of May 14. Realizing the critical need to push the Germans back before they consolidated their gains, Huntzinger planned a two-phased counterattack. In the first wave, he intended to send his reserves to aid General Lafontaine's faltering 71st Division. These units (all of the B Reserve) consisted of two tank (4th and 7th) and two infantry (205th and 213th) battalions, in addition to the infantry unit he had sent the previous day. Behind this initial echelon, Huntzinger received permission to order the 3rd Armored Division (Regular), one of four such units held in reserve by the French Army, and his own 3rd Motorized Division to close the breach at Sedan.[95]

At 0130 hours, Huntzinger gave the order to carry out the first phase of the counterattack. With Guderian's units struggling to protect their exposed infantry by rushing their tanks across the Meuse, the French attack stood a good chance of mauling the Germans. At the very least, Huntzinger calculated, the first wave of his assault would give time for additional units to arrive on the scene to give the Germans a final shove back over the Meuse. Unfortunately for France, events did not take place in the way the commander of the 2nd Army envisioned. The four battalions of this first echelon could not organize themselves for the attack: the 213th considered it too dangerous to march in the dark to Sedan; the 205th encountered the retreating units from the Meuse, which caused enough panic in its own ranks that it could not move forward; the 4th decided to wait for the 205th before advancing; and the 7th claimed that it did not receive the orders to advance in time for an effective attack.[96]

The counterattack that Huntzinger had ordered to take place at 0400 occurred in a haphazard way at 0700 hours.[97] While his forces were gathering themselves for an assault, Guderian's units expanded their precarious bridgehead. The postponed assault, however, had permitted the infantry from the 1st Panzer Division to advance several miles west of Sedan, creating a protective salient for their armor to cross the Meuse. Consequently, when the first wave of the counterattack finally moved forward, the French were surprised to find the Germans several miles away from the Meuse. A determined German defense threw the attackers back. General Huntzinger's initial counterattack had failed to soften the German bridgehead.[98]

Third Failed Counterattack at Sedan. At 0600 hours on May 15, Huntzinger ordered the 3rd Armored Division (Regular) and the 3rd Motorized Division (Regular) to commence the second phase of the counterattack from the town of Stonne. These two units approached the 19th Panzer Corps from the south. Guderian worried that an attack from this direction would outflank his armored forces as they began their critical turn northward from the Meuse toward the 1st French Army. With this concern in mind, he sent the Gross

Deutschland Infantry Regiment south to guard his corps's flank. There they met a sizable French force.

Like the previous counterattack, French units got off to a late start. Although they had earlier received orders to attack at 0600 hours, the commanders of the 3rd Armored and 3rd Motorized divisions informed Huntzinger that at 1130 hours they could not launch their assault until 1500 hours. They told their superiors that their units required additional time to prepare. At 1430 hours, they decided again to delay their assault until 1730 hours.[99]

The Gross Deutschland Infantry Regiment did not wait for the French to attack. They, instead, mounted an assault of their own in the late afternoon. While the French arrayed their armored forces in a thin defensive line around the city, the German infantry charged the northern part of Stonne. A fierce battle ensued, as the city changed hands several times. Skillfully, the German units used their antitank guns to even the score against the French armor. By late evening, the situation appeared grave for the German attackers. Once again, however, the French failed to exploit the opportunity and they mounted a piecemeal assault to push the German infantry back toward Sedan. The slow pace of the operation allowed the Gross Deutschland Regiment to regroup and defeat the second phase of Huntzinger's counterattack.[100] After the war, the German General Hoth reflected that the "French missed a favorable occasion; this counterattack, conducted in a resolute manner, would have transformed defeat into victory."[101]

Fourth Failed Counterattack at Sedan. Because they had encountered stiff resistance from the 3rd Spahi Brigade at La Hornge, the 1st and 2nd German armored divisions stood vulnerable to a counterattack. Hastily, on May 14, General Georges, the deputy French commander-in-chief, created Army Detachment Touchon to plug the widening hole between the French 9th and 2nd armies. General Touchon's corps, consisting of two infantry divisions (the 53rd, a B Reserve unit, and the 14th, a Regular unit), the 2nd Armored Division (Regular), and the remaining light cavalry units from the 9th Army, rushed to the Sedan area north of 2nd Panzer's advance. Georges instructed Touchon to quickly attack the advancing 1st and 2nd Panzer divisions. Here, on May 15, the French had another opportunity to, at the very least, prevent the sixty-two-mile hole in their armies from widening. Time still existed for them to limit the damage.[102]

Once more, the French counterattack sputtered before it reached full steam.[103] The 53rd Infantry Division reached 1st and 2nd Panzer first. Unlike the 3rd Spahis, however, this unit engaged the Germans and then withdrew, after meeting a ferocious assault. Failing to conduct an orderly retreat, the 53rd dissolved as a fighting force. Meanwhile, the 2nd Armored Division faced a series of obstacles that delayed its deployment. A disagreement between Touchon and the staff of the 9th Army kept the 2nd from the forward edge of the battle. After resolving who was actually in charge of the unit, the armored division had trouble mustering all of its tanks. These obstacles effectively thwarted Touchon's attack and hastened the disintegration of the French 9th Army. When he learned of these events, General Georges collapsed sobbing.[104]

Counterattacks at Montcornet, Laon, and Arras: Weak Determination and Flexibility

Casting Touchon's attack aside, the three panzer corps began their northwestern race to the Channel. By the end of May 16, Guderian had advanced fifty-five miles from Sedan and reached Marle and Dercy. The success of their invasion had begun to make the German High Command nervous. After a brief argument with both General Kleist and Army Group A commander Rundstedt about the pace of his progress, Guderian agreed to halt his armored forces on May 17 in order to reconnoiter the location of French reserves.[105] Hitler and his generals worried that the gains made by their armored forces might expose their flanks to French attacks.

Their concerns proved correct, as De Gaulle's 4th Armored Division (Regular) struck Guderian's forces at Montcornet on the same day and again on May 19 at Laon. These hastily cobbled together French soldiers managed to achieve some tactical success, but in "the end these attacks were like flies buzzing around the tortoise head, while never concentrating sufficient forces to threaten its jugular."[106] German forces regrouped and turned back the French attacks, but the assault had created sufficient anxiety among Hitler's generals about the strain on their armored spearheads. The commanders of the three panzer corps, however, did not let the anxieties of their superiors hamper their advance. They pressed onward, and on May 20 Guderian's units reached Abbeville on the English Channel, completing the German encirclement.

The limited success of De Gaulle's efforts finally prompted Gamelin to act. He ordered Georges to prepare counterattacks against the German northern and southern flanks, in the hopes that they could cut off the panzers from the rest of the Wehrmacht. As the planning for this attack got under way, Reynaud used the crisis as an opportunity to reshuffle his cabinet, sending his old nemesis, Defense Minister Daladier, out and bringing the conservative Marshal

Pétain inside the government. The changes did not end with the cabinet. On May 19, Reynaud fired Gamelin and replaced him with the popular General Maxime Weygand. Although Weygand and Pétain had never hidden their disdain for the politicians of the Third Republic, Reynaud thought that bringing both of these men into the government would stiffen France's resolve to fight. In the end, events would prove Reynaud wrong.

In the midst of all of these changes, the Western Allies prepared for some kind of counterattack against the northern and southern flanks of the panzer spearheads. General Weygand had postponed Gamelin's plan for two days until he could review it. The strategy called for British forces from the north to join French units from the south in an assault on Arras. Weygand saw the attack as a chance to re-establish the connection between the Allied armies in the north and the south. While Weygand might have been in a mood for a fight, the generals responsible for carrying out the assault were less enthusiastic. According to British General Ironside, he found generals Billotte and Blanchard

> [in] a state of depression. No plan, no thought of a plan. Ready to be slaughtered. Defeated at the head without casualties. *Très fatigués* and doing nothing. I lost my temper and shook Billotte by the button of his tunic. The man is completely beaten. I got him to agree [to a plan] and Blanchard accepted to take Cambrai. There is absolutely nothing in front of them. They remain quivering behind the water-line north of Cambrai while the fate of France is in the balance.[107]

On May 21, with the strong urging of his government, the commander of the British Expeditionary Force, Lord Gort, sent three of his thirteen divisions south toward Arras to engage the Germans. The recently arrived French 10th Army (Regular) contributed to the counterstroke with two divisions of its own.[108] British tanks made some progress against the Germans, but the lack of coordination between the BEF and the French prevented the Allies from making further gains. French forces arrived in fewer numbers than both sides had agreed were necessary for the operation. The meager contribution from the French meant that the attackers could not exploit any breakthroughs made by British armor. Attacks from the Luftwaffe also had a demoralizing effect on the Allied infantry. With the failure of the counterstroke at Arras, the British began planning for the evacuation of their forces from the Continent.[109]

Weygand Line: Weak Determination and Flexibility

As the three panzer corps reached the English Channel, Army Group B hurriedly marched the remainder of its infantry to prevent an attack from French

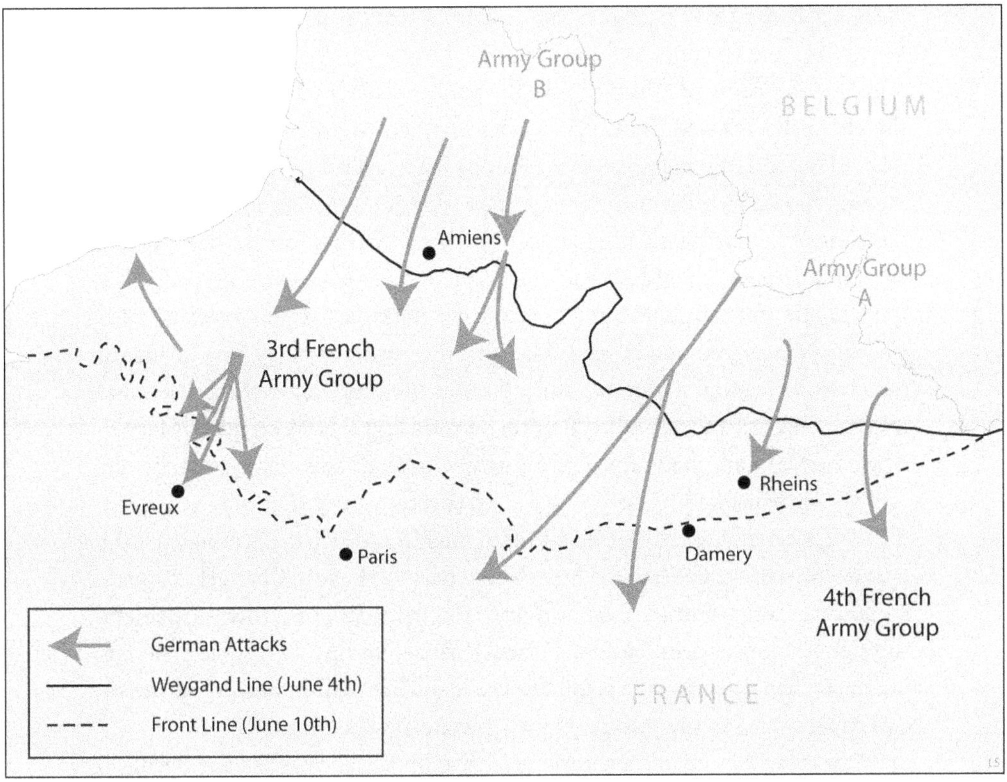

MAP 4.4. Assault on Weygand Line.

forces from the south. By May 22, the armored wedge of German panzers had effectively split the Allied force in half. German forces took positions along the Somme River to guard the armored forces encircling the three remaining Allied armies in the north. As the British, aided by French defenders, staged a miraculous evacuation of their expeditionary force from Dunkirk, the Germans took the opportunity to reorganize and re-equip their panzer corps. Germany's breakthrough along the Meuse had cost its armed forces 60,000 casualties. In contrast, up to this point, France had lost thirty of its divisions with more than 750,000 of its soldiers taken prisoner (see Map 4.4).[110]

To administer the final blow to France (Plan Red), the Germans once more planned to use three armored spearheads to break through the French defenders. The German High Command entrusted Guderian with the task of pushing in the direction of the Swiss border, thereby entrapping the garrison at the Maginot Line. In the center, General Kleist would aim his panzer forces toward

Paris. To the far right, General Hoth sought to strike south and cut off French forces from the Channel.

While the Battle of Dunkirk gave the panzer divisions a much-needed rest, the pause also allowed General Weygand to organize a plan for defending the rest of France. His scheme entailed establishing a defensive boundary from the Somme and Aisne rivers in the west to the Maginot Line in the east. Rather than place the bulk of his remaining forces directly on this line, Weygand decided to employ a defense in depth. When the German forces attacked, he expected his units to give some ground, but as a whole not to break.[111] In his view, this plan gave the French their best chance of keeping the Germans at bay. The balance of forces, however, told a different story. On this occasion, the Wehrmacht could muster three army groups directing 119 divisions against Weygand's 50 French divisions. Germany maintained nearly a three-to-one superiority in manpower.

The German attack began on June 7 and encountered stiff resistance. For two days it appeared that the "Weygand Line" might hold. Then, on June 9, Hoth's panzer corps broke through the French lines near the English Channel at Rouen. Guderian's forces pushed their way past the defenders to create an opening through Champagne. In the center, Kleist's panzer group had trouble penetrating the French defenses. As a result, the German High Command moved Kleist's two panzer corps to support Guderian's breakthrough on the left.[112]

With these two openings in the defensive line, French resistance began to collapse quickly. Guderian's forces drove south over the Plateau de Langres to the Swiss border, cutting off fortress troops at the Maginot Line. At the end of June 11, Kleist widened the breach and crossed the Marne at Chateau Thierry. Weygand, who had suggested the government request an armistice as early as May 29, continued to lobby the French premier to find a way to end hostilities.[113] Reynaud, who had harbored concerns about the morale of both his civilian and military leaders,[114] would have nothing to do with surrender and on June 9 wrote to President Roosevelt declaring: "We shall fight in front of Paris; we shall fight behind Paris; we shall shut ourselves up in one of our provinces, and, if we should be driven out, we shall go to North Africa."[115] Fleeing Paris, the French government relocated to Tours.

While the government remained divided over how much longer to continue the war, the morale of French forces continued to deteriorate. Reports began to surface of acts of sabotage intended to stop the fighting.[116] Similarly, one observer recounted how some French troops began to mutiny rather than continuing the fight:

> In the Hamlet of Tantimont near Germonville, elements of the 23rd and 153rd artillery regiments found themselves encircled by the Germans .... Colonel Charly assembled his officers and told them of his determination to continue fighting. He said that he was aware that "all was lost" but that they were duty bound to fight in order "to save honor." His order was: "Pierce the German lines." The officers were rebellious. Since all was admittedly lost, they argued that they did not see any reason for sacrificing themselves and their men. Colonel Charly was so angry at what he considered cowardice and lack of military honor that at one point he made a gesture as if to reach for his pistol. He thought better of it, and, instead, threatened some of his subordinates with courts-martial for disobedience .... As the ... Colonel passed by his troops, who knew all about the fight between him and the officers, a shot was fired from the dark into his back. Charly dropped dead. A few minutes later, to the relief of the French, the Germans arrived. They ordered the Colonel buried in the village with military honors and then they took all the troops prisoner.[117]

The Decision to Capitulate: Weak Staying Power

The French confronted a bleak state of affairs on June 11, as Churchill flew to Tours to persuade the French leaders to carry on the fight. Reynaud's cabinet remained divided about what to do. A few, like the French premier, wanted to find a variety of ways to resist the Germans. They saw a few options: the government could refuse to surrender and withdraw to France's North African colonies; they could erect some national redoubt, as in Brittany or Bordeaux, and await help from their allies; or they could engage in a guerrilla war against Germany.[118] In their view they needed to choose a course that made a German occupation of France temporary and difficult.

Still others, however, such as Pétain and Weygand, continued to advocate some kind of armistice to save the French Army.[119] For politicians on the Right, such as Pierre Laval, France was about to pay for the sins of the Third Republic. The Left, they contended, had weakened the country's military and had provided ineffectual leadership, and now they saw a chance to begin anew. Germany's victory would give them an excuse to end the Third Republic and begin the process of rebuilding France.[120] Weygand forcefully refused to leave France for North Africa, and he rebutted Reynaud's efforts to convince the cabinet to continue the struggle. In fact, it was the third time France's new commander-in-chief had defied the orders of his civilian superiors.[121]

Fearing a military coup would occur if he remained in power, Reynaud resigned as premier on June 16.[122] With the approval of the French president, Pétain took the reins of government. On the battlefield, the French Army was, in the words of one historian, "now splitting up into fragments, and most of these made little attempt to stand, but merely dissolved in a southerly flow."[123] Many French citizens had also decided the fighting should end. In a few towns, the local populations, led in many instances by their mayors, would prevent the retreating army from destroying bridges. Their determination to keep their country's infrastructure intact eased the German advance southward.[124]

To the relief of most of the country, Pétain's first act was to broadcast a radio address declaring his intention to seek an armistice. Even with his military still fighting, and before consulting the Germans, the new premier decided that it was time to announce his desire to see the fighting end.[125] For the new Right wing–dominated cabinet, the most important goal was not the war with Germany but rebuilding the country as they had sought to do for decades. With Pétain at the head, the "better-Hitler-than-Blum" crowd could now overcome the corruption of the Third Republic and create a government that would best pursue France's interests.

In order to accomplish this task, both Pétain and Weygand believed that they needed to preserve the army. Only with the military intact could they avert one of their greatest fears, a reoccurrence of the Paris Commune of 1871.[126] By saving the armed forces, the new government thought it could prevent the communist agitators they believed contributed to France's weakness from using the country's defeat as a catalyst for revolution. Toward this end Pétain agreed to an armistice with Germany on June 22, ending hostilities, as well as the Third Republic.

Evaluating the Battle of France: An Apathetic French Military

France succumbed quickly on the battlefield for several reasons (see the summaries found in Tables 4.2 and 4.3 at the end of the chapter.) First, French military leaders adopted a strategy that made them vulnerable to Germany's Plan Yellow. When the British Expeditionary Force and the French armies raced to take up positions near Breda, they opened themselves to Germany's three armored spearheads pushing their way through the Ardennes and into the rear of the 9th Army. In addition, France's generals did not anticipate how rapidly the Wehrmacht would exploit its initial breakthrough. The French intelligence service also did a poor job of communicating to the French military Germany's plans, both before and during the conflict. To make matters worse, the British and French could not effectively coordinate their counterattacks once the principal German axis of attack became apparent.

TABLE 4.1: Total Casualties from Battle of France, May 10–June 22, 1940

Battles	French Forces	German Forces	French Casualties	German Casualties	French Per Cap Loss Infliction	German Per Cap Loss Infliction
Plan Yellow	2,776,000	2,758,000	944,500	61,238	0.02	0.34
Plan Red	1,834,000	2,380,000	1,031,000	95,254	0.05	0.43
France Total	2,776,000	2,760,044	1,975,500	156,492	0.05	0.71

Battles	French Loss-Infliction Ratio	German Loss-Infliction Ratio	French KIA	French POWs	German KIA
Plan Yellow	0.06	17	94,500	750,000	10,252
Plan Red	0.11	8.6	24,000	750,000	16,822
France Total	0.07	14.2	118,500	1,500,000	27,024

SOURCES: Michael Clodfelter, *Warfare and Small Armed Conflicts: A Statistical Reference, Vol. II: 1900–1991* (Jefferson, NC: McFarland and Company, 1992); Julian Jackson, *The Fall of France: The Nazi Invasion of 1940* (Oxford: Oxford University Press, 2003); Karl-Heinz Frieser, *The Blitzkrieg Legend: The 1940 Campaign in the West* (Annapolis, MD: Naval Institute Press, 2005); and Martin S. Alexander, "After Dunkirk: The French Army's Performance against 'Case Red,'" *War in History* 14, no. 2 (2007).

More important, the French Army simply lacked a high degree of cohesion. Had French strategy, intelligence, and coordination with its allies been better than it actually was, the end may not have come as quickly as it did. However, an apathetic French military could not have endured a long war with Germany. Most French Army units, but not all, lacked the determination and flexibility to regroup when battlefield events turned against them. On countless occasions, units failed to conduct their missions properly.[127] Even by World War I standards, the French Light Cavalry divisions in the Ardennes failed to relay their information in a timely manner to their respective headquarters.[128] This intelligence might have allowed the French 1st Army to come to the aid of Corap's 9th and Huntzinger's 2nd armies before the situation worsened.

The inability of many units to endure the shock of their first engagements with the German Army also contributed to the French defeat. Several units could not reconstitute themselves for counterattacks. Instead of fighting after the German breakthroughs either at Sedan or the Weygand Line, many French soldiers chose to run, to loot, or even to mutiny. Units suffered from poor morale because of social cleavage in the ranks.[129] What made the situation even worse was the inability of French officers to prevent panic from spreading throughout the ranks.[130] Table 4.1 summarizes the events of May through June 1940. As German casualties can attest, the French military made the Wehrmacht pay a price for its victory. Nevertheless, as the casualty figures show below, French forces fought unevenly. Per capita they inflicted fewer casualties on the German invader, at a surprisingly large rate. Overall, German attackers inflicted per capital fourteen times more casualties than French defenders. Overall, the French Army lacked the determination and flexibility to fight off the highly motivated Wehrmacht.

Moreover, the French military lacked staying power. Without a high degree of regime control, the government could not count on a hard-core group of supporters to keep the country and the armed forces fighting regardless of the strategic circumstance. Instead, the country's political divisions surfaced as the first defeats of May 1940 came to light. These cleavages made France unwilling to suffer the great casualties it had endured in World War I. The same political disputes that had plagued France before 1940 came to the forefront in the midst of the conflict. In general, France's population was relieved to see the fighting stop.[131] There "was a strong sense of preciousness of French Blood."[132]

In particular, politicians on the Right, supported by the military, saw their country's defeat on the battlefield as a golden opportunity to destroy the Third Republic, the government that, in their view, bore the responsibility for weakening France. The country was receiving its due penalty for the Third Republic's transgressions. As French political-military leaders debated strategy, General Weygand declared, "What we are paying for is twenty years of blunders and neglect. It is out of the question to punish the generals and not the teachers who have refused to develop in the children a sense of patriotism and sacrifice."[133] Pétain threatened to resign without an armistice. As historian Julian Jackson has argued, "No Third Republic Premier in June 1940 could have counted in the balance against Pétain once he and Weygand had decided to save the reputation of the army at the expense of the reputation of the Republic."[134] An armistice with Germany enabled them to rebuild France in the way they had strived to do for decades. Rather than waste it by continuing the fight, these leaders concluded that it made better sense to preserve the army as an instrument against communist insurgency. This made sense for French politicians on the Right and their military who feared the Radical Left more than Germany, which many of them admired.

In short, France lacked the will to fight because the country could not unify as it had done in 1914. As one French journalist at the time observed during a conversation following the Battle of France:

"Does it seem to you true to say," he asked, "that the spirit of the French Army and the French People was not as good in 1939 as in 1914 and the will to win was less strong?"

"Many units of the army fought splendidly, but it is true that the war spirit in the French people as a whole was less fiery than in 1914."

"And why? The fate of France was at stake in both instances and the menace greater in 1940."

"That's true, but the France of 1914 was a relatively united country; the France of 1940 was a profoundly disunited country."[135]

A PROFESSIONAL FRENCH MILITARY IN WORLD WAR I

Unlike its performance in May–June 1940, the French Army of World War I demonstrated a remarkable degree of cohesion. After thwarting the German offensive in 1914, most units fought with determination and flexibility for four long years. What accounts for the differences in cohesion between the French military that helped defeat Germany in World War I and the French Army that lost quickly to Germany in World War II?

This brief section uses cohesion theory to explain the variation in French cohesion between the two wars. In contrast to the events of 1940, a professional French Army entered World War I with strong bonds of loyalty within the armed forces. Before 1914, civilian and military leaders shielded the armed forces from societal cleavages that would sap its cohesion twenty-six years later. In a professional military, most units fight with determination and flexibility. However, their staying power diminishes as the probability of wartime victory decreases. Deliberate efforts to create strong bonds inside the armed forces helped it fight with determination and flexibility on the battlefield. Because of these strong organizational loyalties, it was not until April 1917 that significant opposition to the war emerged both at home as well as at the front to briefly undermine the French armed forces. Reeling from the failed Nivelle offensives along the Aisne, several units of the French Army refused to go on the attack. French forces still conducted operations on the defense, and would eventually participate in the British-led offensive at Ypres a few months later. The French kept their soldiers on the battlefield by promising military reforms and by reminding them that with the arrival of the Americans in the war, the balance of power would tilt in favor of an Allied victory. Although the French Army of World War I could not match the staying power of Nazi Germany or the Soviet Union, it showed greater endurance than France in 1940.

Below I describe the domestic political efforts that unified the Third Republic prior to World War I and helped create strong organizational bonds within the armed forces. The following section briefly summarizes the campaign of 1914 on the Western Front, emphasizing how the French Army overcame the shock of the opening months of the war, conducted a fighting retreat, and turned to counterattack at the fateful Battle of the Marne. The third section discusses some of the effects of the war, on French staying power from 1914, through the spring offensives of 1918. Lastly, I outline the performance of the French Army in 1918, with attention to events at the Second Battle of the Marne. I chose these important battles to test cohesion theory's proposition about the battlefield determination and flexibility of a professional military.

On the Eve of World War I

The Third Republic was born out of the bitter defeat of the Franco-Prussian War.[136] After the surrender of Napoleon III at Sedan in the fall of 1870, the rest of the country simply lacked much enthusiasm for continuing the war. Instead, the emperor's embarrassing loss on the battlefield emboldened political reformers who sought to replace the Second Empire with a democracy. As Prussian forces laid siege to Paris, a group of moderate politicians established the Third Republic in Bordeaux on February 12, 1871. The newly formed National Assembly chose Adolphe Thiers to head the government. Worried about the possibility that the continuing war might provide a window of opportunity for more radical elements to seize political power, as they had attempted to do in 1830 and 1848, Thiers quickly negotiated a peace with Prussia, which included relinquishing control over the provinces of Alsace and Lorraine.

After the end of the war, however, Thiers's worst fears were realized. A dispute over wages had angered the largely working-class National Guard in Paris. These units had spent several months defending the capital against a long Prussian siege and grew angry over the way the government treated them. Already well armed, the National Guard seized the city and declared a government of its own on March 26, 1871. The "Paris Commune" controlled Paris until Thiers sent the military into the city to restore order. Bitter street fighting lasted for several days and left 400 dead as well as 1,100 wounded. These events represented a troubling beginning for the new government, and they foreshadowed the political turmoil that would preoccupy the Third Republic throughout its existence.

In the decades to follow, French politics was dominated by a series of disputes that grew out of the widening ideological divide between conservative interest groups and socialist reformers.[137] The Right feared that the program offered by socialist politicians would undermine the economy. Any changes in the economy, they feared, would come at the expense of French business interests and the wealthier classes. They also equated socialist reforms with the same revolutionary elements that had sparked the Paris Commune. A rash of anarchist violence in the 1880s did not help the socialist cause. The Left, in turn, sought reforms that would level the playing field in French society. They pursued a broad agenda designed to improve the lot of the French worker, including a shorter workweek, pensions, and income tax reform. Many of their efforts also concentrated on creating a greater separation between church and state in an attempt to reduce the power of the Catholic Church.

These ideological divides exploded during the Dreyfus Affair of 1894, when a military court convicted a Jewish officer of spying for Germany. The Left accused the French Army of falsely accusing Dreyfus. In response, the Right defended the military and branded the Dreyfusards as unpatriotic. As Alistair Horne explains:

> It was as if some Pandora's box full of all the suppressed and most violent prejudices of the past century had suddenly been opened. Across the nation attitudes polarized and hardened, with the Dreyfusards seen as Republican and progressive, by the intelligentsia; on the other hand, the Army—from whose closed ranks the scandal had sprung—partly Monarchist and largely Catholic, dedicated to hierarchical order, tradition, and obedience. As the crisis deepened, and the Army closed its ranks from external attack, so it seemed to stand further and further apart from the rest of the nation.[138]

On the eve of World War I, however, the French government made an effort to unify the country before what it saw as the impending confrontation with Germany. French politicians decided to set aside their political differences in a drive to bolster societal unity before war. They formed the "Union Sacrée," or the sacred union, a pledge to come together in order to defend France against German aggression. This pact did not resolve the country's political disputes, but merely put them aside until the threat of invasion subsided.[139] Hew Strachan suggests this agreement "reflected not a nationalism that suppressed political divergences, but one that embraced the full range of a liberal society . . . .The Union Sacrée was thus an entirely utilitarian formulation, with the single objective of defending France. Only as the war lengthened would the ideological differences underpinning it become evident."[140]

Creating a Professional French Army: A High Degree of Autonomy

In the years before World War I, the French government also worked hard to prepare its armed forces for war. This was not an easy task, since the military had found itself embroiled in the same political disputes that divided French society.[141] Especially as a result of the Dreyfus affair, the armed forces had lost its standing in society. Given these tensions, fewer and fewer men wanted to make serving in the armed forces a career. Consequently the quality of the officer corps dropped precipitously, since the number of individuals who wanted to embark upon military careers declined. Even more damaging to the relationship between the military and society were the occasions when the government relied on the army to put down strikes. In these ways, societal tensions mirrored the political conflicts that plagued France in the 1930s.[142]

To repair the damage, France's political leaders embarked on a series of policy initiatives in 1911 to improve the military.[143] Known as the "National Revival," the military welcomed these changes, especially since they dovetailed with its own revitalization program started seven years earlier.[144] Against limited opposition from the Left, the government adopted a three-year law for conscription. Not only did the armed forces receive additional manpower and funding, but they also began to improve their training practices. Increasing the amount and the quality of training in the armed forces would prove crucial to sustaining the morale of the army during the war.[145] Over time, the military significantly regained its prestige in the eyes of the public.[146] Together these reforms improved the quality of the French officer corps, an upgrade that would prove particularly fruitful in the very difficult first months of 1914.[147] The "National Revival" spared the French Army from societal divisions, giving it an opportunity to flourish and develop its own internal cohesion. As David Ralston notes:

> The government very early perceived that the army would best be able to carry through its program of reform and to prepare itself for an eventual war with Germany only if it were allowed a very high degree of autonomy and independence within the state. By granting such autonomy, the Republic was relinquishing a vital prerogative: civilian control of the military. But one could argue that since the Republican leaders of France were not prepared to offer a coherent alternate scheme of reform to replace what was being undertaken by the soldiers, too much government supervision over or intervention in technical military matters would only be detrimental to the interests of the army and the nation.[148]

The Campaign in 1914 and the Battle of the Marne: Strong Determination and Flexibility on the Battlefield

When World War I erupted in 1914, Germany had to contend with enemies on its eastern as well as western borders. The German General Staff had long intended to deal with a two-front war by first delivering a knockout blow against France and then moving east to deal with Russia. German strategy, a modified version of the old Schlieffen Plan, called for its military to advance along a sweeping right hook through Belgium that aimed both to envelop the French Army and to encircle Paris from the northwest.[149] Aware of the German approach, France's military leaders nonetheless responded by executing Plan XVII, an attack to liberate the lost provinces of Alsace and Lorraine. The strategy relied upon by the head of the French Army, General Joffre, fit perfectly into the German scheme: while the French extended themselves toward southern Germany, von Moltke's right hook would wheel from the north, cutting across Belgium to deliver the decisive blow in about one month. As French armies would go into the southern side of the revolving door at Alsace, German armies would swing south and west from Belgium into France.

Events did not play out according to either the German or French scripts. The revolving door did not swing as quickly as the Germans would have liked. By the end of August, the French Army had met stiff resistance in Alsace and Lorraine, the area where the Germans had hoped their enemies would overextend themselves. Instead, they found themselves retreating into France, where they would prove useful in the Battle of the Marne. To the north, the three armies of the German right wing had encountered numerous delays in Belgium and raced to encircle Paris. French units meanwhile did their best to conduct a fighting retreat to impede the German advance.

At the beginning of September 1914, the German 1st and 2nd armies, commanded by generals von Kluck and von Bulow, respectively, threatened to reach Paris. To defend the capital, General Joffre ordered his forces nearest the menacing German right wing to retreat westward. These units included, from the left to right, the French 6th Army (directly across from von Kluck's forces), the British Expeditionary Force, the French 5th Army, and the French 9th Army. As the Allied armies withdrew, on September 4 von Kluck altered the script of the modified Schlieffen Plan. Rather than encircle Paris from the rear, he decided to move east of the capital to pursue the retreating French and British forces. Only one day behind them, the German general thought he could quickly roll up the French left. Seizing the capital, he concluded, could wait until the defeat of the French Army.

With visions of a decisive victory in mind, von Kluck moved his units directly east of Paris on September 5. Allied air reconnaissance, however, discovered that this move had exposed the German 1st Army's right flank. Joffre, on the urging of the military governor of Paris, Gallieni, ordered the French 6th Army to counterattack von Kluck's units along its recently exposed flank on September 6. As the bulk of von Kluck's Army fell back to deal with the new attack, a thirty-mile gap opened in between the German 1st and 2nd armies. The French 5th Army wandered into this hole, splitting its units to attack both von Kluck's and von Bulow's armies. At the same time, the BEF also wandered into the same hole, but failed to capitalize on their fortunate position before the Germans reacted. The French 9th Army attacked into the intersection between the German 2nd and 3rd armies.

The battle raged for three days, from September 6 to 9. France's 6th Army barely held its ground under the weight of fierce German counterattacks. Reinforcements brought from Paris via Taxi helped it maintain its position. On September 9, von Kluck received orders to fall back. German forces conducted a fighting retreat forty miles back to the Aisne River. At the cost of 80,000 casualties, the French Army had dashed Germany's hopes of a quick victory in 1914.[150]

Moderate Staying Power: Surviving the Mutinies of 1917

After the Battle of the Marne, France and its allies found themselves in a long, costly war marked by indecisive engagements. Only the Russians endured more agony than the French Army. Over the first three years of the conflict, France averaged 63,000 casualties per month, the highest rate among combatants on the Western Front.[151] Yearly battle deaths from 1914 to 1916 perhaps best capture the pain the French military had to endure in this period: in 1914 France lost 307,000 men; in 1915, 335,000 men; and in 1916, 218,000 men.[152] By the end of the war, 73 percent of all French soldiers mobilized would become casualties; this far exceeds the 41 percent casualty rate of British soldiers inducted into the British Army.[153]

At the outset of hostilities, military leaders promised a quick victory. There was, as Jean-Jacques Becker notes, "the widespread conviction that the war would turn out to be an adventure, cruel perhaps, but of short duration."[154] These hopes were quickly dashed in 1914, and the possibility of victory looked uncertain three years later in 1917. The economic pressures of war took a particular toll on the French public. Resentment grew when the government imposed a larger national income tax in 1916.[155] In January 1917, workers whose wages suffered and who endured higher prices took to the street in the first wave of strikes to afflict the country. Worker strikes spread throughout the country, accelerating in June 1917 when the outcome of the war looked bleak. Only the ascension of Clemenceau later in 1917 seemed to assuage the working class.[156]

These frustrations about the war would also reach the military in 1917. At the beginning of that year, the British and French remained optimistic about the prospect of finally inflicting a decisive defeat on the Germans. Fresh from their grueling defense of Verdun, the French participated in a series of Allied attacks along the Western Front that they hoped would break the back of the German Army. The Nivelle offensives of 1917, named after the French general who designed the attack plans, represented the focal point of the Allied effort. On April 17, a French force of 800,000 men attempted to break through the German lines between Soissons and Reims, a forty-mile front that included the rocky ridges of the Chemin des Dames. Nivelle, one of the heroes of Verdun, virtually guaranteed a French victory. Alerted to the attack, however, the Germans stifled the offensive in four days and inflicted 120,000 casualties on the disillusioned French Army.[157]

The failure of the Nivelle offensives to bring about the victory promised by France's military leaders sparked a mutiny in the armed forces.[158] It appeared that the French armed forces, which had suffered the heaviest losses in proportion to its mobilized strength, had almost reached the breaking point.[159] The turmoil reached its apex on April 29, when sixty-eight divisions of the French Army refused to go on the attack. To restore order, the French military took a number of steps: they replaced Nivelle with Pétain as the head of the army; they arrested more than 23,000 mutinous soldiers; and they executed sixty-one of the instigators. Pétain also quelled the revolt by promising to institute a series of reforms designed to improve the quality of life at the front. More important, he boosted morale by reminding his soldiers that the arrival of American forces would provide the Allies with the additional capabilities that would tip the balance in their favor.[160] By the end of August, the French Army had recovered enough from its disenchantment to participate in end-of-the-year offensives.

Bolstered by the slowly increasing American military presence on the Continent, both French society and its military managed not only to overcome the malaise of 1917 but also to endure the massive German offensives of 1918. According to Jean-Jacques Becker, "There was never more than a brief moment—in the spring of 1917—when morale dropped to a dangerously low level following the sudden crumbling of hopes in a speedy victory. However, once that crisis was over, morale rose and remained adequate even during the worst moments of 1918." By July, Ludendorff's last gamble to win the war for Germany died at the Second Battle of the Marne. With the Germans exhausted and overextended, the Allies embarked on a series of offensives that turned the tide of the war. On the road to victory in the summer and fall of 1918, the French Army had enough stamina to sustain 531,000 additional casualties.[161]

Second Battle of the Marne, 1918: Maintaining Determination and Flexibility on the Battlefield

After the first of his four offensives had failed to achieve a significant breakthrough of the Allied lines, the German quartermaster, General Ludendorff, launched one more attack on the Western Front, on July 15. The German attack took place along both sides of Reims and aimed to reach the Marne River east of Paris.[162] East of Reims, the German 1st and 3rd armies made only minor gains before French forces stopped their advance at 1100 hours. To the west of Reims, however, the 7th German Army smashed its way to the Marne between Chateau Thierry and Epernay. Before the French halted their advance, General Max von Boehn's 7th German Army had established a bridgehead south of the Marne stretching nine miles long and four miles deep. The distant thunder of artillery fire conjured memories of the First Battle of the Marne for the residents of Paris.

On July 17, the French 9th Army, which had moved up from the south to replace the badly mauled 6th Army, halted von Boehn's advance. With this move, the last of Ludendorff's offensives had failed. In the end, they had achieved nothing more than overextending their forces along a wide salient near the Marne. The next day, the Allied commander-in-chief, General Foch, took advantage of the exposed German forces and unleashed a massive counterattack. The French 6th and 10th armies provided the primary efforts for the assault, with the 5th and 9th armies guarding their flanks to the east and south, respectively. With the aid of 350 tanks, the French 6th and 10th armies advanced as far as five miles on the first day.

When the Allies threatened to cut the vital road between Chateau-Thierry and Soissons, Ludendorff began to withdraw his forces from the Marne. On August 2, the Allies liberated Soissons. The next day, German forces conducted an orderly retreat to the Vesle and Aisne rivers. An American attack on August

6 failed to dislodge the entrenched German forces, ending the Second Battle of the Marne. The German spring offensives of 1918 had ended with the Allies seizing the initiative.

CONCLUSION

This chapter explains why French armed forces could fight with cohesion in World War I but could not do the same in World War II. Without organizational autonomy, the French Army of 1940 could not form the internal bonds cohesion requires. In the years prior to World War II, French society remained sharply divided and anxious to avoid another long conflict. For many interest groups in France, including elements of the military, the threat of revolution outweighed their fears of Germany.

These disputes between the Right and the Left infected the military and sapped its will to fight, depriving it of any capacity to endure adversity. An apathetic French Army could not withstand the pressure of early defeats and quickly collapsed. At the critical battle along the Meuse, most units fought without determination or flexibility. They missed ample opportunities to stymie Germany's vulnerable blitzkrieg operations. Similarly, the armed forces lacked staying power. Military leaders pressured civilian leaders to end the war. Douglas Porch describes the two eras this way: "When defeat seemed certain in 1940, a generation of soldiers who had suffered arrogant politicians since 1900 did not pass up the opportunity to pull the rug from under the feet of its tormentors."[163]

In contrast, organizational autonomy and internal bonds bolstered the cohesion of a professional French Army in World War I. (See Tables 4.4 and 4.5.) On the eve of war, civilian leaders gave the military the autonomy to train and to create strong organizational bonds among its members. With these bonds, French units possessed the cohesion to recover from the initial setbacks of 1914 and then survive the agonizing years to follow. As cohesion theory explains, these internal bonds motivate and sustain national militaries, creating a sense of loyalty and trust inside the military. These bonds also imbue soldiers with a sense of duty to country. French forces, therefore, fought with determination and flexibility, turning back the German invasion and going on the offense. They also displayed moderate staying power: except for one brief moment during the mutinies of 1917, France's armed forces managed to endure the war. Even during the mutinies, military officials proved able to restore morale and get French forces fighting once more. The evidence from these two cases shows the explanatory range of cohesion theory. Not only can it explain differences between cases, as the next few chapters will demonstrate, but the theory can also account for within-case variation.

Alternative Arguments

Looking at the same historical evidence, analysts might raise two counter-arguments to the account given here. They might argue that French performance in 1940 reflected an outdated military doctrine. On a different note, they may claim that German strategy, and not weak cohesion, explains why most French units fought without determination and flexibility. Did an outdated military doctrine ensure the defeat of France, and was it responsible for the lack of French cohesion? Although this view remains quite popular, it seems unlikely that doctrine could account for the missing cohesion of the French Army. In fact, the opposite appears more probable: internal divisions, afflicting both French society and its armed forces, prevented extensive training of conscripts and reservists. A defensive doctrine looked like an attractive option for a military without the organizational autonomy to train its armed forces and to generate loyalty and trust among its personnel.[164]

Still further, officers did not display blind adherence to doctrine. French generals in 1940 also understood the need for haste in countering the German advances, but a lack of cohesion, rather than a lack of understanding of the threat, prevented timely counterattacks. More to the point, French units employing methodical tactics performed well against German units in the Gembloux Gap.[165] Some historians even argue that the "methodical battle" actually won World War II. In motivated hands, such as those of the British under General Montgomery, these allegedly outdated techniques proved quite successful.[166] Eugenia Kiesling puts it best: "We will never know whether the doctrines of firepower and methodical battle would have met the case if supported by the kind of training and leadership to which the high command paid rhetorical homage."[167]

Some might suggest that German strategy, not military cohesion, explains why France fought in World War I but not in 1940; French performance varied because German strategy varied. As the reasoning goes, the combination of a flawed Schlieffen Plan and military technology prolonged the war in 1914. German strategy improved in 1940 and, for this reason, allowed Hitler to knock out the French Army quickly.[168] There are three problems with this potential competing explanation.

First, cohesion and the concept of strategy are not mutually exclusive explanations; rather they can complement each other. My argument deals with military cohesion, or the capacity of armed forces to fight with determination and flexibility under stressful battlefield conditions. Strategy deals with how to use force in times of war. Placing stress on an adversary's armed forces is undoubtedly one of the intended objectives of military strategies.

Second, although German strategy in 1940 stressed the French military, it was not without risks. On seven occasions, France's armed forces could have stopped the consolidation and the breakout from the Meuse, but it failed to do so. Nowhere were French forces so outnumbered that they could not have countered the German crossings. In fact, there were ample opportunities for such an attack. Even German histories of the events show that they worried about such an attack because for two days their armor struggled to cross the river in force, while French armored units in the vicinity could have inflicted great damage.[169] Moreover, strategy could not possibly explain the lack of determination displayed by French units in 1940. Nor could it explain why the French military decided to quit the war quickly. In short, strategy and cohesion provide the necessary as well as sufficient conditions for the French defeat.

Third, the amount of stress endured by the French Army in World War I, when German strategy was not at its best, was significantly greater than the pain it experienced in 1940. During the campaign of May to June 1940, the French military suffered 84,000 battle deaths (see Table 4.6). In contrast, the French Army of 1914 sustained 140,000 battle deaths in the first four days of the war (see Table 4.7). The army still retained enough stamina to conduct a month-long fighting retreat back to the Marne. Soon after, it retained sufficient staying power to turn and mount a counteroffensive that came close to inflicting a decisive defeat on Germany. By the end of 1914, the French Army saw 307,000 of its soldiers die in battle. Twenty-six years later, this same military could not overcome the shock of the first series of defeats, whether along the Meuse or the Weygand Line. Nor could French forces sustain huge casualties.

Nationalism. These two cases provide some evidence for the argument that nationalism determines military cohesion. French forces went to war in 1914 unified in the defense of their homeland. They showed determination on the battlefield, as the argument would expect. Similarly, a divided France could not promote nationalism in 1940, and the armed forces fought, according to the argument, with fragile cohesion. A key difference between cohesion theory

TABLE 4.2: Assessing Staying Power: Apathetic French Armed Forces, 1940

Evidence
General Weygand pushed for an armistice on May 29 and became increasingly insubordinate
Senior military leaders, like Weygand and Petain, concluded regime not worth fighting to preserve
President Reynaud resigned fearing a military coup
Increasing acts of indiscipline after the collapse of the Weygand Line

PREDICTION: Weak Staying Power.
Military Collapses Quickly as the Probability of Victory Decreases.

TABLE 4.3: Assessing Battlefield Performance:
Apathetic French Armed Forces, 1940

Battles	Evidence
Dinant	Better equipped French defenders retreated in panic
	French units failed to counterattack and withdrawal in good order
Sedan	55th French Infantry Division could not recover from bombardment
	Better equipped French defenders fled in panic rather than fight
	Many units collapsed when trying to retreat or encountering panicked troops
	Officers struggled to motivate units for counterattacks
	Most German units fought harder than French units
Montcornet, Laon, and Arras	Strong counterattack by elite French 4th Armored division
	Local French commanders in state of shock, hampered willingness to attack
	British forces mounted strong counterattacks
	French units less determined than British units during Arras counterattack
Weygand Line	Most French units fought hard along the Weygand Line for several days*
	Most French units struggled to regroup after German breakthroughs
	Many units succumbed to panic and dissolve rather than fight

PREDICTION: Weak Battlefield Performance.
Only the Best Units Fight with Determination and Flexibility.
(*Indicates evidence against predictions.)

TABLE 4.4: Assessing Staying Power:
Professional French Armed Forces, 1914–18

Evidence
French military recovered from initial defeats and large casualties in 1914 to counterattack
General Petain resolved mutinies by promising fewer costly offensives until Americans arrive
French armed forces recovered from mutinies to defend successfully in spring 1918

PREDICTION: Moderate Staying Power.
Military Collapses Slowly as the Probability of Victory Decreases.

TABLE 4.5: Assessing Battlefield Performance: Professional French Armed Forces, 1914–18

Battles	Evidence
First Marne	Most French units fought after suffering large losses French units improvised to counterattack at Marne
Second Marne	Most French units defended against strong German offensive French units were able to counterattack after stalling German attack

PREDICTION: Strong Battlefield Performance.
Most Units Fight with Determination and Flexibility.

TABLE 4.6: French Casualties: World War II

	Casualties	Initial Forces	Casualties/ Initial Forces	POWs	POWs/ Casualties
1940	2,504,910	3,277,000	76%	1,950,000	77%

SOURCES: Kevin D. Stubbs, *Race to the Front: The Materiel Foundations of Coalition Strategy in the Great War* (Westport, CT: Praeger, 2002); John Ellis and Michael Cox, *The World War I Databook* (London: Aurum Press, 2001); Randal Gray and Christopher Argyle, *Chronicle of the First World War, Vol. I: 1914–1916* (New York: Facts on File, 1990); and Michael Clodfelter, *Warfare and Small Armed Conflicts: A Statistical Reference, Vol. II: 1900–1991* (Jefferson, NC: McFarland and Company, 1992).

TABLE 4.7: French Casualties: World War I

	Casualties	Initial Forces	Casualties/Initial Forces	POWs	POWs/ Casualties
1914	969,000	3,781,000	26%	N/A	N/A
1915	1,598,000	4,978,000	32%	N/A	N/A
1916	1,079,000	4,857,000	22%	N/A	N/A
1917	890,000	4,327,000	20%	N/A	N/A
1918	1,086,000	4,340,000	25%	N/A	N/A
	Total Casualties	Total Mobilized	Casualties/Total Mobilized		
	5,622,000	8,091,000	69%	507,800	8%

SOURCES: See Table 4.6.

and the nationalism claim concerns staying power. The nationalism argument predicts either strong or weak staying power: militaries fight to the death without showing signs of collapse or they collapse quickly. There is no prediction between these two extremes. Only cohesion theory can explain the conditions under which the French Army in World War I began to show signs of collapse and why it would recover.

Small-Group Theory. French armed forces sought to promote primary group bonds within their armed forces by training units together. Units also took on new recruits as a group, rather than receiving individual replacements. In other words, the French Army followed the script for creating bands of brothers within small units.[170] However, the evidence does not support the primary-group argument. The casualty rate for French forces in World War I was too high for these small-unit bonds to form. Individuals could feel loyalty for their country, organization, even fellow soldiers, but not because of a long history of repeated experiences with comrades. The primary group died before it could really play any role in sustaining cohesion.[171] Additionally, for the reasons detailed above, class tensions undermined the training necessary to create these small-unit bonds in the French military of 1940.

Democracy. Lastly, the conventional wisdom about democracies cannot explain why the French Army remained intact from 1914 to 1918, but crumbled in 1940. Because the Third Republic was a democracy, albeit an unstable one, its armed forces should have fought with great determination in both conflicts. Instead, the performance of the French Army varied widely. From 1914 to 1940, the type of government in France remained constant and the will to fight of its armed forces changed over time. The better explanation, then, for why the determination and staying power of these two armies varied is cohesion theory: the French Army lost its strong internal bonds between the two wars.

5 The Soviet Union, 1941

IN THE EARLY MORNING HOURS of June 22, 1941, Nazi Germany launched a surprise invasion of the Soviet Union. For more than five months of Operation Barbarossa, German forces drove a battered Soviet Army to the gates of Moscow. Throughout this ordeal, the Soviet Union lost more than 4.5 million soldiers, casualties that would shatter the cohesion of most militaries. However, the *authoritarian* Red Army refused to break, demonstrating formidable staying power. With their backs to the capital, Soviet armed forces mounted a fierce counterattack. The month-long offensive pushed back the Wehrmacht, erasing many of its gains, and costing the Soviets another 300,000 casualties.[1]

This chapter examines the performance of the Soviet Armed Forces while defending its homeland against the German onslaught. I use *cohesion theory* to explain how and why Stalin's army recovered from the shock of Germany's initial blow to launch a counterattack. I explain how and why a strong state composed of hard-core supporters kept the armed forces intact and fighting. Ideology, the defense of the Russian homeland, and a highly coercive state prevented organized opposition in the armed forces and at home. These factors explain why the czarist Russian army collapsed in 1917 and why the Red Army did not under more dire circumstances.

The chapter begins by explaining the sources of Soviet military cohesion. Next, the chapter examines the key battles Soviet forces fought against Germany's Army Group Center, which sought to capture Moscow. This section also includes a discussion of the Soviet winter counteroffensive. The discussion then turns to an evaluation of Soviet staying power to describe the source of Soviet resilience: a powerful regime capable of motivating its armed forces through a mix of nationalism and coercion. Finally, the conclusion evaluates cohesion theory and compares the Red Army's cohesion with that of Germany and France during World War II.

AN AUTHORITARIAN RED ARMY

The cohesion of the Soviet Armed Forces depended on the regime's strong control over the country's population. Ideological appeals, especially nationalism, and pressure from the regime's hard-core supporters drove most units to fight with great determination on the battlefield. Moreover, regime loyalists relied on the same combination of positive and negative incentives to prevent the armed forces from collapsing even when defeat seemed imminent. Enduring terrible strategic circumstances and enduring more casualties in its second war with Germany, Stalin's armed forces showed much greater staying power than the czar's army in 1917.

While a strong regime might have saved the Soviet Union, the regime's use of terror, however, did prevent the armed forces from training and operating with any autonomy. Stalin's desire to control the military undermined its professionalism. Officers and enlisted personnel could not develop the same bond of loyalty and trust that existed, for example, in the German and French armed forces of World War I. Joseph Stalin's grip on power depended on fear and intimidation, forces that ultimately undermined the military's autonomy to train without political interference.[2] As a result, most Soviet units fought with determination but lacked flexibility, showing little initiative to adapt to changing circumstances on the battlefield. As Stalin relaxed control later in the war, flexibility improved.

The Soviet State: High Degree of Regime Control

The defining feature of Stalin's regime state was its high degree of control over the Soviet Union. Stalin ruled as an autocrat over a vast country of more than 170,000,000 Soviet citizens. Through its ideology and use of political terror, the Soviet state could demand and receive unconditional loyalty from its people. In turn, the regime created an authoritarian military with the staying power it needed to survive the German onslaught of 1941, and the years to follow. These strong regime ties produced a hard-core group of supporters that embraced Stalin's call for unconditional loyalty. The supporters intimidated less enthusiastic citizens, including members of the armed forces, into showing the same level of devotion to the Soviet state.[3]

Soviet ideology during the Stalin era rested on four main pillars.[4] The core of the state's ideological message was a commitment to advancing the Bolshevik revolution of 1917. To this end, Soviet ideology argued that the country was in the midst of its national mission to build a socialist state. The mission required great sacrifice from workers and peasants alike. Plans for industrialization and collectivization of the countryside required great vigilance, not only against foreign enemies but domestic subversives as well.[5] Stalin also erected a cult of personality around his infallibility as a leader. He not only inherited Lenin's mantle as the leader of the revolution, but he also created the image of an all-knowing, ever-present figure in Soviet life.[6] With the population already isolated from the outside world, the strict control of information and a steady diet of propaganda reinforced this message.[7] Finally, when necessary, Stalin would incorporate elements of Russian nationalism to motivate the population. During the first months of the war with Germany, Stalin abandoned standard communist rhetoric about class struggle and instead rallied the nation to defend Holy Mother Russia.[8]

To ensure that the Soviet population embraced the state's ideological message, and to maintain control over the country, Stalin relied on the use of terror. State intimidation and violence against enemies of the revolution was a tradition in Soviet politics. The Bolshevik regime used its secret police to murder 250,000 enemies during the Russian Civil War.[9] Stalin continued this trend during his radical efforts to reform the economy and to collectivize the rural areas of the Soviet Union. During the "Great Terror" of the 1930s, the Commissariat of Internal Affairs (NKVD) imprisoned and murdered thousands that opposed Stalin's economic vision—so-called saboteurs of progress. The forced collectivization of farms in Ukraine caused one of the twentieth century's largest famines, killing a total of 7 million people.[10] Stalin's own paranoia about threats to his rule drove levels of fear and violence upward. Phony charges of treason led to arrests and tortured confessions. These admissions led to further arrests. A cycle of never-ending repression ensued. Membership in the Communist Party did not safeguard individuals from random arrest, imprisonment, and death. Not even the NKVD and its director could escape. Eventually, the terror reached the Soviet Armed Forces, causing great damage to the Soviet efforts to prepare for its eventual war with Germany.

The arrest and execution of Marshall Mikhail Tukachevsky in 1937 marked the beginning of Stalin's purge of the armed forces. Stalin accused Tukachevsky of conspiring with Germany to overthrow the Soviet state. The Soviet dictator, ironically, had promoted him only two years earlier. While these charges were false, Tukachevsky had disagreed with Stalin about the use of political officers and propaganda in the military. After Tukachevsky, Stalin had seventy-one of the eighty-five officers on the Military Council executed. Beyond these military leaders, 45 percent of the senior officers were either fired or executed. A total of 41,000 junior officers suffered the same fate.

By late 1939, Soviet gulags housed 3.5 million enemies of the state. To maintain this aura of fear throughout the population, Stalin practiced killing for the sake of killing, even instructing state security agencies to murder by "random quotas."[11] As Norman Davies observes, "Thousands upon thousands of totally innocent people were shot after being forced to denounce others who would then be shot in turn. And the cycle of false denunciations and murders snowballed until it threatened to paralyze the whole country . . . . Suffice it to say that paralytic fear and coercion drove all the policies whereby the Stalinist state was mobilized to fight the war."[12]

Low Degree of Autonomy for the Soviet Armed Forces

As Stalin's regime asserted absolute control over the Soviet Union, it undermined the military's autonomy to train and, consequently, to build bonds of trust within its ranks.[13] As Roger Reese notes, "Coercion, rather than persuasion began to dominate civil-military relations."[14] Two features of Stalin's police state undermined the military's efforts to create strong internal bonds. The first obstacle to strong organizational ties was the direct control the Soviet state exerted over the military. Soviet military officials had little autonomy to conduct both training and military operations.[15] As a result of these measures, there was "an atmosphere that inhibited the free exchange of ideas and controversy."[16]

Political indoctrination was an important mechanism of control the regime exerted over the military.[17] The responsibility for the political indoctrination of the Red Army fell to the Political Administration of the Revolutionary Military Council of the Workers'-Peasants' Red Army (PUR). To achieve its mission, PUR designed a two-year course of political education in 1925. The program remained in place virtually unchanged until the mid-1930s, when the program was split into three variants, one for literate soldiers, one for the semiliterate, and one for the illiterate. These indoctrination programs continued through, and after, World War II.[18]

Although the most prominent part of the program was militarization, which promoted combat readiness and soldier morale, the second year in service was spent in "political training designed to transform him into a loyal, informed

Soviet citizen and potentially a cadre who might enter the Soviet bureaucracy upon release from the military."[19] Soldiers were educated on government institutions, laws, geography, and Soviet/Russian history, as the Bolsheviks defined it. The "internationalization" training, which also occurred in the second year, attempted to educate soldiers about the foreign policy of the Soviet state. They learned to identify nations bordering the Soviet state and to distinguish enemies from friends. Additionally, thousands of soldiers enrolled in the International Organization for Aid to Imprisoned Fighters for the Revolution (MOPR) in an effort to forge symbolic ties with foreign workers; leaders of foreign governments sympathetic to the Soviet Union or of foreign revolutionary parties that paid frequent visits to Red Army units.

Political officers, or commissars, also played a part in the state's effort to control the armed forces.[20] They took charge of the political education during training. Commissars not only taught but also monitored what took place within units. On the battlefield, the Soviet state would sometimes give them dual command alongside a unit's regular officer. This ensured that soldiers remained loyal to the state while at war.[21] As Roger Reese describes:

> Commissars, not commanders, were charged with keeping morale up and discipline firm. Since the officers were not responsible for their men's morale and discipline, they took none of the usual measures to ensure the health and welfare of their men. This lack of involvement weakened commanders' bonds with their men, diminished their feeling of responsibility, and consequently lowered the level of respect accorded them by their men.[22]

On the battlefield, commissars played an instrumental role in motivating common Red Army soldiers fighting on the Eastern Front. They were a brutal and effective instrument with which the Stavka, the Soviet High Command under Stalin, imparted its demands for unwavering loyalty and complete sacrifice for the Soviet motherland. Comparing the Red Army's ability to fight during World War II with the collapse of the czarist army of World War I, David Glantz highlights the role of commissars:

> In contrast [to the czar's army of 1917], Stalin's army remained a coherent, viable, and reliable military force throughout its war in spite of even greater defeats and losses [than what was suffered during World War I] because an armada of commissars and political workers exercised stringent political control and military tribunals and other security organs used harsh punitive measures ruthlessly and effectively to enforce discipline within its ranks.[23]

The second obstacle to strong organizational bonds in the Soviet military was the regime's use of coercion and intimidation. In addition to the purges, which cut down a wide swath of officers, the regime enforced draconian measures of punishment. The Soviet Military Code warned deserters that they risked death if caught. Soldiers choosing to surrender rather than fight would bring punishment to their families.[24] On the battlefield, the NKVD would administer punishment to personnel failing to carry out their duties.[25] One report suggests that the NKVD killed 200,000 soldiers during four years of war. At Stalingrad, they shot 15,000 men of the Red Army.[26]

Coupled with the purges of the Soviet officer corps, the military lived in a climate of fear and suspicion. Under these circumstances, soldiers professed a strong loyalty to the state and worried about expressing any other. "They possessed a primary loyalty to the Communist Party and at best a secondary loyalty to the Red Army, if they identified with the army at all."[27] Moreover, there existed an atmosphere of "suspicion, intimidation, and irresolution" that would hamper units on the battlefield.[28] The purges also hampered the quality of the Soviet armed forces, leaving it with politically reliable officers that were also mediocre and lacked experience.[29]

Following its defeat to Finland in 1940, the Red Army, under Timoshenko's leadership, attempted to "redress the balance in the Red Army which had swung in favor of the Political Administration."[30] An order issued on May 16, 1940, directed Red Army troops to "be trained only in the things necessary for war and only in what they would be required to do on the battlefield."[31] Additional reforms focused on increased obedience to the officer corps by rank-and-file soldiers and discipline within the army.[32] The Red Army leadership raced to put the revised military program into effect as the threat of war with Germany grew larger.

However, as John Erickson observed, "With Timoshenko's reforms just beginning [in 1940], in no sense could the Soviet Union be regarded as ready for war. Three years of degeneration [during the purges] could not be re-couped in a few weeks."[33] Between 1937 and 1939, nearly 35,000 officers had lost their positions. Some 90 percent of the military district commanders lost their jobs to subordinates. Although no more than 7.7 percent of officers lost their jobs, and the purge did manage to clear some "dead wood" among the officer corps, the turnover still created problems.[34] Not only did the purge instill the fear of "political error" among the officer corps, the repression also hindered training and preparation.[35] Even as the Soviet General Staff attempted to institute reforms to learn from the Soviet experience in Poland and then Finland, officers felt reluctant to embrace these changes too quickly. As David Stahel explains, "There was also an undercurrent of fear about being associated with the new reforms. Engaging too quickly with them or demonstrating too much initiative might prove dangerous if the pendulum should swing again and the men now advocating the changes should suddenly become counter-revolutionary."[36]

Still further, the purge interfered with recruitment, training, and supply, while also severely undercutting morale. Shortfalls in materiel also compounded the deleterious effects of the purge. Part of the prewar reforms included an expansion of the Soviet Army that in the short term created undermanned units. Meanwhile training suffered, since newly promoted officers could not draw on experience to properly train fresh recruits. A poorly educated officer corps compounded these difficulties.[37] Many units also deployed without their required equipment, because of shortages or repairs. Soviet armored forces represented a particularly glaring deficiency. Out of the 23,767 tanks it could field, only 1,861 included the more modern, formidable KV heavy tanks or T-34 medium tanks.[38]

Causal Logic: An Authoritarian Red Army Fought with Strong Staying Power but with Moderate Battlefield Performance

The combination of a regime with a high degree of control over Soviet society and a low degree of autonomy for the armed forces created an authoritarian Red Army. According to cohesion theory, Soviet armed forces will display strong staying power: avoiding collapse even when victory looks unlikely. A high degree of control permits a state to keep its military fighting through a mix of positive and negative incentives. In the Soviet case, Stalin motivated the armed forces and society with appeals to Russian nationalism as well as "healthy doses of harsh discipline."[39]

Additionally, most of its units will fight with determination on the battlefield. The same set of incentives that keep the armed force from collapsing as a whole also motivate units in combat to fight hard. Lastly, most units will lack flexibility on the battlefield. Without the autonomy to train, the armed forces will lack the bond of trust that enables officers to take risks and to show initiative when confronting difficult circumstances on the battlefield, such as breaking out of encirclements. Below, I discuss first the determination and flexibility of Soviet units in key battles of 1941, and then I describe the staying power of the Red Army.

OPERATION BARBAROSSA: JUNE 1941 TO JANUARY 1942

Germany's attack on the Soviet Union caught Stalin and most of the Soviet Army by surprise. The early morning assault on June 22, 1941, inflicted great damage on the Soviet Air Force, giving the Luftwaffe an advantage it maintained for the rest of the year, weather permitting. Germany invaded the Soviet Union with more than 3,000,000 soldiers as well as an additional 500,000 troops from its allies. To oppose them, the Soviet Union had arrayed along its western frontiers 2,900,000 of the 5,000,000 personnel in its large armed forces. This discussion focuses primarily on Army Group Center's drive to take Moscow.

Frontier Battles of Bialystock-Minsk: Determination but Lack of Flexibility

The Battle of Bialystok-Minsk, also known as the Frontier Battles, took place during the opening stage of Operation Barbarossa, from June 22 to July 9, 1941. It pitted Field Marshal Fedor von Bock and the 1,800,000 men of Army Group Center against the 627,300 soldiers of Lieutenant-General D. G. Pavlov's Western Front. The capture of Bialystok-Minsk would open an approach to Smolensk, and then, the road to Moscow. The battlefield stretched from the Soviet border to the east side of Minsk. During operations, Soviet forces surprised the Wehrmacht by fighting with unexpected tenacity. In the end, German armored forces encircled and then destroyed the defending Soviet armies. The Red Army lost 417,000 as killed and wounded, with another 290,000 becoming prisoners of war (POWs).

The first day of the assault began with a massive air and artillery bombardment (see Map 5.1). Three days later, with the Red Air Force decimated, the Luftwaffe gained control of the skies.[40] In the confusion, German forces overran several border posts before guards could assemble. Other Soviet units fought to their last round, delaying the Wehrmacht for a few hours.[41] Stalin and General Timoshenko, now commander of all Soviet forces on the central front, reacted by ordering a general counteroffensive. Although the battlefield remained chaotic and they lacked information, Soviet officers obeyed these orders. They understood that any deviation from a directive would result in severe punishment. With this knowledge undoubtedly in mind, Pavlov organized a counterattack to prevent the encirclement of Bialystock. Soviet forces at the time lacked effective communication, air cover, and logistical support. Nevertheless, they mounted an attack and then fell into a German antitank ambush.

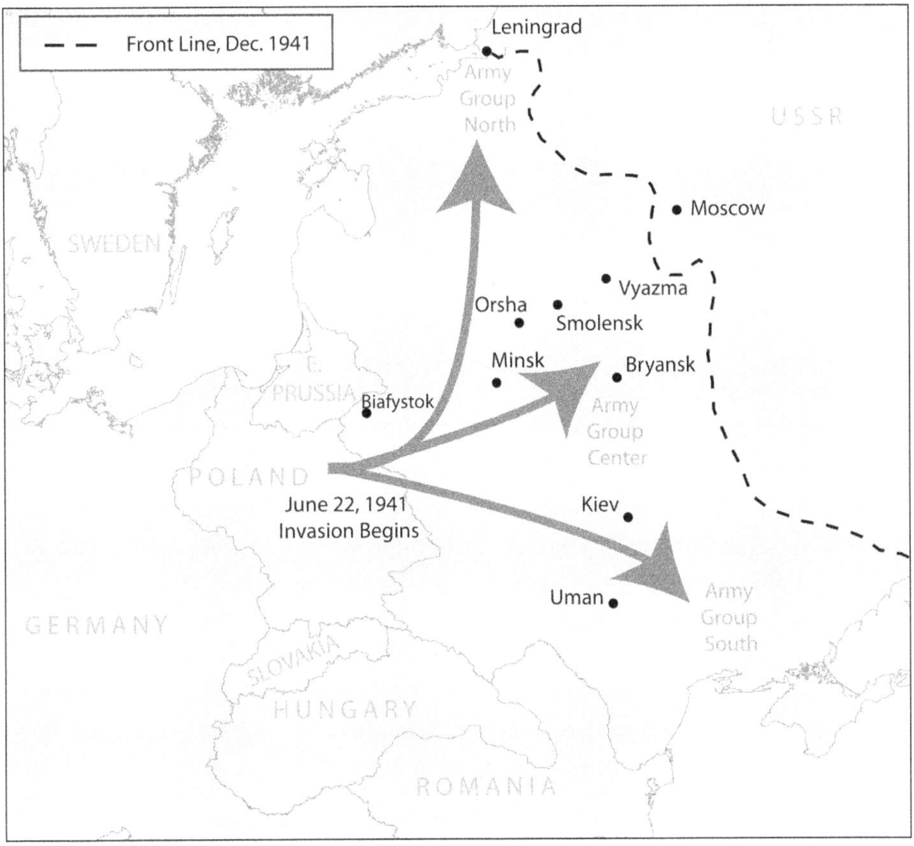

MAP 5.1. Operation Barbarossa.

On June 25, the Germans fought off Pavlov's disorganized and chaotic counterattack.[42] His failure opened the way to Minsk for the German panzer forces. Five days later, on June 30, Von Bock's 2nd and 3rd panzer groups enveloped the 10th, 3rd, and 13th Soviet armies west of Minsk. Despite frenzied counterattacks, the German forces completed the encirclement on July 9. This maneuver effectively destroyed Pavlov's Western Front. Stalin responded to this defeat by having Pavlov executed on July 22, which reminded the officer corps that the penalty for failure was death at either the hands of the NKVD or the German invader.[43] During the offensive, Army Group Center had advanced 360 miles and inflicted 417,000 casualties on Pavlov's forces.[44]

These results demonstrated that Soviet forces were ill-prepared for the German attack at Bialystok-Minsk. Even though the battle served as the first real example of a Soviet failure, it did foreshadow the Red Army's ability to resist in dire situations. On the positive side of the ledger, Soviet units fought with great determination. More often than not, "encircled Red Army units were fighting hard, which forced the Germans to divert troops to deal with them rather than pushing on as fast as possible to seize key objectives and keep the withdrawing elements of the Red Army off balance."[45] The undermanned fortress at Brest, for example, managed to fight for a month before German forces overtook them.[46] On the negative side of the ledger, Soviet forces showed little flexibility in their defense of Briansk and Minsk.[47] This paralysis most likely reflects the advantage Germany gained from tactical surprise. The shock of the invasion caused chaos among the Soviet defenders by severing an already sclerotic command and control.

The battle also revealed the Wehrmacht's difficulty with assembling enough forces to complete encirclements. This problem allowed many Soviet troops to escape and to fight in the future. At the same time, Hitler worried that the panzer groups advanced too fast and too far without supporting infantry. He increasingly ordered them to pause, which inadvertently allowed the Red Army to regroup. This hesitation occurred again after the Battle of Smolensk. For Soviet military leaders, the execution of high-level officers for failing to "adequately" fight off the Germans along the Western Front left a lasting impression. Soviet commanders would show even greater ferocity in future battles.

The Battle for Smolensk: Determination but Lack of Flexibility

The initial surprise of Operation Barbarossa severely hampered the Red Army's defenses. By July 3, Army Group Center had destroyed the Russian forward elements of the Western Front. German forces stood on the Dvina and Dnepr rivers, ready to resume their exploitation once supplies, infantry support, and Hitler's nerve had caught up to the victorious panzer groups.[48] The Battle of Smolensk, from early July to August 1941, involved a series of intense struggles between Army Group Center and several Soviet armies. Capturing Smolensk would open the roadway to Moscow. Approximately 820,300 German soldiers faced off against 581,600 Soviet defenders.[49]

The opening engagements of the battle occurred on July 10, when General Heinz Guderian's 2nd Panzer Group approached Smolensk from the south and encountered the Soviet 13th Army near the Dnepr River. The 2nd Panzer Group punched through the 13th Army, encircling several of its divisions near Mogilev. These encircled units resisted for two weeks but failed to halt the German advance.[50] Meanwhile, on July 13, the Soviet 19th Army arrived by rail to stop 2nd Panzer Group's drive to Smolensk. As soon as it dismounted, the 19th Army launched an attack. They offered a determined effort but failed to halt the 2nd Panzer's drive to Smolensk. By the evening of July 13, elements of 2nd Panzer Group had pushed to eleven miles south of Smolensk, and the 19th Soviet Army began to dissolve.[51]

Fearing the loss of Smolensk, the Soviets conducted a series of counterattacks all along Army Group Center's front. To the south, the 3rd and 21st Soviet armies struck the southern flank of 2nd Panzer Group. To the north, 20th Army launched a series of bloody assaults against 3rd Panzer Group near Vitebsk. Once again, these efforts slowed but did not stop Bock's forces from moving closer to their objective. At the time, these attacks appeared uncoordinated. Historians today, such as David Glantz, recognize that these attacks were part of a larger Soviet effort to stop Army Group Center from reaching Smolensk.[52]

The 3rd Panzer Group, commanded by General Hermann Hoth, advanced northeast of Smolensk on July 14, driving the Soviet 16th Army out of the city. For two days, the Germans hesitated, trying to decide if they should encircle and destroy nearby Soviet units or press forward to Moscow. Finally, on July 16, Army Group Center cleared Smolensk of the remaining Soviet defenders. To the northeast of the city, the 16th Soviet Army, now a ragtag collection of shattered units, defended its position against attacks by Hoth's 7th Panzer Division.

For nearly a week, a cat and mouse game ensued between Army Group Center and the Soviet defenders of Smolensk. German forces would surround the Soviets, only to see them escape. On July 21, Army Group Center made an attempt to annihilate these retreating Soviet units. Near Smolensk, the 2nd and 3rd panzer groups encircled Soviet 16th and 20th armies.[53] To free these trapped forces, Stavka attempted another counterattack with the four armies: the 29th, 30th, 28th, and 24th. This counterstroke placed immense pressure on the panzer units, inflicting frightful casualties on both sides.

The Soviet efforts to defend Smolensk eventually failed, not because of a lack of determination. Instead the defeat stemmed from logistical problems and, more important, an inability to coordinate counterattacks for maximum effect. These last set of difficulties reflected deficiencies in Soviet training. Stavka, worried about the ability of its commanders, tried to orchestrate every counterattack from its headquarters, away from the battle. Red Army officers, of all the ranks, suffered from insufficient operational training to coordinate large-scale counterattacks and to make adjustments when necessary. Later in the war, Soviet units would improve their ability to conduct these types of operations.[54]

On July 27, the 2nd and 3rd panzer groups trapped most of the 16th, 19th, and 20th Soviet armies east of Smolensk. Once again, exhausted and dangerously overextended, the German armored forces managed to encircle several Soviet armies. In response, the 20th Army led an attack to break out from the German trap. With the aid of nearby Soviet units, the encircled armies managed to escape one more time.[55] The 3rd Panzer Group possessed the capabilities to surround these large formations but lacked enough supporting infantry to destroy them. By late July, another series of Soviet counterattacks began to slow Army Group Center's advance. Again, 3rd Panzer Group tried to destroy the 16th Soviet Army. Almost immediately, however, the determined Red Army defenders stymied the panzer divisions, with the help of a large artillery barrage and a few KV-1 tanks. By the end of the month, Hitler and his generals ordered Army Group Center to halt for a rest and refit.[56]

At the end of the battle, the momentum of the German Blitzkrieg had been severely slowed, and there had been approximately 298,000 German casualties. The defense of Smolensk cost the Soviets 344,926 casualties.[57] The Soviet ability to slow the German advance stemmed from the multiple counterattacks—though poorly coordinated—that placed pressure on overextended panzer units. The lack of sufficient infantry support for the panzer divisions became glaringly apparent during the battle. Moreover, high casualties on both sides illustrated how Stavka's divisions slowed down the German invasion through bitter fighting.

The struggle in and around Smolensk hindered the German advance for the first time during Operation Barbarossa. Franz Halder, Chief of the German General Staff, admitted, "It is increasingly clear we underestimated the Russian colossus . . . . These forces are not always well-armed and equipped and they are often poorly led. But they are there."[58] Throughout the invasion the Wehrmacht had battered and beaten Soviet troops. At Smolensk the Soviet military put on a stubborn defense. At one point the 18th Panzer Division had only twelve tanks still in operation.[59] "When they ran out of bullets and shells—as was all too often in the early stages of the war—they fought with knives or bayonets."[60] Throughout the battle the Soviet forces remained moderately effective while taking large casualties. A relentless number of Soviet counterattacks impeded the Wehrmacht's advance. Sadly, as a result of poor coordination, weak fire support, and an almost total lack of logistical support, the attacks failed to do more than delay the Germans.[61]

In the end, the tenacious Soviet resistance did, however, prompt the Germans to abandon any thoughts of a quick follow-up attack toward Moscow. "This limited victory gave the Red Army a great morale boost and a precious month to reorganize for the defense of Moscow."[62] Hitler also ordered the bulk of the panzer forces removed from Army Group Center and sent to the Northern and Southern Army groups. There they would contribute to new operations. This last deployment order denied them the rest and repairs they required to retain combat effectiveness. Perhaps it also contributed to the German failure to take Moscow three months later.[63]

The Battle for Kiev: Determination but Lack of Flexibility

Germany scored its greatest battlefield victory on the Eastern Front at Kiev. The fighting took place from August 21 to September 26, 1941, with Army Group South opposed by nearly the entire Soviet Southwestern Front. Approximately 300,000 Germans were pitted against 676,000 Soviet forces.[64] The two forces fought around the Ukrainian capital of Kiev in the Dnieper Bend. Two panzer groups encircled the city from the east, trapping the Russian defenders. The Soviets suffered 616,000 casualties, with many marched into cruel captivity. German casualties numbered around 100,000.[65]

After Smolensk, Hitler began to shift his focus away from Moscow. The German leader now looked southward, hoping to capture Soviet industry and croplands. Army leadership, led by Chief of the German General Staff Franz Halder, instead wanted to capitalize on the Smolensk victory and push on to Moscow. Nevertheless, Hitler prevailed and "possibly saved the Soviet capital."[66] The industrial and agricultural heartland of the Ukraine became one of his primary targets. In Hitler's view, these objectives stood well within reach of the Wehrmacht before the harsh winter would hit. "Throughout July and August, German forces had driven Southwestern and Southern Front forces deep into the Ukraine, encircling most of the three armies (6th, 12th, and 18th) near Uman, besieging Odessa, and reaching the outskirts of Kiev and the southern Dnepr by 30 August."[67] Hitler removed many of Army Group Center's panzer forces to assist in the taking of Kiev. He argued that by turning toward the south, Guderian's 2nd Panzer Group could cut through the exposed flanks of the Soviet forces defending the Ukrainian capital. General von Kleist's 1st Panzer Group could link up with Guderian's forces by attacking northward. Stalin, meanwhile, believed that the Germans planned to concentrate their next moves against Moscow and Leningrad. He, therefore, left the Soviet defenders vulnerable in the Kiev region.

Reluctantly following Hitler's orders, on August 24, 2nd Panzer Group turned south to assist Army Group South with the capture of Kiev.[68] In late August, Stalin finally recognized German intentions. He ordered Stavka to prevent the capture of Kiev at all costs. As part of these efforts, the newly formed Briansk Front, under General Eremenko, launched an attack on Guderian's 2nd Panzer Group as they drove south toward the city. Soviet military leaders initially wanted General Kirponos, the Southwest Front commander, to join in the attack from Kiev. Stalin argued against this two-prong attack, asserting that Eremenko's Briansk Front could deal with Guderian. Unfortunately for Kiev's defenders, Stalin was wrong. Guderian's panzer group deflected the attack and met Kleist's forces east of the Ukrainian capital in early September.[69]

As German forces began to complete the encirclement of Kiev, Kirponos pushed for a withdrawal. To preserve his forces, he pleaded with Stavka for permission to move his forces eastward from Kiev, out of the impending encirclement. Soviet Marshal Budenny, the commander of the Southwestern Direction, overseeing both Kirponos and Eremenko, and his military council member, Nikita Khrushchev, also recommended to Stalin that Soviet forces evacuate the city or risk destruction. As Budenny warned, "Delay with the withdrawal of the Southwestern Front will lead to losses in men and a large quantity of equipment."[70] Moscow replied by replacing Budenny with Marshal Timoshenko to emphatically deny any request for withdrawal.[71]

Meanwhile, Eremenko's Briansk Front mounted a fierce counterattack on the 2nd Panzer Group, as it tried to complete the encirclement of Kiev. If they could not stop Guderian, they would at least buy time for Kirponos to escape the coming German trap. While Eremenko lost 100,000 soldiers to prevent the encirclement, Stalin continued to forbid any retreat from Kiev. The Soviet dictator admonished the "panicky" response of Kiev's defenders, recommending "the maintenance of extreme coolness and steadfastness on the part of all commanders at all levels."[72] On September 16, Army Group South completed the armored noose around Kiev as well as the entire Southwestern Front. The Wehrmacht surrounded a total of four Soviet armies: the 5th, 21st, 26th, and 37th.[73]

The Battle for Moscow: Determination but Lack of Flexibility

By December 1941, the Wehrmacht had advanced 540–720 miles into the Soviet Union. Their presence at the gates of Moscow threatened the survival of Stalin's regime. In its autumn push eastward, dubbed Operation Typhoon, German forces had become bogged down in seasonal rains and mud. By mid-November, the ground had hardened, permitting Bock's Army Group Center to resume its drive forward. Soviet losses by this time were terrible. The Red Army had begun the summer with 5 million soldiers and entered the fall with 2.3 million, with 45 percent of its population behind enemy lines.[74] With the Germans at the outskirts of Moscow, one Soviet officer worried, "Hitler is going to push us back to Siberia."[75] German gains, however, had come at a considerable price to the Wehrmacht. By the assault on Moscow, German forces had lost a quarter of their strength from the beginning of the invasion in June.[76]

The Battle of Moscow took place in two phases. In the first phase, Army Group Center led the assault on the Soviet capital, from November 15 to December 4. The second phase consisted of Marshall Zhukov's counteroffensive, which took place from December 6 to January 7. In the initial phase, approximately 800,000 German troops faced off against 760,000 Soviet soldiers.[77] The balance of forces changed dramatically in the Red Army's favor in the second phase, as 1,021,700 Soviet soldiers threw Army Group Center back.[78]

The engagements in both phases took place from Kalinin to the north of the capital to Tula in the south. German planners viewed the capture of these two cities as the crucial step to enveloping Moscow.[79] By the battle's conclusion, the German offensive was abruptly stopped and resulted in 155,000 Wehrmacht casualties.[80] The counteroffensive cost the Russians 370,995 casualties, but they did push German forces back several hundred kilometers. Zhukov's counteroffensive erased Germany's gains during Operation Typhoon.[81] Soviet success derived from well-placed concentration of troops, fresh divisions, overextended and poorly equipped German forces, and Army Group Center's lack of preparation for a counterattack. The NKVD also provided blocking formations behind attacking Soviet divisions to stop any retreat that was without orders.

With the ground frozen hard, Bock's Army Group resumed its offensive on November 15, 1941. To meet the attack, Marshal Zhukov anchored his defenses on the Mozhaisk line. Zhukov also conducted a few spoiling attacks anticipating Bock's renewed advance. In one of these attacks the Soviet 44th Mongolian Cavalry Division charged across an open, snow-covered field near Klin.

The German 106th Infantry Division cut down the 2,000 cavalrymen and their horses without suffering any casualties. Horses proved useful for getting troops to battle in winter, but "too vulnerable to maneuver in a pitched battle."[82]

In mid-November, the 3rd Panzer Group struggled to seize the highway from Kalinin through Klin to Moscow. German armor pushed forward slowly against fierce resistance, taking Klin on November 24.[83] Then, four days later, it seized a bridgehead across the Volga-Moscow Canal, 35 kilometers from the Kremlin. In the south, the newly renamed 2nd Panzer Army forced its way forward in an attempt to encircle Tula from the east, another stepping-stone to Moscow. Counterattacks, coupled with freezing temperatures and shortages of fuel, ammo, and functioning vehicles stopped the advance. "Dogged Soviet defenses did as much to stop the Germans as did bad weather and poor supply lines." Guderian repeatedly requested an end to the offensive, but Hitler refused to give his consent.[84] The 4th Panzer Group closed to within 20 kilometers of Moscow on November 28.[85]

By early December, Army Group Center had nearly reached its objective. The weather, however, turned against Bock's forces. On the night of December 4, the temperature dropped to minus 34 degrees Celsius. The cold caught German forces unprepared, as soldiers and equipment fell victim to the extreme winter conditions. More than 133,000 cases of frostbite weakened German forces.[86] The Wehrmacht did not bring the proper winter clothing to deal with the change in weather. German planners had optimistically, perhaps naively, predicted the invasion of the Soviet Union ending before winter. At the very least, they believed that the Wehrmacht would have occupied one of the country's major cities. They did not envision the German Army bivouacked in the Russian countryside. Nor could the German military's logistics cope with the demands of supplying a large army, several thousand miles away, without enough railway lines.[87]

The Soviet counteroffensive began at first light on December 5 in the Kalinin Front's sector, north of Moscow. For three days, Zhukov's forces mounted a fierce attack, creating confusion in the German units, which started to retreat. Stavka concentrated on shattering the German pincers north and south of Moscow. Zhukov told his forces to avoid frontal attacks. Instead, he instructed them to assault the retreating Germans from the flanks and the rear.[88] Using "shock groups," the Soviets punched holes through the Wehrmacht's front lines. After ten days, the Red Army had retaken Klin, and then Kalinin by the end of the month. Reaching these objectives dislodged the northern German pincer.

Soviet units liberated Tula in a few weeks, and by the end of the month they had retaken Kaluga, clearing the southern pincer threatening Moscow.[89] Hitler ordered his forces to stand their ground, a directive that probably prevented a collapse of German defenses.[90] By the end of Zhukov's counterstroke, the Red Army had pushed the Wehrmacht back 225 kilometers.[91]

Strong Soviet Staying Power

How and why did the Red Army remain intact and avoid collapse under the pressure of Operation Barbarossa? The answer lies in the sources of Soviet staying power. The heavy casualties endured by Soviet forces posed a grave threat to the regime in Moscow. Soviet forces, for example, suffered twenty casualties to every German soldier lost in battle. A large swath of European Russia fell into German hands during the initial invasion.[92] Facing these losses, Stalin rightly worried that the invasion could signal his doom, either through a German victory or at the hands of disgruntled Soviet citizens that might take the opportunity in the midst of the chaos to revolt.[93] Nevertheless, by exerting strong control over the country, the Soviet state prevented the collapse of the military and safeguarded its survival. Stalin relied on both positive and negative incentives to keep the Red Army fighting. As Roger Reese puts it, "[T]he Stalinist state succeeded in mobilizing, through inspiration and coercion, just enough of the population for military service to ensure its survival."[94]

Ideological appeals represented one way the regime motivated the Soviet nation. Not only did Stalin and his government reference Soviet ideals to inspire the populace, but they also called upon the Russian people to defend their homeland against a ruthless invader.[95] This message included the rehabilitation of traditional Russian heroes, such as Alexander Nevsky and Mikhail Kutuzov.[96] For the younger generation, appeals to Soviet principles made good sense, since they knew nothing other than collectivism and service to the state.[97] Germany's brutal treatment of the Soviet people and captured soldiers also motivated the Red Army to continue the struggle.[98] German atrocities raised the stakes of a Soviet defeat.

Coercion provided a second mechanism Stalin's regime used to prevent the collapse of the Soviet war effort. As the country's plight grew more perilous, the Soviets answered by increasing terror at home.[99] At the front, blocking formations helped keep the Red Army fighting.[100] These units aimed to deter desertions, picking up and punishing soldiers that had abandoned their positions. In some cases, they shot individual deserters.[101] The control of information helped the regime keep morale from cascading too low after sequential defeats. A network of informants and secret police searched for potential enemies of the state within the armed forces. To bolster discipline, Stalin issued Order 270, declaring the summary execution of deserters as well as officers that refused to lead from the front.[102] Ruthless in its discipline, the NKVD executed an estimated 98,000 soldiers in 1941 and 1942.[103] One former soldier noted, "An extreme situation calls for extreme measures."[104]

As a result of these measures, Stalin's regime sustained the Red Army through its crisis in 1941 and through the end of the war. The government managed a massive evacuation of industry from the threatened western parts of the country eastward, out of reach of the German invader. To bolster the ranks of the armed forces, the government enlisted large numbers of volunteers. During the defense of Moscow, for example, the regime raised a volunteer force called the *opolchenie*, from intellectuals, old men, and students aged seventeen to fifty-five.[105] Although the armed forces suffered nearly 1 million desertions and surrenders in 1941, these numbers reflected problems in tactics rather than disloyalty to the regime.[106] The Red Army's ability to survive the horrific events of Operation Barbarossa reflected the power of the Soviet state. As David Glantz argues, "The fact that neither collapse nor coup took place bore mute testimony to the ruthless effectiveness of Stalin's regime."[107]

Soviet Units Lacked Flexibility during Operation Barbarossa

While the power of Stalin's regime gave the Red Army its staying power, this same degree of control hurt the battlefield performance of Soviet units. Without the autonomy to train and to conduct operations free from the interference of political officers or regime officials, Soviet units fought with little flexibility, experiencing problems when surprised, when needing to change battlefield plans, or when encircled. Primarily, the climate of fear and distrust promoted by the regime's security forces discouraged initiative in the military. A rigid command and control structure also deprived lower-level officers of opportunities for flexibility.[108] On several occasions during Operation Barbarossa, Stalin intervened to prevent his local commanders from withdrawing. Soviet officers also feared giving accurate battlefield reports, worried that a wrong answer might bring imprisonment, or worse.

Insufficient training also bore responsibility for inflexible units. With little confidence in the training of these officers, the Soviet armed forces discouraged initiative, convinced that subordinates could not adapt on the battlefield. With poor training, many officers could not improvise during combat.[109] In particular, they lacked the training to break out of encirclements. This reality contradicted Stalin's instruction that only "deficient" officers led their units into encirclement and deserved punishment.[110] Max Hastings summarizes the paralysis that this fear created in Soviet units in one particularly suitable anecdote:

> In an army in which fear played so large a part, many officers were reluctant to accept orders by telephone. They demanded written instructions, which could be preserved and produced if matters went awry. "Orders were never a matter for discussion," said Lieutenant Alexandr Sergeev. Even when German artillery was registering on Captain Vasily Krylov's Katyusha battery, it was unthinkable for him to shift position without a direct order. Individual initiative was discouraged.[111]

CONCLUSION

The case of the Red Army during Operation Barbarossa provides compelling evidence in support of cohesion theory. A high degree of regime control, in the form of Stalin's repressive government, held the Soviet armed forces together under extremely difficult circumstances. The Red Army exhibited strong staying power in its recovery from the surprise attack of June 1941. Over the subsequent months, Stalin's army regrouped from several prodigious defeats. It then showed great endurance by mounting a successful counteroffensive at the gates of Moscow. Table 5.1 illustrates the Soviet willingness to suffer large casualty rates in battle.

The highly coercive Soviet state supplied the Red Army with its staying power. Domestic resistance to Stalin's policies at home or at the front proved both too dangerous and difficult for any serious opposition to the Soviet war effort to emerge. Compared with Reynaud's France, Stalin could rely on a highly coercive government to maintain domestic and military support for the Soviet war effort. The Soviet dictator would not have tolerated a General Weygand, refusing to keep fighting. Stalin's military effort also proved more resilient than the Russian armies led by Czar Nicholas II in World War I. Tables 5.4 and 5.5 illustrate the willingness of the Red Army to endure far greater casualties than its czarist predecessor.

On the battlefield, Soviet soldiers fought tenaciously to defend their homeland. Officers and enlisted personnel alike also knew that disobeying orders risked punishment for them, or their families. The presence of commissars and the NKVD was a constant reminder of Stalin's grip on power. One can imagine how the NKVD could have also motivated French units to fight in 1940 at places

TABLE 5.1: Soviet and German Casualties, June 22, 1941, to January 1942

Battles	Soviet Forces	German Forces	Soviet Casualties	German Casualties
Bialystock-Minsk, June 22–June 29	1,989,900	3,206,000	747,870	92,120
Smolensk, July 3–Aug. 5	581,600	820,300	344,926	298,000
Kiev, Aug. 23–Sept. 26	676,000	300,000	616,000	100,000
Moscow Defense, Sept. 30–Dec. 5	1,250,000	1,200,000	658,279	120,996
Moscow Offensive, Dec. 5–Jan. 7	1,021,700	800,000	370,995	103,600
Total for Period	7,052,700	3,206,000	4,808,904	831, 050

Battles	Soviet Per Cap Loss Infliction	German Per Cap Loss Infliction	Soviet Loss-Infliction Ratio	German Loss Infliction Ratio
Bialystock-Minsk, June 22–June 29	0.05	0.23	0.20	4.97
Smolensk, July 3–Aug. 5	0.51	0.42	1.22	0.82
Kiev, Aug. 23–Sept. 26	0.07	2.05	0.54	29
Moscow Defense, Sept. 30–Dec. 5	0.11	0.54	0.20	4.99
Moscow Offensive, Dec. 5–Jan. 7	0.10	0.46	0.2187	4.63
Total for Period	0.12	1.34	0.09	11.17

SOURCES: Christopher Bellamy, *Absolute War: Soviet Russia in the Second World War* (New York: Vintage Books, 2007); David Glantz and Jonathan House, *When Titans Clash: How the Red Army Stopped Hitler* (Lawrence: University of Kansas Press, 1995); Michael Clodfelter, *Warfare and Small Armed Conflicts: A Statistical Reference, Vol. II: 1900–1991* (Jefferson, NC: McFarland and Company, 1992); David M. Glantz, *Before Stalingrad: Barbarossa, Hitler's Invasion of Russia, 1941*, 2nd ed. (Stroud, Gloucestershire: Tempus Publishing, 2003); Bryan Perrett, *The Battle Book: Crucial Conflicts in History from 1469 BC to the Present* (London: Arms and Armour, 1992); Robert Kershaw, *War without Garlands: Operation Barbarossa 1941–1942* (Surrey: Ian Allan Publishing, 2009); Earl F. Ziemke and Magna E. Bauer, *Moscow to Stalingrad: Decision in the East* (Washington, DC: U.S. Army Center of Military History, 1987); David Glantz, "The Soviet-German War 1941–1945: Myths and Realities: A Survey Essay," paper presented as the 20th Anniversary Distinguished Lecture at the Strom Thurmond Institute of Government and Public Affairs, Clemson University, South Carolina, October 11, 2001.

like Sedan. In their tenacious defense, Soviet units readily sacrificed the lives of soldiers. As one Soviet officer put it, "Of course the Red Army was reckless with the lives of our men . . . . Nobody knew how many died, and who cared, anyway?"[112] (See Tables 5.2 and 5.3 for summaries of Soviet staying power and battlefield performance.)

TABLE 5.2: Assessing Staying Power: Authoritarian Soviet Armed Forces, 1941

Evidence
Soviet armed forces fought and inflicted casualties despite defeats, large casualties, and desertions
Stalin's regime used mix of ideology and coercion to motivate armed forces and public
Red Army recovers from substantial defeats to mount massive offensive near Moscow

PREDICTION: Strong Staying Power.
Military Collapses Only When Adversary Possesses Crushing Material Superiority.

TABLE 5.3: Assessing Battlefield Performance: Authoritarian Soviet Armed Forces, 1941

Battles	Evidence
Bialystok-Minsk	Despite surprise, most units defend and counterattack
	Defenders of Brest Fortress hold out for one month
	Units lack flexibility to adjust to German tactical surprise
Smolensk	Soviet units inflict large casualties on German attackers
	Experience trouble coordinating counterattacks
Kiev	Red Army units mount numerous counterattacks to break out of encirclements
	Coordinating breakout efforts difficult due to Stavka's direct control
Moscow	Tenacious defense of the capital includes newly raised units of volunteers
	Soviet's mount counterattacks but cannot exploit initial victories

Most Units Fight with Determination but Lack Flexibility.
PREDICTION: Moderate Battlefield Performance.

A high degree of regime control, however, created one problem for the Soviet armed forces. Stalin's reign of terror and anxiety about military opposition prevented the Red Army from obtaining the autonomy to train. The military found it impossible to create bonds of loyalty and trust within its ranks. Strict control by the state and Stalin's purges created an atmosphere of distrust that prevented their formation. Indoctrination hampered training for warfighting. On the battlefield, this sometimes caused Soviet forces to fight with less discipline than militaries that possess these bonds. Without tight control, units would panic, and they displayed little flexibility, deviating from established battle plans. In the eyes of the Soviet state, such difficulties were likely considered a small price to pay for the absolute obedience of the Red Army.

With respect to the existing alternative arguments, cohesion theory also offers a superior interpretation of this case. Clearly, this historical episode contradicts the claim that militaries from democracies fight harder than the armed

TABLE 5.4: Russian Casualties: World War I

	Casualties	Initial Forces	Casualties/ Initial Forces
1914	1,800,000	6,553,000	27%
1915	2,450,000	5,047,000	49%
1916	2,400,000	3,048,000	79%
1917	693,000	2,350,000	30%
	Total Casualties	Total Mobilized	Casualties/ Total Mobilized
	7,412,000	15,378,000	48%

SOURCES: Kevin D. Stubbs, *Race to the Front: The Materiel Foundations of Coalition Strategy in the Great War* (Westport, CT: Praeger, 2002); John Ellis and Michael Cox, *The World War I Databook* (London: Aurum Press, 2001); Randal Gray and Christopher Argyle, *Chronicle of the First World War, Vol. I: 1914–1916* (New York: Facts on File, 1990); and Michael Clodfelter, *Warfare and Small Armed Conflicts: A Statistical Reference, Vol. II: 1900–1991* (Jefferson, NC: McFarland and Company, 1992).

TABLE 5.5: Soviet Union Casualties: World War II

	Casualties	Initial Forces	Casualties/ Initial Forces
1941	4,308,094	4,197,000	102%
1942	7,080,801	6,124,000	137%
1943	7,483,647	6,389,000	117%
1944	6,503,204	6,930,000	101%
1945	2,823,381	6,330,000	44%
	Total Casualties	Total Mobilized	Casualties/ Total Mobilized
	28,199,127	34,476,000	82%

SOURCES: See Table 5.4.

forces of nondemocracies. Stalin's repressive and decidedly undemocratic regime gave the Red Army its wartime staying power as well as its determination in combat. The small-group theory also sheds little light on Red Army performance. Although the Soviet military tried to recruit and to train regiments from small localities, the high turnover rate caused by the enormous casualties prevented bonds from forming among small groups. Fear of Soviet informants also created distrust among the ranks.[113]

The best alternative argument to cohesion theory is the claim that nationalism motivated and created a cohesive Soviet military. Stalin, after all, married Soviet ideology to a patriotic desire to defend the Russian homeland. He drew on past Russian heroes as well as Soviet ideals to bolster the desperate defense of the Soviet Union. While nationalism certainly played a role in the Red Army's cohesion, especially its staying power, it does not explain the conditions under which these appeals work. Under less dire circumstances, the czar's government promoted Russian nationalism during World War I, only to experience a revolution in 1917. No such opposition emerged in the Soviet Union. As cohesion theory explains, a high degree of regime control creates a hard-core group of government supporters who fight to the death and pressure the armed forces to do the same.

6 North Vietnam, 1965–73

WHY DID THE North Vietnamese Army (NVA), the military arm of an undemocratic government, remain a cohesive fighting force in the face of numerous casualties during its war to unify Vietnam?[1] What motivated the NVA during its costly, prolonged stalemate and eventual victory over the United States and its allies? Over the course of twenty years, from 1954 to 1975, communist-led North Vietnam suffered 1,298,000 casualties.[2] The North Vietnamese stood firm against an allied coalition whose ground forces maintained superior airpower, firepower, and manpower. At the same time, they endured successive American bombing campaigns that consisted of 527,000 sorties used to drop 6,162,000 tons of munitions, more explosives than U.S. bombers dropped in World War II.[3]

This chapter examines the cohesion of the North Vietnamese Army. I argue that the Democratic Republic of Vietnam fielded *a messianic* army capable of fighting with a high degree of cohesion. As cohesion theory would predict, the sources of NVA's staying power were a high degree of regime control and the military's organizational autonomy to train. This combination enabled most North Vietnamese units to fight with determination and flexibility on the battlefield while enduring a long war, even when victory seemed uncertain. Over the eight years the countries engaged in significant fighting (1965–73), this messianic military demonstrated an ability to sustain casualty rates that were significantly higher than those of the United States. Although they lacked the skill constantly displayed by their U.S. opponents, they more than compensated by displaying a fierce tenacity in combat, taking risks and being willing to incur large casualties in engagements. Through this battlefield determination, the

North Vietnamese credibly demonstrated their willingness to endure great pain to achieve their war aims, a level of pain the U.S. ultimately concluded it did not want to match. The chapter does not recount the entire history of the North Vietnamese struggle against the United States. Instead, to test the explanatory power of cohesion theory in this case, I examine two representative engagements to detail the NVA's battlefield performance. The chapter also highlights the blood price North Vietnam was willing and able to pay for its ultimate victory to illustrate its staying power.

The chapter's discussion consists of four parts. First, I outline the sources of the NVA's cohesion. The armed forces of the Democratic Republic of Vietnam derived much of their motivation from a regime with a high degree of regime control over the population. Hanoi won unconditional loyalty by appealing to nationalist desires to liberate the country and by communist promises to improve the quality of life of its citizens. The regime could also enforce their demands for loyalty through a coercive grip on the country's political institutions. These national demands for unconditional loyalty echoed the same demands the North Vietnamese Army made on its members. Strong internal bonds took root inside the NVA because the regime permitted its military organizations the autonomy to train for warfighting. Even regime indoctrination efforts reinforced this focus on enhancing battlefield performance. North Vietnam's messianic military could, therefore, fight with determination and flexibility on the battlefield, while also displaying enough staying power to outlast the United States.

In the second section, I discuss the NVA's performance at the Battle of Ia Drang in the fall of 1965. There North Vietnamese forces experienced one of their first major engagements with the United States. Near Chu Pong Mountain along the Cambodian border, American forces, largely elements of the 1st Cavalry Division, attacked several regiments of North Vietnamese in the midst of preparations for an offensive of their own. La Drang represents a good case because it provides solid evidence to evaluate the battlefield performance of North Vietnamese and U.S. forces—the subject of this chapter—in a series of detailed engagements. Although they lost the battle, the NVA displayed what would remain the defining features of their cohesion: a willingness to endure great casualties to wear down their opponents. The next two sections describe the effect of the Tet Offensive of 1968 on North Vietnamese society and illustrate how the military retained its staying power through the following year.

A MESSIANIC NORTH VIETNAMESE ARMY

A large reason for the North Vietnamese victory was the cohesion of its messianic military. With a high degree of control, the government in Hanoi could motivate the population and the armed forces through its ideology, which married communism with nationalism. Without significant political opposition, the regime successfully imposed their ideological demands for unconditional loyalty on the population. Another source of the armed forces' motivation came from the norms of loyalty and trust instilled by its officers during training.

A High Degree of Regime Control

The roots of North Vietnam's ideology grew out of the nationalist movements launched in the 1890s to free the country from the shackles of French domination. From the turn of the century until World War II, the French faced a series of challenges to their rule of Indochina. Leaders, from Phan Dinh Phung to Phan Boi Chau, mounted several failed attempts to liberate Vietnam. These efforts to free their country from French colonialism did not achieve much noticeable measure of success until 1945. With the departure of the Japanese at the end of World War II, Ho Chi Minh's Marxist Viet Minh movement briefly seized control of the country before the French returned to regain power over their colony.[4]

For the next nine years, the Viet Minh engaged in a long struggle with the French for control of Vietnam. Winning support in the north and in the countryside of the south, Ho Chi Minh's forces undermined French efforts to re-establish their grip over the region. The communist insurgents garnered backing from large segments of the population by appealing to nationalist sentiments calling for Vietnamese self-determination. Ho Chi Minh succeeded where other rebellions had faltered because his movement promised to free the country from French rule and the oppression of the corrupt Vietnamese upper class that had long collaborated with the colonists.[5] After their defeat at Dien Bien Phu in the summer of 1954, the French departed. Pending an election to decide who would rule the country, the Geneva Accords divided Vietnam in half, with the communists controlling the north.[6]

Flush with the success of their movement, Ho Chi Minh began to consolidate his sway in the north by championing an ideology infusing nationalist rhetoric into a communist vision for future social reform. Foremost, he pledged to liberate the remainder of the country from all foreign influence. Because the nation had a long history of fighting against external invaders, citizens had become accustomed to struggles for independence.[7] It came as no great surprise, therefore, when the Democratic Republic of Vietnam began to sponsor a communist insurgency movement in the south at the conclusion of the Geneva Accords. It was the growing threat posed by the Viet Cong (VC), as well as the prospect of an invasion of the south by North Vietnam, that prompted more American military intervention in the early 1960s. Good to their word, the regime in Hanoi was undaunted by American involvement. As the United States increased its military presence in South Vietnam, the North Vietnamese stepped up their efforts to unify what they saw as their country.

Many Vietnamese, however, viewed the nationalist movements of the past as simply efforts by one set of narrow interests to replace the French and enrich themselves in the process. To this end, the communists offered more than simple nationalism. Their promises of economic and political equality also resonated with a population long under the domination of, first, foreign governments, and later, corrupt landed elites.[8] In pursuit of economic and political equality, Ho Chi Minh's regime sought to convince the people of North Vietnam of the need to see their personal interests in terms of the nation's interests.

The Democratic Republic of Vietnam won the allegiance of its citizens by doing more than simply offering empty rhetoric. Early on, communist officials made good on their assurances to improve the economy and the general welfare of their citizens.[9] Once the French had departed, the party turned its attention to extending famine relief to rural areas neglected during the war.[10] This move helped offset the brief anger caused by the brutal land reform efforts introduced to restart the struggling economy. While the Viet Minh solidified their control over the countryside, they also made headway in urban areas by permitting some measure of private business to continue. Ho Chi Minh's version of communism resembled Lenin's New Economic Policy more than it did Mao's collectivization strategy.

At the same time, the government folded many of the country's existing institutions under the growing umbrella of the Communist Party, creating a variety of avenues to promote and to enforce its ideology among the population. The regime also oversaw an overhaul of the educational system, providing an opportunity to inculcate the country's youth with their ideological beliefs.[11]

Causal Logic: High Regime Control Generated Strong Military Staying Power. Not only did the ideology of the North Vietnamese regime bolster its civilians, but it also reinforced and sustained the determination of the country's armed forces. Throughout the conflict, the North Vietnamese Army could maintain a high level of morale.[12] Even after suffering costly tactical defeats, such as the Battle of Ia Drang, the communists still managed to send highly motivated forces to the battlefield.[13] Hanoi, perhaps swept up in its own rhetoric, concluded that North Vietnam was destined to emerge victorious. From this perspective, the north was willing to outlast the United States in a bloody stalemate. This was, after all, a country led by the same leader who once told the French: "You can kill ten of my men for every one I kill of yours, but even at those odds, you will lose and I will win."[14]

The regime's ideology appealed to the entire nation's desire to oust foreign invaders, to unify the country, and to improve, once and for all, the economic plight of everyone in Vietnam. In effect, both the government and its soldiers pursued a cause worthy of individual sacrifice. As Brian Jenkins described it:

> The genius of the North Vietnamese people is their tenacity. It is also their most terrible weapon. Hanoi's apparent determination to go on fighting reflects convictions that in their eyes seem correct—so correct that the alternative of not fighting may be inconceivable. Confucian doctrine, imported from China centuries ago, permeates the arguments put forward by the Vietnamese Communists. Terms such a "just cause," and, "legitimate government," dominate the speech of their leaders. Vietnamese Communists firmly believe that they possess the "Mandate of Heaven" to rule all of Vietnam and, therefore, must emerge victorious eventually.[15]

Average North Vietnamese expressed these beliefs during the war. "I never think the front could lose the war," one infantryman noted. "This is a war of the Vietnamese people fighting against American imperialists. As long as the people still exist and as long as the American people are still in Vietnam dominating the Vietnamese people, the war will go on." Another NVA fighter added: "I think the war will go on a long time and the Vietnamese people will certainly win it."[16] Young recruits especially expressed this commitment to the war: "I am leaving camp tonight. My heart is filled with love for my homeland. I pledge to achieve victory before returning to my homeland." Many soldiers echoed this devotion to country and the party's ideology. As one soldier wrote home:

> We are in the prime of our lives, a time that should be devoted to serving society, contributing all our strength to the Party and to the Fatherland. We belong to the generation that is advancing speedily forward and that is scoring many achievements on the battlefield. I am determined to sacrifice my personal life and emotions to make my parents happy, and also to be able to continue marching on the Revolutionary path that I am following.[17]

Perhaps the best evidence of the NVA's staying power was its ability to rebound from devastating defeats and continue to pressure American forces. The Tet Offensive of 1968, for instance, grew out of a belief by the North Vietnamese political and military leadership that if they could land a large enough blow against the United States they might drive their opponent from the battlefield. While they guessed incorrectly about their ability to defeat American units in combat, they did manage to turn American public opinion against the war, even though the NVA and its Viet Cong allies suffered terrible losses. In spite of the damage of the Tet Offensive, Hanoi concluded that it had to put additional pressure on the United States. With an eye on political events in the enemy's homeland, the North Vietnamese regrouped from their battlefield defeats to mount additional costly attacks through the rest of 1968.[18]

High Degree of Military Autonomy

When it ascended to power, first in 1945 and then nine years later with the departure of the French, the communist regime harbored great concerns about fashioning an independent military. Like many revolutionary governments, they worried that once created, the armed forces might attempt to seize political power and end the communist experiment.[19] This concern about the loyalty of the armed forces led the regime to establish an imposing Communist Party presence in its armed forces. Party officials placed representatives at every level of the NVA's organization, from every squad to the high command. With the intention of monitoring as well as controlling the armed forces, the party oversaw all military operations.[20]

This overt presence of the regime inside the military began to change once the Communist Party solidified its legitimacy as the government of North Vietnam. Moreover, the prospect of war with the United States reinforced the need to create a military organization free to train and prepare its soldiers for war. Under these circumstances, the government's ideology fell into the background, reinforcing what soldiers learned from their training about the importance of working as a team to accomplish national goals. The party's ideological message of communism and nationalism moved from being a method of controlling the military to a source of motivation for soldiers, and most saw these ideas as a rationale for a long war.[21] Under these conditions, the NVA obtained the autonomy to focus on training.

Like any military organization, the North Vietnamese Army sought to instill its own principles of unconditional loyalty and trust. Officers bore the responsibility for training, leading in battle, and for indoctrinating soldiers. In addition, the organization encouraged its officers to watch over men in order to maintain their fighting spirit. In training, as well as combat, officers constantly remained on the lookout for men who showed more than the usual signs of loneliness, lack of morale, or general unhappiness. Officers also tried to encourage a spirit of camaraderie by interacting closely with their men. As one captured soldier attested, "We even wrestled with our company commander!"[22]

The normal North Vietnamese soldier received vigorous training. Although basic training typically lasted two months, the NVA recruit did not have the same training materials, weapons, and supplies that his American counterpart did. Training focused on getting soldiers accustomed to working in groups, and to survive in the tough terrain of the region without a steady stream of supplies.[23] To bolster a soldier's morale, the military created an institution known as the three-man cell, whereby soldiers trained and fought in proximity with two of their comrades. The three-man cell represented the military's scheme for ensuring that men formed tight friendships within small groups of soldiers. It also became another mechanism for the armed forces to reinforce its norms of loyalty and trust, as well as the importance of working in groups.[24]

Causal Logic: A High Degree of Autonomy Created Strong Battlefield Performance. This mixture of difficult training and an officer corps committed to promoting norms of unconditional loyalty and trust bolstered the will to fight of the North Vietnamese Army. In its long, bloody struggle with the American military, the NVA saw few defections. From 1963 to 1969, it experienced only 1,700 defections, less than 1 percent of its total forces.[25] The morale of the North Vietnamese Army also remained high, despite countless bombings, defeats in battle, and long periods of boredom. Perhaps the truest test of the NVA's mettle was its ability to sustain enormous casualties and still continue the fight against a superior American enemy. As one analyst at the time put it, "Everyone who has spoken with the NVA—if only for a few minutes—vouches for the excellence of their morale, which is not accidental."[26]

THE BATTLE OF LA DRANG (OCTOBER 27 TO NOVEMBER 26, 1965)

In the summer of 1964, communist leaders in the north decided to change their wartime strategy. Up to that point, they had relied primarily on the efforts of the Viet Cong both to foment unrest in the south and to fight units of the Army of the Republic of Vietnam (ARVN). For several years they had disingenuously claimed that VC insurgents had sprung up in the south without aid

MAP 6.1. South Vietnam.

from Hanoi. The Viet Cong, they said, represented an indigenous movement sparked by the corruption of the regime in the south. Although they had decided as far back as 1959 to take charge of the war in South Vietnam, their regular armored forces, the North Vietnamese Army, needed several years to ready themselves for incursions into enemy territory. Over time, North Vietnamese regulars began to augment the ranks of the Viet Cong, causing some tension with their communist brethren in the south (see Map 6.1).

At the same time, the leadership in the north planned to utilize its freedom of movement inside Laos and Cambodia, moving along the Ho Chi Minh Trail, to begin inserting forces into South Vietnam. One of the first objectives the north sought, in the fall of 1965, was to dominate Route 19, a road that ran between the Cambodian border and the city of Qui Nhon in the east. Seizing this route would permit them to cut the South Vietnamese republic in two. To capture this critical highway, communist forces began to deploy in large numbers in the Ia Drang Valley near the Chu Pong Mountain. Once in place, the NVA units staged a number of probing attacks in October against the Republic of Vietnam to lure ARVN forces from their base at Plei Me, a few miles southwest of Pleiku.[27]

After these assaults against the Army of the Republic of Vietnam, the American commander-in-chief, General Westmoreland, ordered the U.S. 1st Cavalry Division to deploy near Pleiku.[28] Westmoreland tasked the newly formed airmobile 1st Cavalry Division to locate the base of operation used by the North Vietnamese and to destroy it. These NVA forces included the 32nd, 33rd, and 66th regiments under the command of Brigadier General Chu Huy Mân. The three units of communist regulars consisted of approximately 3,000 soldiers. Their opponent, the 1st Cavalry, included eight maneuver battalions controlled by three brigades. In addition, the American division contained four artillery battalions, an air cavalry squadron, an engineer and signal battalion, and an aviation group. The division's true strength, more importantly, was its ability to quickly transport its light forces by air. They could establish a perimeter in neutral or hostile territory, opening the way for additional forces to link up with them.[29]

Initial Battles: Evidence of Determination and Flexibility

Initially, American forces had trouble finding the NVA. Then, on November 1, 1965, the 1st Squadron of the 9th Cavalry, commanded by Lieutenant-Colonel John B. Stockton, seized a hospital five miles west of Plei Me. The facility belonged to the 33rd North Vietnamese Regiment. They killed 135 communist fighters in a quick, one-sided engagement. They also captured 44 soldiers as well as a huge cache of weapons and other supplies.

A few hours later, in the evening, NVA forces mounted a counterattack to throw the Americans back. Against Stockton's squadron, the North Vietnamese staged a furious assault. Because they threw themselves with much determination into the heart of the prepared American defenses, the communist units suffered horrible casualties. Reinforced by another battalion, the Americans forced the depleted 33rd Regiment to withdrawal. In this second damaging encounter of the day, the NVA lost 250 of its soldiers.[30]

For the 33rd Regiment, these brief engagements cost them dearly. Overall, the regiment saw 40 percent of its men killed, including two of three battalion commanders. Because the regiment had offered fierce resistance, by the end of the engagement it could muster only one hundred men. Still further, in the midst of the fighting, the 33rd had left behind nearly all of its antiaircraft guns and mortars. In contrast, U.S. forces suffered lightly, with 11 deaths and 51 wounded.[31]

On November 3, Stockton received orders to search for additional communist units in the Ia Drang Valley. After further reconnaissance, Stockton's units discovered a heavy NVA presence, from the 66th Regiment, near the western side of the Chu Pong Mountain. Close to the Cambodian border, the American forces sent three platoons to establish ambushes, covering three separate trails, to stop any North Vietnamese units from moving eastward. Along one trail, during the early hours of the evening, the Americans soon observed a reinforced company of the 66th Regiment, with close to one hundred soldiers, moving to the west. They stopped 120 yards from the Americans, who lay in wait for them. After a long rest, the North Vietnamese unit continued their march.

To make the most of their ambush, the Americans allowed the lead column of the NVA units to pass through the fire zone. Once the heavier elements of the company entered the trap, the Americans fired eight claymore mines simultaneously. A few seconds later, they followed up these detonations with rifle and machine gun fire. After inflicting heavy casualties on the NVA company, the three American platoons pulled back to form a defensive perimeter. Almost immediately, a large North Vietnamese force launched a massive counterattack that threatened to overrun Stockton's platoons. Again, reinforced by another company, the Americans eventually beat back the communist assault.

From interrogations of their prisoners, captured over the space of two days, the 1st Squadron garnered further intelligence regarding the Ia Drang Valley. The area, indeed, represented a staging ground for North Vietnamese forces. Since October 1965, the 1st Cavalry Division had searched for the NVA along the Cambodian border, where they made their trouble-free incursions into South Vietnam from the Ho Chi Minh Trail. Over the next two weeks, American forces would continue their hunt for the core elements of the NVA forces near Ia Drang.[32]

The Fight at Landing Zone X-Ray: NVA Determination and Flexibility

The heaviest engagements in the Battle for Ia Drang began on November 14, 1965, when Lieutenant-Colonel Harold G. Moore's 1st Battalion, 7th Cavalry assaulted Landing Zone (LZ) X-Ray near Chu Pong Mountain. New intelligence reports had hinted at the possibility of an NVA base of operations farther inland from Cambodia than had been anticipated. American units, thus, correctly began to suspect the presence of the North Vietnamese wedged in between the eastern side of Chu Pong and the Ia Drang Valley. Colonel Moore's forces sought to make contact with the enemy. Then, the remainder of the division would conduct a follow-on attack against the difficult to find North Vietnamese units.[33]

Moore's unit consisted of four understrength companies, a total of 450 men for the entire battalion. With a shortage of helicopters, the battalion would need a total of four hours to transport all four companies to the landing zone. This situation presented a great danger to the Americans. Unwittingly, they had stumbled upon the staging ground for the 66th Regiment, the strongest NVA unit in the valley. Commanded by Lieutenant-Colonel La Ngoc Châu, the 66th neared 1,600 soldiers in total strength. As Moore's forces began to secure the landing area, they presented a vulnerable target for the superior Vietnamese defenders. Seeing an opportunity to keep his attackers off balance, Châu pounced on the Americans as they arrived.[34]

As soon as the first platoons of Moore's understrength battalion began landing, his units began to fan out in order to secure the perimeter of the small landing zone. As soon as they had arrived, one of Moore's sergeants spotted a lone North Vietnamese soldier sitting on the ground. He quickly grabbed the startled man and took him prisoner. After questioning the captive, he learned that there were "three battalions on the mountain who want very much to kill Americans but have not been able to find any."[35] At the time it was 1050 hours, and Moore would have to wait for more than an hour to secure the LZ.

By 1210 hours, Moore had two companies on the ground with him at Landing Zone X-Ray. Even though two companies had yet to arrive, Moore judged it better to go on the attack instead of waiting for the NVA to begin their assault against his forces. Rather than permit the North Vietnamese to threaten the critical landing area, Moore decided to send his two companies north and west toward the smattering of gunfire that had begun to increase during the morning. He sent his Bravo Company, under the direction of Captain John Herren, in the direction of Chu Pong.[36]

Twenty minutes later, the smattering of gunfire from the mountain had grown in intensity. Herren ordered his two platoons to advance in a line with his third platoon covering the rear as a reserve. At the same time, his company's mortars would provide as much support as they could with their meager forty rounds. After a few minutes, the 1st Platoon, on the far left, encountered a column of NVA soldiers casually moving down the mountain in the direction of the landing zone. The platoon opened fire on the advancing North Vietnamese and soon found itself embroiled in an intense fight.[37]

As soon as he received word of contact with the enemy, Herren sent his 2nd Platoon, commanded by Lieutenant Herrick, to move up along the left side to support the neighboring platoon. En route to reinforce the 1st Platoon, Herrick's unit spotted additional NVA forces moving along farther to their right. Herrick received permission to pursue them with the warning not to get drawn too deeply inside the North Vietnamese forces. Herren worried that his 2nd Platoon would become separated from the rest of his units. Meanwhile, the 3rd Platoon reached the right side of the 1st Platoon and joined the fight.

Herrick formed up the two squads of his platoon side by side and ordered them across a waist-deep stream toward a ridge on the opposite side. Once they crossed the stream they found themselves attacked by fifty North Vietnamese who threw themselves down from their trail toward the Americans. While one of Herrick's squads engaged the enemy forces, another swung around to the right to outflank them. Even though the NVA unit had taken cover amid the dense vegetation and the tall anthills, the flanking maneuver surprised them. From this position, the second squad killed scores of the North Vietnamese before another unit came running from the same direction. Fortunately for the Americans, the NVA ran into the same L-shaped ambush and suffered massive casualties.[38]

While 3rd Platoon found itself in the midst of a one-sided firefight, Herrick's 2nd Platoon moved farther away from the rest of Bravo Company. Soon the 2nd Platoon discovered it had been cut off from the remainder of the American forces. Complicating their situation, an entire NVA battalion came rushing down the mountain in their direction. Quickly, the other two platoons formed up and attacked along a line to reach their isolated sister unit.

To the left of this attack, Captain Nadal's Alpha Company pushed through a dry creek bed north and west of the landing zone. It was there that his 3rd Platoon met a NVA force of 150 men running toward them. As Nadal recalled, "The enemy on the mountain started moving down rapidly in somewhat uncoordinated attacks." Hurriedly, the soldiers of the 3rd Platoon dropped their packs and formed in a line of two squads to attack the larger North Vietnamese unit. Within minutes most of the platoon's lead squad fell in the fierce firefight, including their commander Lieutenant Taft.[39]

The remainder of the platoon stood firm under the North Vietnamese onslaught. When additional NVA units arrived, they attempted to outflank the platoon from the left. Their flanking maneuver ran them straight into 2nd Platoon, Alpha Company, also rushing to provide reinforcements. As one staff sergeant put it, "Fifty NVA came right across my front and were cut down almost immediately and they did not turn and return our fire."[40] After suffering massive losses, the badly mauled communist units withdrew across the creek bed.

Moore's battalion received a needed boost at 1430 hours when his last two companies finally arrived at Landing Zone X-Ray. To cover a gaping hole in his defensive perimeter, Moore sent the fresh Charlie and Delta companies to push out south from the LZ. After settling into position, Charlie Company came under attack from the 7th Battalion of the 66th NVA Regiment. Since the North Vietnamese thought they had found an undefended gap in Moore's lines, the Americans caught them unprepared. With the element of surprise on their side, Charlie Company quickly threw the 7th Battalion back.

After Charlie Company's swift victory, Moore completed the final airlift to bring the remainder of his battalion to the landing zone. By 1520 hours, enough forces had arrived to enable him to form a reserve platoon that could plug gaps in his widening perimeter. As the 1st Battalion regrouped for another attack against the North Vietnamese, the division's artillery began pounding the NVA units on the mountain. For two hours they shelled Chu Pong while supporting air forces dropped napalm.

At 1620 hours, Captain Nadal's three platoons prepared to make another attack to reach the isolated 2nd Platoon, Bravo Company. While advancing fifty yards in the direction of the cut-off platoon, the North Vietnamese launched another counterattack. Once more communist forces raced toward the Americans with seemingly little regard for the machine gun fire their opponents directed against them. With their bayonets fixed, the soldiers of Alpha Company fought off the initial NVA charge and moved forward. Throughout the evening they met further resistance and decided that they could not reach Lieutenant Herrick's beleaguered platoon. The action had cost Alpha Company thirty-one enlisted men and three officers.[41] Moore informed the 2nd Platoon that they would have to hold out until the following morning.

In the evening, the North Vietnamese forces made a costly mistake. Shortly after 2000 hours, two American resupply helicopters approaching Landing Zone X-Ray spotted the signal lights of NVA units moving down from Chu Pong. The North Vietnamese were making another push to dislodge the U.S. 1st

Battalion, 7th Cavalry. With the aid of the lights, however, the Americans could see that the communist column spanned an area 300 yards wide and a half-mile long. Their route took them in the direction of the cut-off 2nd Platoon, Bravo Company and the northern part of the LZ.

American forces used the North Vietnamese error to call in supporting artillery strikes. Throughout the night, the shelling took a heavy toll on the communist forces.[42] Secondary explosions erupted when the American barrage hit the NVA ammunition depots. During the night, several North Vietnamese units broke through the artillery barrage to attack the 2nd Platoon. Even then, however, the platoon leaders still managed to call in artillery to prevent the NVA from getting close to the unit's remaining twenty men.

Events continued to worsen for the North Vietnamese defending the Ia Drang Valley. Although they had planned another assault on the landing zone at 0200 hours on the morning of November 15, 1965, they could not organize themselves for the operation. Adding to their problems, American forces now used a steady stream of air strikes, coupled with the constant barrage from the division's artillery, to disrupt the 66th NVA Regiment. Their nighttime counterattack had to wait until the dawn.[43]

At first light, Moore sent a series of patrols out to probe for the location of NVA units. They shortly discovered the forces of the 66th Regiment streaming toward the landing zone through the waist-high grass. American Charlie Company took the brunt of the assault. In the attack, they suffered heavy casualties, losing their company as well as two platoon commanders. One soldier involved in the engagement remembered, "The enemy was all over, at least a couple hundred walking around for three or four minutes; it seemed like three or four hours. They were shooting and laughing and giggling."[44]

Repeating the code word "Broken Arrow," Colonel Moore's forward air controller radioed for support. American air forces knew that one of their units stood on the verge of collapse. They swiftly began a series of air strikes driving the NVA reserves and follow-up forces from the area near Charlie Company. Once more the division's artillery drove the North Vietnamese back. However, a few of the communist units were already too close to the American ground forces to fall victim to the barrage of steel. To reinforce Charlie Company, then, Moore sent a platoon from another unit. After losing two soldiers, they managed to plug the gap in Charlie Company' sector.[45]

Two hours later, the Alpha Company of the 2nd Battalion, 7th Cavalry arrived in their helicopters to further reinforce American units in the area. With the additional forces, the cavalry troopers pushed the North Vietnamese back and extended the perimeter of the landing zone. In several minutes, they killed scores of NVA soldiers. During the morning engagement, Charlie Company had lost 111 men.[46]

To the left of this fierce engagement, the newly arrived H-15 Main Force Viet Cong Battalion prepared to attack Delta Company. Fortunately for the Americans, this fresh communist unit mounted its assault through an area where Delta Company could clearly see its approach. Through the evening the company had prepared for this kind of attack, placing nine machine guns, each with a crew of four, in front of their positions. Undaunted, the Viet Cong ran toward the Americans and fell victim to the superior firepower of Delta Company. One machine gunner alone killed one hundred VC soldiers. After experiencing terrible losses, the Viet Cong unit withdrew.[47]

In the afternoon, the last elements of the 2nd Battalion, 7th Cavalry marched to Landing Zone X-Ray to bolster Moore's forces. Their arrival coincided with a B-52 assault against Chu Pong Mountain, tying down additional NVA units. Under the cover of this bombardment, the remainder of Captain Herren's Bravo Company and the 2nd Battalion moved slowly to the stranded 2nd Platoon. At 1500 hours they pushed through a circle of NVA units and reached the trapped Americans. During their ordeal, the 2nd Platoon had lost eleven of its twenty-nine members. Seven of its soldiers were wounded.[48]

That evening, the 1st Battalion, 7th Cavalry evacuated its wounded from the area. Moore then prepared for the unit's second night in the landing zone. The Americans knew that a still sizable communist force remained nearby. They believed that the NVA would soon mount yet another counterattack. Instead, the North Vietnamese gave them four uninterrupted hours in which to prepare for an assault. Along the area formerly occupied by Charlie Company, American forces dug foxholes capable of holding three men. In front of these positions, they placed their machine guns. With the additional time, they also sent a reconnaissance squad out front to set trip-flares and booby traps.

Before daybreak the NVA forces began their next assault. The Americans became aware of their presence when the North Vietnamese set off the trip flares set the night before. Approximately one hundred soldiers of the NVA 7th Battalion, 66th Regiment mounted a frontal attack against the prepared American positions. Within ten minutes they had fallen victim to a hail of machine gun and artillery fire. Twenty minutes later, the North Vietnamese rushed down the mountainside in even greater strength. Fresh with replacements, they charged the American units. Once more they succumbed to the storm of metal fired at them from the air and the ground.

Although outnumbered, the Americans maintained their positions. In the lull before another wave of North Vietnamese approached, a group of Moore's men, led by Sergeant Setelin, went forward to set an ambush. At 0500 hours, trip flares alerted the Americans to another attack, and Setelin's troops held their ground allowing them to outflank the NVA soldiers as they came forward. Once the communist forces came into the open ground, the Americans, along with Setelin's group, opened fire. As the sergeant later recalled, "They stepped into the open area. The flares were burning, they were lit up, and it was easy. We opened up and picked them off."[49] Within one half-hour, the assault ended with another NVA retreat.

In one last gasp, the North Vietnamese threw all of their forces into the fray at 0627 hours. This assault was even less organized than the previous ones as waves of NVA soldiers came down the Chu Pong straight into the American defenses. One American described the action as "a shooting gallery."[50] Those North Vietnamese that did not close with the Americans fell under another shower of artillery and mortar fire. The fight ended in fifteen minutes. Casualty figures suggest a lopsided engagement: the Americans suffered six wounded and none killed; the NVA, in contrast, lost hundreds.[51]

With the North Vietnamese counterattacks ended, elements of the newly arrived American 2nd Battalion, 7th Cavalry began to push their way past the perimeter moving eastward. Aided by air strikes, the remaining units of the 1st Battalion, 7th Cavalry pushed out in the opposite direction, solidifying American gains. They met little resistance from the NVA units, which by this time had retreated in disarray. By the afternoon, the 1st Battalion and its attached companies left the landing zone via helicopter.

In their effort to drive the American forces from Landing Zone X-Ray, the North Vietnamese experienced horrific losses. According to American estimates, the NVA lost 634 soldiers in the close combat. Artillery fire and air strikes killed another 1,215 communist soldiers. Moore's battalion, and those units that had provided reinforcements, suffered 79 dead and 121 wounded.[52]

Evaluating the Battle of Ia Drang

Through a combination of airborne assaults and aerial bombardments, the 1st Cavalry Division secured the Pleiku area. Their operations in the Ia Drang Valley disrupted the NVA from gathering enough strength to move along Route

19 to the sea. Not only had the American effort prevented the NVA from cutting South Vietnam in two, it had also demonstrated the effectiveness of air mobility warfare. In total, American forces lost 305 men. In contrast, the NVA lost 3,561 soldiers in a failed effort to establish a staging ground for an ambitious invasion of the south.[53]

The Battle of Ia Drang demonstrated the key characteristics of the North Vietnamese Army during its war to liberate the south. NVA units fought with great determination, showing a willingness to suffer disproportionately larger casualties than their American forces without losing their ability to fight. To dislodge the 1st Cavalry Division from the area near the Chu Pong Mountain, it mounted numerous counterattacks, but to no avail. North Vietnamese forces also fought with flexibility: they were capable of altering their battlefield plans and recovering from defeats quickly to mount counterattacks. While the NVA forces demonstrated great determination in the defense of their staging ground in the Ia Drang Valley, they made numerous mistakes, such as pausing to kill American wounded, launching numerous frontal assaults without attempting to turn their opponent's flank, and exposing themselves to American air attacks by using flares at night. In the end, however, the NVA exhibited its most important trait: namely, an insensitivity to large casualties that would eventually help North Vietnam to emerge victorious against the United States.[54]

NORTH VIETNAM AND THE AFTERMATH OF TET

The North Vietnamese attempted to score a decisive victory and, thereby, turn the course of the war with the Tet Offensive of 1968. They had hoped that the series of coordinated attacks by NVA and their Viet Cong allies would spark an uprising in the south. With revolution finally spreading throughout South Vietnam, they reasoned, the United States would find itself in an impossible strategic situation. This combination of simultaneous conventional and guerrilla operations would overwhelm the American military and end the war.

Instead, the North Vietnamese lost each battle of the Tet Offensive and they suffered enormous casualties in the process. From the perspective of the battlefield, Tet had ended in a resounding defeat for communist forces. Not only had American forces badly mauled NVA units along the Demilitarized Zone, but also the losses suffered by the Viet Cong had rendered it a nearly useless military instrument. The communists had failed to achieve any of their stated objectives. Moreover, from January to March 1968, they had incurred more than 58,000 battle deaths.[55] It was the American reaction to Tet, however, that turned a communist operational defeat into a strategic success.

In the aftermath, the North Vietnamese public did not view Tet as the triumph that it would later become. Government leaders had promised that the offensives of 1968 would end the war and the costly expenditure of lives that came with it. One Hanoi newspaper noted that the people and armed forces "have lost confidence in the higher echelon leadership and in the revolutionary capability of the people. They think that our assessment of enemy capabilities is inaccurate, our strategic determination is erroneous, and we have to lower our requirements to prolong the war. They have become doubtful of victory, pessimistic, and display a shirking attitude."[56]

Recognizing that opposition to the war might surface, the North Vietnamese government made a concerted effort to rally the military and the nation as whole. They reminded their citizens that the Viet Cong suffered the bulk of the losses, and that the North Vietnamese Army remained intact, ready to continue the struggle. After the election of Richard Nixon to the American presidency, the government in Hanoi also confidently argued that the United States had lost the stomach for war. Still further, rather than gloss over the possibility of a quick victory, the communist leadership reminded the North Vietnamese public that "victory would not come suddenly but in a complicated and tortuous way." They honestly predicted that the country could no longer count on "victory in a short period of time." To bolster public support, Hanoi told its citizens that they would win because they could endure more pain than the Americans could. The fact that the North Vietnamese promised to continue the war, while the new U.S. government guaranteed that it would end it, shows the strength of Hanoi's control over its citizens.

Support for the war eventually rebounded among the population and armed forces. Even as the war to liberate the south continued to take a toll on manpower and resources, the regime could count on the staying power of the NVA to remain firm. A mix of factors kept the North Vietnamese military fighting. One element of motivation was the fear of punishment. A RAND Corporation study based on interviews of NVA prisoners of war found, "Fear of family punishment is not imaginary. Families of ralliers [defectors] and deserters do suffer and many ralliers and POW report having personally seen this. The families' food rations are reduced or eliminated entirely, and they are ostracized. If they are working on a government job, they are fired. Thus, propaganda designed to arouse homesickness is double-edged, as it calls the NVA soldier's attention to the fact that the family would suffer."[57]

Another part of this commitment stemmed from the American bombing campaign against North Vietnam, which seemed to validate the regime's warnings about the threats from foreign invaders. The more people were killed by American bombings, the more the country supported the regime's goal of pushing forward to victory.[58] To further motivate the population, the government inundated the cities and rural hamlets with a steady diet of propaganda. With slogans like "All for the Front Lines," Hanoi created the impression that the entire nation needed to work together to defeat foreign aggression. The nationalist appeals of regime seemed to inspire its citizens.[59] As one Vietnamese psychologist reflected years later: "Even the lowliest peasant is deeply nationalistic and in times of war, the sentiment can border on xenophobia."[60] In the words of one soldier, "We all wanted to see our loved ones again, but we all knew that the country was being invaded, that many in the South were being dominated by Americans . . . so all we could do was to fight the war."[61]

The regime's efforts to promote nationalism and communism reinforced the training personnel received in the NVA. As Konrad Kellen describes it:

> The analyst found particularly remarkable (as emphasized in the text) the degree to which men do not simply "mouth" what they have been told, but seem to have fully absorbed and assimilated it, rendering it in their own terms, illustrating it with their own examples and experiences. Thus, what may have begun as indoctrination has become sincere conviction, opinion, and emotion, and may, therefore, be regarded as virtually impossible to dislodge. The men polled here—all except one—are unlikely to change their views or to lose their readiness to make the necessary sacrifices commensurate with those views. They can perhaps be killed, but they probably cannot be dissuaded either by words or by hardships. As found in previous analyses, all of the elements of strength and cohesion, impressive by themselves, tend to reinforce each other. The enemy soldier (both the private and low-level cadre, VC and NVA) trusts his leaders, likes his political officer, gains strength from criticism/self-criticism and the three-man cell, draws pride from his military successes, is encouraged by what he sees as the unalterable support and sympathy of the people, and relies heavily on what he insists is the righteousness of his cause. This parallels the findings of some of the 1965, 1966, and 1967 RAND studies on the same subject.[62]

THE 1969 CAMPAIGN AND OPERATION DEWEY CANYON

Operation Dewey Canyon represents an ideal setting in which to assess NVA determination and flexibility on the battlefield, but also their staying power as it recovered from the terrible defeats of the previous year. In January 1969, North Vietnamese forces had begun to regroup from their costly offensive during Tet a year earlier. Part of this recovery involved a change of tactics—namely, avoiding large engagements with the American military and its superior firepower. At the same time, U.S. forces, now under the leadership of Creighton Abrams, had decided to pursue the NVA units into their sanctuaries. Many of these outposts were sprinkled throughout the A Shau Valley, a hilly region near the border with Laos. Typically, NVA forces resupplied these sanctuaries through Laos, and from there they would launch attacks against the coastal cities of South Vietnam.[63]

Abrams wanted to mount a series of preemptive assaults to stop the North Vietnamese forces moving from Laos into the A Shau Valley along both routes 548 and 922.[64] To sweep the area in an operation known as Dewey Canyon, the Americans relied on the 3rd Marine Division. In 1968, they had engaged and defeated the NVA's 320th Division three times. Now the North Vietnamese replaced this exhausted unit with six independent regiments: the 138th, 270th, 84th, 31st, 27th, and the 126th Naval Sapper. In the southern region of the valley, the NVA deployed three battalions of the 812th Regiment. Operating throughout the A Shau Valley, the North Vietnamese deployed 36,800 troops, half of which were combat units.

The American plan called for the Marines to seize a series of abandoned outposts near Laos and to use these as staging grounds. On January 18, 1969, the Marines reopened several of these unused fire support bases and placed their 105-millimeter artillery in areas where they could shell NVA units sneaking into South Vietnam from Laos. On February 2, the North Vietnamese struck first by shelling Fire Support Base Cunningham with their artillery pieces located safely in Laos. The barrage killed five Marines and left another five wounded. American military leaders denied U.S. units permission to cross into Laos and destroy the NVA guns.

Operation Dewey Canyon: Evidence of Determination and Flexibility

On February 11, 1969, three Marine battalions moved into the Da Krong area of the A Shau Valley. There they attacked a North Vietnamese redoubt known as Base Camp 611, which was close to the Laotian border. For several days, the Marines attacked well-fortified NVA positions. The communist defenders, however, refused to give any ground. During the evening hours, they would send out patrols to probe the American positions. Some NVA soldiers would even tie themselves to trees so that they could fire and throw grenades down on the Marines. Eventually the NVA withdrew, and the encounter at Base Camp 611 proved inconclusive.

Nine days later, along a ridgeline that overlooked Route 922, a company of Marines encountered a convoy of NVA forces moving out of the A Shau Valley into Laos. The North Vietnamese had apparently caught wind of Operation Dewey Canyon and had decided to withdraw to the safety of the Laotian countryside. The Marines discovered that they could not call in artillery fire onto the retreating communists because their guns were too far out of range to do any damage. Captain Winecoff's Hotel Company decided to move into Laos in order to set an ambush for the NVA forces.

It would take the Marines two days to prepare for their attack. On February 22, Captain Winecoff's three platoons waited for Vietnamese units to approach. Early that morning, they encountered a large convoy of trucks and heavy tracked vehicles. The Marines sprang their ambush against the surprised North Vietnamese units. In less than one half hour, the three platoons destroyed most of the vehicles and then quickly dashed back across the border into South Vietnam.

Farther inside South Vietnam, the other Marine battalions cleared out the area near Route 548. During one phase of the operation, Lieutenant Wesley Fox's Alpha Company stumbled upon the NVA unit, dug in along a ridgeline, protected by steep hills and a dense covering of tall trees. Both the terrain and the poor weather precluded any artillery and close air support. Thus, on February 23, the three platoons of Alpha Company had little choice but to engage the enemy directly. Once again, the North Vietnamese put up a tough defense. In the fierce fighting, all three of Lieutenant Fox's subordinate officers were killed or wounded. Eventually the Marines took the hill at the cost of ten men. Almost everyone in the company was wounded once in the engagement. The North Vietnamese lost 105 of its soldiers.

By the time Operation Dewey Canyon ended on March 19, 1969, the Marines had suffered 130 battle deaths and 920 wounded. In contrast, the North Vietnamese saw 1,617 of its soldiers killed in the fighting. They also lost 1,233 individual weapons, 16 artillery pieces, 73 antiaircraft guns, 26 mortars, 92 trucks, more than 807,000 rounds of ammunition, and 222,000 pounds of rice. More important, American units had severed their routes from Laos to the coastal cities of South Vietnam. For this reason, American military leaders reasoned, the spring of 1969 passed more peacefully in Hue and Da Nang than in previous years.

CONCLUSION

Although the United States withdrew the bulk of its combat forces in 1973, it took the North Vietnamese military two more years to finally unify the country by force. The prize they had sought decades earlier had come at a terrible price. The regime's high degree of control over the population and the autonomy of its military to train created a messianic army. It was capable of enduring a long and costly conflict. The regime could count on a motivated set of recruits, little opposition on the home front, and a loyal army. The armed forces could send units into battle filled with soldiers that were motivated by ideology as well as bonds of loyalty to their fellow soldiers. (See Tables 6.2 and 6.3.)

Even after the terrible losses and defeats of the Tet Offensive, the North Vietnamese recovered and continued to endure additional battlefield defeats at the hands of American forces. They would consistently tolerate higher casualties than the United States without losing their will to fight (see Table 6.1). Some estimates calculate that the North Vietnamese lost 500,000 of their fighting men between 1962 and 1975, with some approximations suggesting that the total was as high as 850,000 soldiers.[66] Motivated by an intense nationalism that was reinforced by communist ideology, the typical Vietnamese soldier was willing to continue risking his life in order to eject the American invader. A decade later, Vietnamese general Giap recounted his military's efforts this way:

> We sought to break the will of the American people to continue the conflict.... We were waging a people's war ... a total war in which every man, every women, every unit, big or small, is sustained by a mobilized population.... America's sophisticated weapons, electronic devices, and the rest were to no avail. Despite its military power, America misgauged the limits of its power. In war, there are two factors—human beings and weapons. Ultimately, though, human beings are the decisive factor. Human Beings! Human Beings![67]

Table 6.1: American and North Vietnamese Casualties, 1965–73

	United States	North Vietnam
1965	1,369	35,000
1966	5,008	55,524
1967	9,378	88,149
1968	14,589	181,149
1969	9,414	156,954
1970	4,221	103,638
1971	1,381	98,094
1972	300	131
1973	237	5,625

SOURCE: Michael Clodfelter, *Warfare and Small Armed Conflicts: A Statistical Reference, Vol. II: 1900–1991* (Jefferson, NC: McFarland and Company, 1992), 1292–1300.

Table 6.2: Assessing Staying Power: Messianic North Vietnamese Army, 1965–73

Evidence
North Vietnamese recovered from devastating defeat of the Tet Offensive
Soldiers motivated by combination of communism, nationalism and fear of punishment
NVA units mount additional offensives after the Tet defeat

PREDICTION: Strong Staying Power.
Military Collapses Only When Adversary Possesses Crushing Material Superiority.

Table 6.3: Assessing Battlefield Performance: Messianic North Vietnamese Army, 1965–73

Battles	Evidence
IA Drang	NVA units attack and counterattack despite heavy casualties
	NVA units counterattacked after surprise
Dewey Canyon	North Vietnamese units capable of defending soon after Tet defeat
	NVA units respond to Marine offensive with probing attacks and tactical changes

PREDICTION: Strong Battlefield Performance.
Most Units Fight with Determination and Flexibility.

This kind of calculation made perfect sense for a messianic military. After the war, the American analyst Colonel Harry G. Summers remarked to a North Vietnamese officer, "You know you never defeated us on the battlefield." To which the communist officer replied, "[T]hat may be so, but it is also irrelevant."[68]

7 The United States, 1968–72

FROM 1960 TO 1975, the United States, one of the most powerful countries in the world, tried unsuccessfully to compel one of the weakest nations, the communist-led Democratic Republic of Vietnam, to recognize the independence of its neighbor, South Vietnam. In the beginning, American involvement mostly entailed advising its ally in Saigon on how best to conduct a counterinsurgency against the Viet Cong (VC) infiltrating from the north. Eventually, the United States began to assume a greater role in the civil war between North and South Vietnam. With the backing of the Gulf of Tonkin Resolution of July 1964, President Johnson increased the American military presence in South Vietnam from 23,000 to 485,000 soldiers by the end of 1967.[1]

Over the next few years, the U.S. armed forces went from advising to conducting counterinsurgency operations, fighting regular units of the North Vietnamese Army (NVA), and mounting an ambitious series of bombing campaigns against North Vietnam.[2] All the while, the American government and public believed that the regime in Hanoi would ultimately capitulate under the pressure.[3] This optimism, however, would change with events in 1968.

The Tet Offensive of late January to March 1968 marked a key turning point in the course of the Vietnam War. After the campaign, the Viet Cong ceased to function as an effective combat force in South Vietnam. The NVA, moreover, suffered two serious setbacks against American forces at Hue and Khe Sanh. In operational terms, the entire offensive resulted in a massive defeat for communist forces.

While the North Vietnamese and their Viet Cong allies lost most engagements on the battlefield, the resilience of the communist war effort suggested that the two sides had become locked in a stalemate. Regardless of the superiority in material resources and personnel the United States military poured into Vietnam, a decisive and favorable end to the war seemed elusive. Prior to the Tet Offensive, American leaders had promised that victory was in sight. After Tet, breaking the deadlock appeared increasingly unlikely, and the American public concluded that their country's involvement in the war had become too costly. The American public lost their appetite for war, voicing their opposition in protests and at the ballot box. At the same time, U.S. armed forces pursued a different strategy, reducing the number of forces and preparing South Vietnamese units for the eventual American withdrawal. These conditions slowly eroded the staying power of the U.S. military.

This chapter argues, according to *cohesion theory*, that the United States deployed a *professional* military to fight the Vietnam War. This professional military grew out of a combination of weak regime control over American society and strong organizational autonomy for the armed forces to train. As such, U.S. armed forces displayed strong battlefield determination and flexibility, but their staying power slowly deteriorated when the war became a stalemate and American political leaders began a drawdown of personnel to end the country's commitment to the war. U.S. units continued to fight hard in combat, but incidents of dissension and disorder began to fill the ranks, especially among support units in the rear areas.[4]

I also use cohesion theory to explain why the U.S. military did better than primary group theory would expect. Because the armed forces relied on individual and not unit replacement policies, primary group bonds, according to this widely accepted view, would not form and the armed forces should lack cohesion. My theory, instead, can explain why the U.S. military fought with determination and only gradually lost its staying power. In other words, U.S. armed forces displayed more cohesion than primary group theory would predict.

A PROFESSIONAL AMERICAN MILITARY

When the United States entered the Vietnam War, it relied on a professional military to do the fighting. These armed forces came from a country where the government, or regime, exerted little control over society. Government existed to protect the individual liberties of its citizens, not to demand unconditional loyalty from them, as in North Vietnam. It was, in fact, this fierce commitment to individualism that united Americans. U.S. military planners recognized the strong individualism that American society instilled in its citizens and, therefore, potential recruits. To compensate for this tradition of individualism, American policy-makers created an autonomous military capable of training recruits to think as a team instead of pursing their own personal interests. In other words, the American military sought to create ties within its ranks that it knew did not exist within the broader society. With freedom from outside interference to train, American organizations worked hard to turn proud individualists into soldiers, sailors, airmen, or marines who believed in unconditional loyalty to their country as well as to one another.

The military cohesion of American forces in the Vietnam War was determined by two factors: weak governmental control over society but strong autonomy for military organizations to conduct training. Training and strong organizational bonds within the armed forces created units capable of fighting with determination and flexibility on the battlefield. However, with a government at home that could not exert strong control over the population, as the regime in Hanoi did in North Vietnam, the U.S. military suffered from moderate staying power. Consequently, when the Tet Offensive confirmed suspicions that the war had become a stalemate, U.S. forces slowly lost their staying power. At home, the population was less willing to support wars in which victory no longer appeared possible, persuading American policy-makers to draw down forces and look for an exit to the conflict. Within the armed forces, personnel began to question why they were fighting. The combination of stalemate and withdrawal relentlessly undermined the staying power of the American military.

Weak Regime Ties

Throughout its history, the United States has always espoused a government where individual citizens could pursue their own interests mostly free of state interference. As champions of individualism, Americans have always fought to prevent undue government involvement in politics or the economy. In this liberal society, individuals view the world through the lens of their personal interests, with the state primarily existing to protect their right to live under these conditions. Because of these liberal beliefs, historians have argued, more collectivist ideals, such as socialism, have never found a place in the American political tradition.[5] These views influence how the country's citizens perceive military service. The citizens of liberal societies generally believe that their government should not call upon them to risk their lives for their country unless the country's survival is at stake.[6] As Michael Walzer suggests, "Indeed, the great advantage of liberal society may simply be this: that no one can be asked to die for public reasons or on behalf of the state."[7]

These liberal views about the relationship between the state and the people, however, have not prevented Americans from giving their lives in service to their country. The United States lost 578,000 soldiers in World War I, World War II, and the Korean War combined.[8] Although the American public enthusiastically backed its country's participation in World War I and World War II, its unconditional support did not extend to U.S. involvement in Korea. When the Korean War began to appear deadlocked, support for the intervention effort dropped precipitously, until a cease-fire ended hostilities.[9] From Korea, U.S. policy-makers learned the limits of public support for wars: the American people did not want their soldiers to become involved in stalemated conflicts. However, they also took away one additional lesson: if left unchecked, communist aggression would spread from one country to the next. The Korean experience would draw the United States into Vietnam, where it suffered another painful stalemate.

With the conflict over Korea in mind, American leaders tried to construct a consensus for fighting the Cold War. They attempted to arrive at a list of rationales for American military intervention. President Eisenhower, for example, justified American involvement in places like Korea with the Domino Theory: "You have a row of dominos set up. You knock over the first one, and what will happen to the last one is a certainty that it will go over very quickly. So you could have the beginning of a disintegration that would have the most profound consequences."[10] In other words, the Cold War might turn hot in order to prevent the spread of communism from one country to another. Whatever reasons they might provide for sending troops into battle, American leaders realized after Korea that they could not count on the public's wholesale support for the kind of limited interventions that came to characterize the Cold War.

President Lyndon Johnson's administration became involved in Vietnam out of a concern that instability in Southeast Asia would erode American credibility. In Johnson's view, the United States could not claim to back the creation of a democratic South Vietnam and then abandon its ally in the face of a communist insurgency. Concerned with demonstrating its resolve to stop the spread of communism, the Johnson administration continued aiding the South Vietnamese government, a commitment it took over from the Kennedy presidency. By 1965, President Johnson had augmented the 7,100 military advisors Kennedy had sent to South Vietnam with 184,300 American soldiers.

Initially, Kennedy had voiced some apprehension about becoming entangled in the Vietnamese civil war. Before his untimely death, Kennedy realized that the United States could very quickly become embroiled in a difficult war. He warned Arthur Schlesinger, "The troops will march in; the bands will play; the crowds will cheer; and in four days everyone will have forgotten. Then we will be told to send more troops. It's like taking a drink. The effect wears off, and you have to take another."[11]

Concern about the conflict in Vietnam escalating also worried Johnson when he assumed the presidency. In the mid-1960s, the United States not only faced the possibility of a widening war in Southeast Asia, it also had to confront civil rights struggles in the south and increasing demands for social reform. To resolve some of these tensions, and to follow Kennedy's legacy, Johnson proposed his Great Society programs. Johnson wanted his presidency remembered for the progress he made on domestic as well as foreign policy. Looming large in the background, however, was the worsening conflict in Vietnam. Fearing that political conservatives would use the downfall of the South Vietnamese government as an opportunity to derail his ambitious domestic legislation, Johnson slowly increased American involvement in Southeast Asia.[12]

The largest increase in American forces in South Vietnam came in 1964. After U.S. naval forces skirmished with North Vietnamese patrol boats in the Gulf of Tonkin, Johnson won congressional approval to do whatever he deemed necessary to protect American forces in the region and support South Vietnam. The Joint Chiefs of Staff urged Johnson to invade North Vietnam, predicting, "There seems to be no reason we cannot win if such is our will—and if that will is manifested in strategy and tactical operations." Johnson, however, believed the cost of invasion too high, since it entailed sending 700,000 to 1 million men into what the military told him might become a seven-year conflict. He instead sought victory on the cheap and decided to follow the advice of Air Force general Curtis Le May, to conduct a bombing campaign that he hoped would bring Hanoi to its knees.[13] While deciding against a larger deployment, Johnson did allow American forces to grow to 485,600 by the end of 1967 for the purposes of helping the South Vietnamese fend off growing incursions from the north and to provide security for U.S. operations.

This desire to spare America a costly war not only influenced Johnson's military strategy for the war, but it also affected his decision-making in terms of what kind of sacrifices he would demand from the public.[14] In his efforts to avoid imposing pain on the American people, he refused to raise taxes or to call a national emergency.[15] More important, contrary to the advice of members of his administration, Johnson decided against mobilizing the reserves and the National Guard. Johnson worried that the country would oppose this move and undercut the support he had garnered for his Great Society programs.[16] This move contradicted the military's conventional wisdom at the time, viewing these forces as essential to preparing the military for war.[17] In addition, the president prevented the military from extending the time individual soldiers spent on a tour of duty. Because of these changes to manpower policy, the forces sent to Vietnam would consist largely of inexperienced conscripts.[18]

Strong Organizational Autonomy for Training

Unlike Hitler's Germany, the United States could not rely on a strong, coercive government to create military cohesion. Instead, the armed forces relied on training to establish strong bonds within its ranks. Free from government interference and domestic political cleavages, American officers used training to instill norms of loyalty and trust among their personnel. Almost all militaries attempt to promote these norms as means to improve performance on the battlefield. As Jonathan Shay explains:

> Any army, ancient or modern, is a social construction defined by shared expectations and values. Some of these are embodied in formal regulations, defined authority, written orders, ranks, incentives, punishments, and formal task and occupational definitions. Others circulate as traditions, archetypical stories of things to be emulated or shunned, and accepted truth about what is praiseworthy and what is culpable. All together, these form a moral world that most of the participants most of the time regard as legitimate, "natural," and personally binding. The moral power of an army is so great that it can motivate men to get up and out of a trench and step into enemy machine-gun fire.[19]

In these respects, the American military during the Vietnam War was no different. A sense of homogeneity, or the belief that every member acted together for the team in pursuit of common goals, represented one of the most important values the armed forces attempted to inculcate in its members. The U.S. military recognized the need to remove the individualist tendencies recruits had acquired as civilians.[20] To instill the value of teamwork, all branches of the military put soldiers through a rigorous course of training. This experience would establish the value of working in groups and set the stage for soldiers to form strong bonds with one another.[21] Soldiers learned to think of each other as a group first and as individuals second. By emphasizing a common fate, the military, after the Korean War, became a model institution for racial integration.[22]

Another principle championed by the American military was trust among its members.[23] Officers attempted to create a climate of trust among enlisted men as well as officers. Whether in training or actual combat, officers demonstrated competence and expected their men to do the same. Not only did officers hold high expectations for their soldiers, but they also shared the same hardships as their men, often leading them into battle instead of remaining in the rear.[24] This sense of trust and mutual obligation, acquired in difficult training, often created tight bonds between enlisted men and their officers.[25] Charlie Moskos describes these bonds between soldiers as a social contract, with each man agreeing to perform at his highest potential in order to ensure each other's survival.[26] The American military also advanced homogeneity and trust as worthy values by using merit as a standard for promotions.[27]

Moreover, this atmosphere of trust also permitted the American military to push their officers to show greater flexibility on the battlefield. Changes in military technology made initiative in combat a virtue. According to S. L. A. Marshall: "As more and more impact has gone into the hitting power of weapons, necessitating ever widening deployments in the forces of battle, the quality of the initiative in the individual has become the most praised of the military virtues."[28] Trainers hoped that by teaching the value of initiative, and trusting their soldiers' judgment, they would improve the chances of officers adapting to the exigencies of the modern battlefield.

THE TET OFFENSIVE: U.S. ARMED FORCES FIGHT WITH DETERMINATION AND FLEXIBILITY

In the summer of 1967, the North Vietnamese leadership decided to change its strategy for conducting the war.[29] Up to that point, the war had not gone well for communist forces. After more than two years of confrontations with U.S. forces, they had concluded that the NVA and the Viet Cong could not compete with American superiority in mobility and firepower. Some in the leadership, such as General Nyugen Chi Trahn, argued for a reduction of VC operations in the south replaced by a greater effort by the North Vietnamese Army. This view envisioned a long, drawn-out war to wear down the American will to fight.

General Vo Nyugen Giap offered a different approach. He argued for one massive offensive that would inflict a defeat on the Americans and South Vietnamese on the scale of his victory at Dien Bien Phu. Giap's proposal reflected the concept of the "General Offensive" found in communist doctrine. Unlike

Trahn, who sought a more conventional war of attrition, Giap wanted to make use of the Viet Cong as well as the North Vietnamese Army. He envisioned a plan whereby a coordinated attack by both the NVA and the VC would generate a "General Uprising" in South Vietnam. Those sympathetic with the communist cause would use the instability caused by the massive offensive, he reasoned, to overthrow the government in Saigon.[30]

With Trahn's death in July of 1967, the leadership in Hanoi decided to execute Giap's plan as soon as possible. Even though the general claimed that his forces needed two more years to prepare for it, the Politburo, convinced the population in the south was supportive of its aims, believed the time for decisive action had arrived. Giap's strategy rested on three assumptions about what would happen once the offensive unfolded. First, he believed that the Army of the Republic of Vietnam did not have the stomach to fight and would fold once the combined attacks by the Viet Cong and the NVA began. Second, the population in the south would follow the script of communist doctrine and conduct a "General Uprising," toppling the South Vietnamese regime. Finally, he remained convinced that the shock of the assault would erode the American will to fight.[31]

To offset their relative disadvantage in firepower and mobility, Giap's strategy depended upon securing the element of surprise. The North Vietnamese, seeking to catch their adversaries unprepared, set the time for their offensive to coincide with the celebration of Tet on January 31, 1968, the start of the lunar New Year and the most important holiday in Vietnam. Lest they fall into enemy hands, Giap also kept the details of his plans from his field commanders till the very last possible moment. This last precaution would create trouble for him once the offensive began. Before the offensive, the North Vietnamese also conducted a series of diversionary attacks against Army of the Republic of Vietnam (ARVN) and U.S. forces. Giap hoped to draw his opponents into the countryside, away from the urban centers the Viet Cong sought to seize in the opening moves of the offensive. In addition, NVA forces began a massive buildup near Khe Sanh, across the Laotian border, in an effort to convince the Americans that the U.S. Marine base there represented their near-term objective. Giap played on Westmoreland's fear that Khe Sanh might become another Dien Bien Phu, when in fact he sought to capture the city of Hue to the south.

Some American commanders, however, began to detect the outlines of North Vietnamese designs. Showing flexibility, U.S. units began to adjust their defensive plans. In particular, Lieutenant-General Frederick C. Weyand started to view the peripheral communist attacks as the opening stage to some larger attack to come in the future. Noticing an increase in VC radio traffic around Saigon, he requested and received permission from Westmoreland to station additional U.S. forces closer to the capital. As a result of his foresight, twenty-seven American battalions, rather than the fourteen originally planned, stood nearby and ready to come to Saigon's aid once hostilities finally broke out.[32]

Giap intended his Tet Offensive to begin on January 31, 1968. However, his desire for secrecy created confusion about the start date among some of his field commanders. Some could not decide if the start date coincided with the old lunar calendar of South Vietnam or the new calendar issues by Hanoi. Consequently, twelve cities in the central region of the south, including Pleiku, Da Nang, and Na Trang, came under attack by the Viet Cong at 0015 hours on January 30, an entire day earlier than Giap had planned. These premature assaults canceled the pre-Tet cease-fire and alerted both ARVN and U.S. forces. The lack of coordination cost Giap the element of surprise he desperately needed for a chance at victory.[33]

At 0130 hours on January 31, the remainder of the scheduled Tet attacks took place with assaults against 105 of South Vietnam's cities. The Viet Cong's attempts to seize the American embassy in Saigon became one of the most newsworthy events of the entire war. The American media quickly reported on the bitter fighting taking place there, sending reports back home almost hourly. Although American units secured the embassy by the next morning, the presence of communist forces inside the compound seemed to the American public to signal the vulnerability of the military. Except for Khe Sanh, Hue, and the area surrounding Saigon, the fighting ended in a few days.[34]

The Battle of Hue: January 30 to February 25, 1968

The longest and bloodiest of the battles of the Tet Offensive occurred in the cultural and intellectual center of Vietnam, the city of Hue.[35] A zigzag moat and an outer wall surrounded the city's imposing citadel, with the Palace of Peace at its center. Inside the city were two Allied military installations: the headquarters of ARVN's 1st Division and the U.S. Military Assistance Command Vietnam (MACV) compound.

After receiving word of attacks against cities farther south, Brigadier General Ngo Quang Truong put his forces on alert. The move would prove critical in preventing the communists from completely taking over Hue from within the city's walls. Inside, Hue supporters of the North Vietnamese had been planning for several months to help the NVA seize the city during the Tet Offensive. Two days before the actual attack, elements of the 12th Viet Cong and Hue City Sapper battalions began to slip into the city to prepare for the assault. Early in the morning of January 31, ARVN patrols spotted several battalions of North Vietnamese regulars approaching Hue from the west. Under the cover of dense fog, these units reached Hue unencumbered. A few hours later, they began to shell the headquarters of the ARVN's 1st Division with 122-mm rocket fire.

Two regiments conducted the attacks on Hue. The 6th NVA Regiment, commanded by Lieutenant-Colonel Nguyên Trong Dan, launched its assault from the west, with the citadel as its objective. The 4th Regiment, led by Lieutenant-Colonel Nguyên Van, made its way to Hue from the south and east. Although ARVN forces delayed the 4th's approach to the city with an ambush, the NVA units eventually mounted an attack on the southern part of the city, including the MACV compound. By the morning, near 0800 hours, the NVA held much of Hue south of the river, all of Gia Hôi, and the southern half of the Citadel. When they reached the Palace of Peace, communist forces defiantly hoisted the Viet Cong flag.

Meanwhile, ARVN units still held the northern part of the Citadel and approximately 200 Americans as well as Australian advisors still guarded the MACV compound. Fighting with great determination, these holdouts complicated communist plans for seizing Hue. After receiving a distress call from the Americans within the MACV compound, the Marine Corps base at Phu Bai dispatched a relief column to rescue them. The original force, however, proved too small to make much progress into the city.

With reinforcements, the Marines finally reached the compound and received orders to link up with Truong's encircled ARVN forces. North Vietnamese forces thwarted these initial attempts to reconnect American and ARVN units. Over the next few days, the 1st Marine Division continued to gradually send additional forces to push out the large NVA presence but failed to accomplish the task. The NVA battalions and Viet Cong offered fierce resistance to the American counterattack.

Once inside the city, North Vietnamese forces augmented their stockpile of manpower and weapons. When communist units reached the city's jails, they freed 2,500 inmates, 500 of whom joined the attackers. The NVA forces also seized an ARVN depot stocked with U.S. weapons and ammunition. Moreover, their main supply line into Hue from the A Shau Valley remained open, keeping them equipped and well armed. A few days later, to bolster the attacking force, five battalions joined the nine that originally had begun the assault.

In the meantime, concerns about damaging the city of Hue prevented American forces from taking advantage of their superiority in mobility and firepower. The U.S. effort to push the NVA out of Hue would be hampered by these concerns until February 12, when the ARVN I Corps commander finally relented and authorized American units to use whatever force was necessary to clear the city. Moreover, the U.S. High Command underestimated both the size and the intention of the NVA attack well into the battle. General Westmoreland maintained that communist forces aimed to overrun Khe Sanh to the northwest near the Demilitarized Zone (DMZ). Accordingly, for several weeks he husbanded his reserves for units in that area, cautious not to devote too much effort to Hue, since he believed that the main NVA effort would come elsewhere. As a result, Westmoreland misjudged the size of the North Vietnamese blocking forces that encircled the city, hampering his attempts to free the city.[36]

On February 2, the 1st Cavalry Division attempted to cut communist supply lines into Hue. Six miles northwest of the city, the division commander sent 2nd Battalion, 12th Cavalry to secure a landing zone. Almost immediately, the unit came under heavy attack from a North Vietnamese force in the area. Simultaneously, a sister battalion, the 5th Battalion, 7th Cavalry approached from the west and sought to link with the 2nd, 12th. This force also met a fierce assault from NVA forces, preventing it from reaching the other battalion until February 9, 1968. Westmoreland had guessed incorrectly. His intelligence had placed the NVA 304th, 325C, and 324B divisions near Khe Sanh. In reality, however, they stood nearby and posed a formidable obstacle to U.S. 1st Cavalry's efforts to slice North Vietnamese supply lines to Hue.

By the second week of February, Westmoreland finally read North Vietnamese intentions correctly. To cut off Hue, the American commander committed six battalions to the mission. The reinforced 3rd Brigade of the 1st Cavalry Division attacked from the north and the west. Two battalions of the 101st Airborne Division joined the assault from the south. Inside Hue, three battalions of U.S. Marines had joined the fight and cleared the southern bank of the city by February 10.

The American defenders began to gain the upper hand on February 10. By that time, the 1st Battalion, 5th Marines crossed the river and joined ARVN forces in the fight for the citadel. Units from the Republic of Vietnam now had nearly eleven battalions struggling to recapture the center of Hue. Even with these reinforcements, however, NVA forces fought stubbornly to maintain their gains. Bitter house-to-house fighting carried on for two weeks. In the midst of this close combat, both sides made use of tear gas to secure buildings.

On the outskirts of Hue, fierce fighting between the NVA units and the 1st Cavalry also continued. With air support, the 1st Cavalry eventually managed to slice the communist supply line to the city on February 21. With their enemy cut off from its supplies, the Allied defenders finally pushed their way into the Citadel. On February 24, ARVN units swept into the old Imperial Palace, only to discover that communist forces had withdrawn. The Battle of Hue had ended.

Predictably, both U.S. and North Vietnamese forces fought tenaciously for control of Hue. Each side suffered in the fight. American forces, including the U.S. Army and Marines, had suffered 216 dead and 1,364 wounded. ARVN losses amounted to 384 dead and 1,830 wounded. The North Vietnamese attackers experienced even more casualties, with 5,000 dead, 89 captured, and innumerable numbers of wounded combatants. With as many as 5,700 people executed, the civilian population endured even more terrible losses. Fifty percent of Hue was in ruins, causing 116,000 of the city's 140,000 inhabitants to lose their homes.[37]

The Siege of Khe Sanh: January 21 to April 1, 1968

Some of the most prolonged fighting of the Vietnam War took place during the NVA's siege of the U.S. Marine Base at Khe Sanh.[38] Because the action entailed a more conventional operation than most in Vietnam, it differed from the urban fighting that characterized most of the Tet Offensive. The American compound near the DMZ represented one of several outposts that served as infiltration barriers between a divided Vietnam. Westmoreland intended bases like Khe Sanh to alert his units rather than completely stop communist incursions on their own. They also provided a staging ground for American attacks against NVA forces in the area. From April to October 1967, for instance, elements of the 3rd Marine Amphibious Force (MAF) cleared the hills surrounding Khe Sanh of North Vietnamese units.[39] In this effort, the Americans lost 160 dead and 700 wounded, but defeated an entire NVA regiment as well as a large artillery emplacement.[40] By the end of the spring of 1967, the 26th Marine Regiment had replaced the 3rd MAF. At the same time, Westmoreland concluded from the success of the recent operation that American forces could hold out against superior numbers of North Vietnamese units, provided they could call upon adequate air support and aerial resupply. He thus won the support of the

Johnson administration for enlarging the American commitment to Khe Sanh, providing the NVA with an inviting target.

Initially, Giap had intended his attacks on Khe Sanh to divert the attention of American units from his actual objectives, the seizing of several major cities in the south. To this end, the NVA began to mass troops near the outpost, along the DMZ and the Laotian border. These units included the 325C Division, which had participated in the Battle of Dien Bien Phu, the 304th Division, two regiments of the 320th Division, an armored regiment, and two artillery regiments. North Vietnamese forces totaled nearly 30,000 soldiers. Over the course of the siege, the North Vietnamese attack would appear less and less like a diversion and more like an ambitious effort to inflict a decisive conventional defeat on American forces.[41]

On the evening of January 2, 1968, a Marine patrol killed five NVA officers near the base at Khe Sanh. The incident reinforced Westmoreland's growing suspicion that the North Vietnamese sought to attack the Marine compound. The seizure by communist forces of the only major road to Khe Sanh, Route 9, reinforced his hunch about the possibility of an NVA offensive against the Marines. In the next few weeks, Westmoreland reinforced the base from the air. By the middle of January, 6,000 Marines, accompanied by numerous reporters hungry for a news story, defended the Khe Sanh plateau and its four surrounding hills—950, 881, 861, 558—each named for its respective height. Colonel David Lownds, who commanded the 26th Marine Regiment, divided his forces in two, with one half defending the compound and the other devoted to defending the circle of hills. Fire support for the American forces was provided by 105-mm howitzers and mortars.[42]

The American defenders knew that the North Vietnamese stood poised for an attack but did not know their positions. Lownds took the risk of dispatching some of his forces on a patrol to gather information on their opponent's intentions. To prompt the NVA to give away their nearby positions, Lownds sent a large patrol to the area near Hill 881 on the morning of January 20. The patrol consisted of three platoons. As they advanced to where they thought they could establish contact with enemy forces, an artillery barrage preceded them. Lownds hoped that the artillery assault would prompt the North Vietnamese to reveal their locations.

Instead, the veteran NVA units waited until one American platoon stumbled upon them. This first unit, led by Lieutenant Thomas Brindley, found itself under attack by rocket-propelled grenades as well as close-range rifle and machine gun fire. Brindley's point man fell immediately, and several of his platoon members were wounded. A second platoon rushed to the scene and outflanked the NVA ambush. Brindley then ordered his unit to rush the enemy positions. His platoon took additional casualties, and Brindley died leading the charge. With fighter-bombers dropping napalm on the North Vietnamese, the Marines pushed the NVA back and secured the area near Hill 881.[43] They also received information from one of their prisoners about major attacks planned for the next day on hills 881 and 861.

Adjusting to the surprise encounter, the Marines quickly went to work fortifying the area with barbed wire, claymore mines, and trip-flares. According to schedule, the NVA attacked Hill 861 at 0300 hours on January 21. They overran the American positions, but the Marines counterattacked successfully at 0500. Thirty minutes later, the North Vietnamese responded with a massive artillery and rocket bombardment against the main Khe Sanh compound. In the process they hit the primary American ammunition depot, which left the defenders with barely enough ordnance to return fire during the day. The event made for spectacular headlines in the press about the beleaguered American defenders. C-130 cargo planes, however, would later manage to resupply the Marines with twenty-four tons of artillery shells. At 0630, the NVA mounted an attack against the base, but American air and artillery barrages drove them back. Despite the setbacks of the day, the Marine defenses would hold together.

From January 21 to February 5, 1968, the North Vietnamese made a series of attacks against Marine positions on Hill 861. During these assaults, the Americans pressured the enemy from the air with attacks by fighter-bombers and B-52s against communist command centers in Laos. On February 5, NVA units overran a portion of Hill 861, killing seven Marines in the process. Undaunted, American forces retook the position using tear gas as well as air and artillery support. At the end of the day, however, the Marines needed to engage in hand-to-hand combat to retake the hill.[44]

On February 21, the North Vietnamese overran the Marines on Hill 861 once more. This time the Americans lost twenty-one soldiers and suffered twenty-seven wounded. NVA forces would press the attack on the Marine compound for another two weeks. Unable to break through the American defenses, communist forces began to withdraw on March 6. Seeing an opportunity to secure the area around their base, American units pursued their attackers. By April 1, they completed operations against the NVA in the region. They also reopened Route 9 connecting the Marine outpost with the rest of American forces.

The Siege of Khe Sanh ended up becoming a costly diversion for the North Vietnamese. Although the attacks did initially draw American attention away from events at Hue, the NVA failed to capitalize on the distraction. Through a disciplined defense of the outpost, the Marines were able to prevent the communists from turning Khe Sanh into another Dien Bien Phu, as both Westmoreland and Johnson feared it might become. The Americans experienced 1,261 killed in action and 5,302 wounded during the siege and the actions to secure the area. The North Vietnamese attackers suffered even greater losses, with an estimated 10,000 to 15,000 of their soldiers dying for Giap's diversion.[45]

AMERICAN STAYING POWER: SOURCES OF EROSION

Early in the U.S. involvement in Vietnam, Johnson worried that American support for the conflict might erode over time. The president realized, "The weakest chink in our armor is public opinion."[46] Any military effort undertaken by a democracy, as B. H. Liddell Hart notes, "rests on a popular foundation—that for the supply of men and munitions and even for the chance of continuing the fight at all, it depends on the consent of the 'man in the street.'"[47] Opposition to the war began, in fact, to mount in 1965 with Johnson's bombing campaigns. In April of 1965, 12,000 students staged a large march against the war in Washington, DC. Picking up on the possibility of growing dissent, newspapers, such as the *New York Times*, warned against the prospect of "lives lost, blood spilt and treasure wasted, of fighting a war on a jungle front 7,000 miles from the coast of California."[48] Months before the Tet Offensive, a majority of the American people started to question the wisdom of their country's involvement. For many, including those in the armed forces, Vietnam, like the Korean conflict before, lacked the clarity of purpose associated with World War II.[49]

Against this potential wellspring of opposition, Johnson struggled from 1965 to 1967 to maintain a consensus in favor of the war. As one student of the Johnson presidency noted, "Throughout the United States' involvement in Vietnam, the President fought to maintain this public support. He knew that the American people disliked long and interminable wars."[50] When American forces first landed in South Vietnam in large numbers in 1965, many in the United States believed they would defeat the Viet Cong and their North Vietnamese allies quickly.[51] However, by 1967 the American bombing campaign had escalated without much in terms of results: sorties increased from 25,000 in 1965 to 108,000; bomb tonnage increased in the same period from 63,000 to

226,000. The United States had deployed nearly a half-million troops, dropped more bombs than it had in all theaters of World War II, and was spending $2 billion per month on the war.[52] For many, because American troops did not mount any offensives deep into North Vietnamese territory, the war began to look fruitless, with communist forces constantly replenishing their ranks and avoiding the decisive engagement U.S. military leaders so desperately sought.

With the Tet Offensive, the tide of American public opinion began to turn against the war. In the months following these highly visible engagements, the United States suffered monthly casualty rates higher than had been experienced in both World War II and Korea.[53] For many Americans, the war after Tet came to represent a costly venture with no end in sight.[54] As one study of American support for the war put it, "The only indisputable fact about public opinion during the Korean and the Vietnam conflicts is that as the two wars continued, public support for the original decisions to intervene decreased substantially."[55] Most notably, after the Tet Offensive, every time U.S. casualties increased by a factor of ten, support for the Vietnam War decreased by nearly 15 percent.[56] One Gallup poll noted that Johnson's already low popularity dropped to 36 percent in March 1968, with only a mere 26 percent polled believing the war should continue.[57]

The Johnson administration began to understand that the wheels had come off the limited public bandwagon of support for the war. McGeorge Bundy, the president's special assistant for National Security Affairs warned, "If one thing is more clear than another it is that we simply are not going to go on at the present rate."[58] He went a step further and explained, "What has happened is that a great many people—even very determined and loyal people—have begun to think Vietnam really is a bottomless pit."[59] The new secretary of defense, Clark Clifford, suggested, "I see more and more fighting and more and more casualties and no end in sight."[60] Even Secretary of State Rusk, who had scolded reporters on television about their criticizing the war only months before, noted, "The element of hope has been taken away by the Tet Offensive. People don't think there is likely to be an end."[61]

The American press also voiced concern over the course of events. Editors at the *Wall Street Journal* proclaimed, "The American people should be getting ready to accept, if they haven't already, the prospect that the whole Vietnam effort may be doomed."[62] Even former congressional advocates of American intervention in Southeast Asia started to criticize the war, and Johnson's Great Society legislation began to suffer as a result.[63] Other prominent observers, such as George Kennan and Walter Lippmann, began to voice their doubts about the need for the war to continue.[64]

What angered the American public, in particular, were the Johnson administration's assurances before Tet that the United States stood on the verge of winning the war. A gap appeared between the government's earlier pronouncements and what Americans saw as reality in the press. Only one month before the Tet Offensive, Secretary of Defense McNamara had firmly proclaimed, "There is no military stalemate."[65] Westmoreland expressed a similar opinion when he noted, "I am very, very encouraged. We are making real progress."[66] Even after the Vietnamese offensive Westmoreland continued to suggest that victory was near, since American forces now killed communist soldiers faster than North Vietnam could replace them.[67] In short, the events of Tet appeared to undermine the credibility of the Johnson administration and its military leaders. This fact did not escape officials in government. Deputy Defense Secretary John McNaughton recognized it when he reported, "A feeling is widely and strongly held that the 'Establishment' is out of its mind . . . . Related to this feeling is the increased polarization that is taking place in the United States with seeds of the worst split in our people in more than a century."[68]

In the summer of 1968, Johnson gathered a group of former diplomatic officials, the so-called Wise Men, who included former Secretary of State Dean Acheson, to offer advice about the conduct of the war. They warned the president that "endless inconclusive fighting" was "the most serious single cause of domestic disquiet." As a whole, they recommended that Johnson quickly find a strategy that reduced American casualties.[69] What worried them most was the growing protest movement in the United States calling for an immediate end to the war. The president, they argued, would find it difficult to continue the war under the weight of the protests.

According to George Herring, the effect of the broadening peace movement

> challenged the rationale of the war and indeed of a generation of Cold War foreign policies. It exposed error and self-deception in the government's claims, encouraging distrust of political authority. It limited Johnson's military options and may have headed off any tendency toward more drastic escalation. Perhaps more important, the war movement caused fatigue and anxiety among the policymakers and the public, thus eventually encouraging efforts to find a way out of the war.[70]

The antiwar movement had started its efforts to end the war long before the Tet Offensive, culminating with a series of marches all through 1967 that drew hundreds of thousands of participants in a wide range of cities from Washington, DC, to New York, to San Francisco.[71] The various movements had unified under the common goal of bringing home American soldiers from Southeast Asia, an appeal that resonated across political affiliations.[72] Westmoreland's request for 206,000 additional troops after the Tet Offensive only galvanized additional support for the peace movement.[73] Coupled with a growing crime rate and rioting in American cities stemming from outrage over the assassination of Martin Luther King, the demonstrations of those opposed to the war added to the impression that the United States stood on the verge of a serious domestic crisis.[74]

The ranks of the protesters became increasingly filled with moderates. As *Time* magazine noted at the time, "So many of the folks—far from being professional liberals or agitators or youths simply trying to avoid the draft—were pure, straight middle-class adults who had simply decided in their pure, straight middle-class ways, that it was time for the U.S. to get the hell out of the war in Vietnam."[75] More startling, polling numbers tended to underestimate the number of lower-middle-class voters opposed to the war.[76] The war, more than any other conflict since the Civil War, seemed to divide an American society already undergoing massive social change.[77]

For American policy-makers, the numerous draft-eligible men and members of the military who sympathized with the peace movement became one of the most troubling aspects of the public protests against the war. The large demonstration in October of 1967 against the selective service system, or what was known as the draft, represented only a harbinger of what would happen in the United States after the Tet Offensive.[78] Throughout the 1960s, the number of deferments individuals could acquire to avoid military service increased, creating the impression that the system established for the draft was unfair. A number of men eligible for the draft enlisted, not out of a spirit of volunteerism but out of the belief that enlisting would allow them to avoid serving as infantrymen in Vietnam. Their understanding of how the system worked was accurate: in 1968 and 1969, 88 percent of the infantrymen serving in Vietnam were draftees.[79]

After the Tet Offensive, American draft policies inspired even more protests.[82] During large numbers of rallies, men eligible for conscription burned their draft cards, conducted sit-ins at induction centers, and participated in numerous burglaries to destroy military records.[83] As the war continued, the number of desertions within the military began to increase. Between 1966 and

TABLE 7.1: American Force Levels and Killed in Action, 1965–71

	1965	1966	1967	1968	1969	1970	1971
U.S. Forces at the End of Year	184,300	385,300	485,600	536,100	475,200	334,600	156,800
U.S. Killed by Year	636	6,008	9,377	14,589	9,414	4,221	1,381
U.S. Killed in Action to Date	636	6,644	16,021	30,610	40,024	44,245	45,626
Percentage of Total American Forces Killed in Action per Year	0.3%	1.5%	1.9%	2.7%	1.9%	1.2%	0.8%

SOURCE: Harry G. Summers, Jr., *The Vietnam War Almanac*, 2nd ed. (Novato, CA: Presidio Press, 1999), 33–59.
NOTE: U.S. forces reached their highest level in the spring of 1969 with 543,400.

TABLE 7.2: Communist and American Losses, 1968

Killed in action	During Tet	After Tet
Americans	3,895	10,694
Communist Forces	58,000	36,556

SOURCE: Summers, *The Vietnam War Almanac*, 44–48.

1973, the Defense Department counted 503,926 acts of desertion.[84] In 1968, the U.S. military began to take notice of growing drug problems and racial strife within its ranks. These same problems would reduce the fighting effectiveness of the armed forces from 1971 to 1972, the closing years of involvement in Vietnam.[85] One soldier was arrested for carrying a sign in front of the White House reading: "120,000 American Casualties—Why?"[86]

These large social forces would ultimately influence Johnson's subsequent strategy for conducting the war in Vietnam. To demonstrate his eagerness for peace and to quell domestic opposition at home, Johnson instituted a pause in the American bombing campaign.[87] He also decided against seeking re-election as president. At the same time, the Defense Department commissioned a series of studies looking at ways to recast Westmoreland's search-and-destroy strategy against the communists in order to reduce American casualties.[88] Although Westmoreland requested more than 200,000 reinforcements after Tet, Johnson decided to augment the American presence by a smaller fraction of troops.[89] His efforts to de-escalate the war did little to reduce American casualties. Fighting in the Dong Ha corridor along the Demilitarized Zone saw an increase in casualties after Tet, as the United States lost 327 soldiers in fierce engagements.[90]

The stalemate in Vietnam helped propel Richard Nixon, who claimed to have in mind a plan for "peace with honor," into the presidency in 1968. His innovative plans for ending the war, however, tended to reflect a continuation of American policy, sparking even more opposition at home to U.S. involvement in Vietnam. In an effort to gain leverage against the North Vietnamese by cutting their supply lines into the south, the Nixon administration approved a series of incursions into Cambodia in 1970. Although the president endeavored to keep the invasion secret, once information about his escalation of the war reached the American people, public protests mounted.[91] Under great domestic pressure, Nixon was forced to reduce the American force levels in Vietnam (see Tables 7.1 and 7.2). By 1973, the United States had ended its direct involvement in Vietnam according to the dictates of a peace accord with the North Vietnamese reached in Paris.[92]

AFTER TET AND THE BATTLE FOR HAMBURGER HILL: U.S. FORCES STILL DETERMINED AND FLEXIBLE

During the spring of 1969, the United States undertook a series of operations to keep pressure on the North Vietnamese base camps that began to pop up along the border with Laos. The area of concern was the A Shau Valley, a region of rugged terrain whose proximity to the Laotian border allowed the NVA to quickly shuttle their forces in and out of South Vietnam. From the A Shau Valley, the North Vietnamese could prepare for attacks against American forces at Khe Sanh and the city of Hue. American forces entered the valley in an attempt to cut off NVA before they could conduct these assaults on the coastal provinces. It was during this brief operation that the American units spent several days ejecting tenacious NVA forces from Ap Bai Mountain, or Hamburger Hill.[94]

Along with two units from the 1st Army of the Republic of Vietnam, five infantry battalions of the 101st Airborne under Major General Melvin Zais led the American assault into the A Shau Valley. They planned for each of the units to comb the area. When a battalion met enemy units, according to the strategy, they would call for reinforcements that would promptly arrive using the division's helicopters. American planners wanted to prevent communist forces from retreating into Laos before they could destroy it.[95] In the first few days of the operation, U.S. forces experienced only light contact with the enemy. Documents captured by the U.S. 3rd Battalion, 187th Regiment, however, suggested that the 29th NVA Regiment lurked somewhere in the region.

On May 11, Lieutenant-Colonel Honeycutt sent his 3rd Battalion to scour the area near Ap Bia Mountain. When his Bravo Company made heavy contact with the North Vietnamese, he called in Cobra helicopter gun-ships for support. They conducted a heavy artillery rocket attack that, unfortunately, hit the battalion's command post. The "friendly fire" incident badly damaged the unit's communications equipment, making it difficult for the battalion headquarters to control its companies. For the next three days, American units would struggle to coordinate their assaults against the North Vietnamese forces dug in along Ap Bia Mountain. To make matters worse, the Americans had underestimated the enemy forces they had encountered. What they did not know yet was that they had stumbled upon several elements of the 29th NVA Regiment, known as the "Pride of Ho Chi Minh."

For three days, Honeycutt's units attempted to dislodge the North Vietnamese defenders. The difficult terrain and weather only complicated their task. Unlike most NVA units in 1969, however, the communist forces on Ap Bia Mountain refused to withdraw in the face of superior American fire power. Instead, they offered stiff resistance. The willingness of the North Vietnamese commander to take heavy casualties convinced the Americans that they had a tough fight on their hands.

With the help of the 1st Battalion/506th Regiment, Honeycutt coordinated a two-prong attack on Ap Bia Mountain. Because the 1st Battalion had trouble approaching the NVA positions from the south, Honeycutt's plan had to wait until May 17, when all units were in place. In the meantime, American news reporters had caught wind of the difficulties faced by the 101st Airborne Division in the A Shau Valley. Over the next few days, the American media began to arrive to cover what seemed one more example of another fruitless engagement in war that the country's political leaders had said would soon end.

On May 17, the 1st Battalion attempted an assault from the south but made little progress and remained 1,500 meters from the top of the mountain. A day later, both battalions finally managed to coordinate an assault. The Americans hoped to overwhelm the NVA in order to prevent them from concentrating their fire. They got their wish as the battle degenerated into fierce hand-to-hand combat. A roaring thunderstorm forced the American forces to fall back, ending the assault.

The American units conducted their final attack on Ap Bia Mountain on May 20, 1969. After ten days, and under the rapt attention of the press corps, the 101st Airborne finally brought its full combat power to bear. A ninety-minute artillery and aerial bombardment against the NVA defenders preceded the ground attack. The U.S. Air Force flew 272 attack sorties in which they dropped more than 1 million pounds of bombs and 152,000 pounds of napalm. This massive display of firepower destroyed the 7th and 8th battalions of the NVA's 29th Regiment. When American forces finally seized the mountain, they discovered 630 North Vietnamese dead. The engagement had cost 56 American lives.[96]

To the American public, the purpose of fighting for what was later called "Hamburger Hill" remained unclear. Although the lengthy engagement had rendered an entire NVA regiment unsuitable for further combat, American forces later abandoned the mountain that they had spent several days attempting to take. Put differently, American units had won yet another battle, but the effort did not appear to bring the war closer to a satisfactory conclusion. The costly engagement would lead to a further congressional reappraisal of American strategy in Vietnam. More important, the public expressed outrage over the American casualties expended for Ap Bia Mountain. The Battle for "Hamburger Hill" demonstrated that, at least in this engagement, U.S. forces had not lost their will to fight. Beginning in 1969, however, the American military's discipline began to fracture.

THE EROSION OF AMERICAN STAYING POWER: 1969–72

While the American public voiced vigorous disapproval of the U.S. involvement in Vietnam, the country's military continued to perform effectively on the battlefield. As American citizens protested the war at home, the military demonstrated great discipline in the field against the Viet Cong and its NVA allies. Even though communist forces saw dissension in the United States as a reason to pursue additional offensives after Tet, American forces still proved capable of inflicting defeats upon North Vietnamese units. With a new commander, General Creighton Abrams, American units held their ground and prevented both the NVA and the Viet Cong from translating the opposition in the United States into significant battlefield gains in South Vietnam.[97]

The American military performed quite admirably, even though it could not completely separate itself from events occurring at home. Although President Nixon claimed that he was taking steps to end the U.S. military commitment in Vietnam, the reduction of forces would occur over a period of three years. In the process, Nixon sought to create conditions favorable to turn the war over to the South Vietnamese Army, a strategy known as the "Vietnamization of the War." To this end, American forces had to give their South Vietnamese allies a period of security that would enable them to improve their own capabilities.

In conducting these missions, American forces would still fight with enough determination to win engagements against North Vietnamese units. Unfortunately, the long process of withdrawal would also begin to weaken American staying power. U.S. forces already understood that the public had turned against the war. For the individual soldier, then, risking his life for a war that the country no longer wanted to win seemed to make less and less sense. At a time when American society was growing increasingly opposed to the country's involvement in Vietnam, the rationale for fighting appeared vague and uncertain. Under these conditions, the morale of American forces began to deteriorate.

The decline in the U.S. military's discipline took a sharp turn downward beginning in 1969 and ending with the significant drawdown of American combat units in 1973. Even though the U.S. commitment was ending, this was still a period when American forces defended the countryside from communist incursions as well as conducted offensive operations to disrupt NVA sanctuaries. In a climate in which few soldiers wanted to spend a tour of duty risking their lives in a war no one believed was winnable, the decay in determination manifested in a variety of forms. As Kyle Longley explains:

> The divisions in the country mirrored those that developed in the military as soldiers, especially after 1968, increasingly opposed the war and some became protestors both within the military and after they left it. The protests took multiple forms, from the most extreme of desertion in a combat zone to random acts of disobedience such as drawing peace signs on helmets in a combat zone. For the military protestor, unlike the civilian, the act required more courage and willingness to accept harsher punishments. The acts began early on, but intensified as the war escalated and the military drafted more young men who questioned their service. Ultimately, the dissent within the military helped undermine morale and discipline.[98]

Military officials began to notice an increase in drug use among their personnel. A Defense Department study found that in 1971, 50.9 percent of U.S. soldiers had used marijuana, 28.5 percent had used other narcotics such as heroin and opium, and 28.5 percent had used other psychedelic drugs.[99] Most observers, however, have also noted that drug use increased throughout the American armed forces and was not unique to those units stationed in Vietnam. More troubling for U.S. officials were the growing racial tensions within the ranks. Race relations in the military, an institution that took pride in its egalitarianism, began to mirror the turmoil at home.[100] African-Americans increasingly felt that they were fighting and dying for an unjust society at a disproportionate rate. Unfortunately, the facts supported their suspicions. From 1965 to 1968, the rate of African-Americans killed in every army division was higher than the percentage of African-Americans in the unit. This situation, fortunately, would change by 1969.[101] The tensions, however, would continue to afflict American units. In the words of one report from MACV:

> Unusual psychological pressures were placed upon US military personnel in the Republic of Vietnam (RVN) during 1970 .... US troops were being withdrawn, with some loss of a sense of mission by those remaining. Criticism from home of US policies and actions in Vietnam helped create dissidence. Activities of undisciplined soldiers attracted public attention. Many troops were bored or restless.[102]

Members of the American military also began to lose respect for authority. From 1968 to 1971, the desertion rate among U.S. forces more than doubled (see Table 7.3).[103] Soldiers sometimes described their plight with the phrase "The unwilling, led by the unqualified, doing the unnecessary, for the ungrateful."[104] At the same time, the number of combat refusals increased, as men became simply unwilling to undertake risky missions. One soldier explained, "We're getting out anyway. Why should I be the last man killed in a no-good war?"[105] In 1968 military officials convicted 82 of their men for insubordination, but that number jumped to 131 in 1970. Many of these convictions occurred in cases in which soldiers simply decided not to carry out missions. Similarly, medical personnel began to see a jump in the number of men who wounded themselves to avoid combat.[106]

Even more troubling were the number of "fragging" incidents that came to plague the American military. The term "fragging" refers to officers who are attacked, usually, as the name indicates, with a grenade. Many of these incidents took place when soldiers thought their officers ordered them to conduct reckless missions. In 1968, military officials recorded 126 attacks on officers. "The Defense Department reported 788 fraggings from 1969 to 1972, while the Marines investigated more than twenty fraggings in an eight-month period in 1969. Others estimate that more than 1,000 officers and non-commissioned officers may have been killed at the hands of their own men."[107] (See Table 7.3).

U.S. military leaders recognized that the combination of withdrawals and the lack of a clear path to victory undercut the staying power of its armed forces. General Creighton Abrams concluded, "I need to get this Army home to save it."[108] Nevertheless, even as discipline decayed, U.S. units still fought with de-

TABLE 7.3: Discipline Problems in American Forces

	1968	1969	1970	1971
Desertion Rates (Per 1,000 Average Enlisted Monthly Strength)	29.1	42.4	52.3	73.5
Convictions for Insubordination	82	117	131	N/A
Fragging Incidents	N/A	126	271	333

SOURCE: Guenter Lewy, *American in Vietnam* (Oxford: Oxford University Press, 1978), 94–107.

termination and flexibility on the battlefield. As Gregory Daddis explains, "Despite a deterioration of performance in some areas, suggestions of a complete disintegration seem overstated. As General Bruce Palmer recalled, leaders still held their units together and carried out their missions 'right up to the end.'"[109]

CONCLUSION

Although the American military demonstrated great skill in winning every major battle of the Vietnam War, it could not overcome the draining effect that opposition at home had on its will to fight. As cohesion theory predicts, a professional military will show great determination and flexibility in its battlefield performance because of the strong bonds within the armed forces developed through training (see Tables 7.4 and 7.5.). Without a strong regime capable of coercing the population and military, however, U.S. forces would fight with only a moderate level of staying power. The more victory appeared unlikely, the more domestic opposition mounted to slowly undermine military staying power, through protests at home and through a steady withdrawal of the U.S. forces. Even though North Vietnamese forces suffered a terrible defeat, the Tet Offensive suggested that the conflict would continue as a costly deadlock. The year 1968, then, represented a turning point in the war because American society proved no longer willing to tolerate additional casualties in a conflict that seemed unending. Pronouncements by American military and political leaders that victory was near compounded the problem by raising expectations that the war would soon end. When the true, protracted nature of the conflict surfaced, the public no longer thought the war was worth the fight. As *New York Times* correspondent C. L. Sulzberger put it: "We lost the war in the Mississippi valley, not the Mekong valley. Successive American governments were never able to muster the necessary mass support at home."[111]

The effect of opposition at home did not prevent armed forces from winning battles, but it gradually did force the military to cope with growing racial

TABLE 7.4: Assessing Staying Power: Professional U.S. Military, 1968–72

Evidence
Domestic opposition to the war undermined U.S. staying power
Replacements brought public disapproval of the war with them to the front lines
U.S. drawdown undermined morale, some personnel concluded war not worth the individual risks
U.S. units still fought hard as dissension grew in form of combat refusals, fraggings, and crime

PREDICTION: Moderate Staying Power.
Military Collapses Slowly as the Probability of Victory Decreases.

TABLE 7.5: Assessing Battlefield Performance: Professional U.S. Military, 1968–72

Battles	Evidence
Tet Offensive/Hue	U.S. units take Hue at the cost of heavy casualties
	Most U.S. units regrouped and counterattacked after surprise
Khe Sanh	U.S. units take risks to uncover enemy positions
	Recovered from unexpected encounter with NVA units to defend
Hamburger Hill	U.S. units ejected tenacious NVA defender from Ap Bai
	U.S. mounted persistent assaults against unexpectedly determined defender

PREDICTION: Strong Battlefield Performance.
Most Units Fight with Determination and Flexibility.

tensions in the ranks, desertions, insubordination, and "fraggings." Soldiers did not understand why they should risk their lives in a war that not only seemed unending but also appeared to be a war the American public no longer wanted to fight. By the end of the war, the lack of discipline in the American military had grown worse. As one officer put it in 1971, "Our army that now remains in Vietnam is in a state approaching collapse, with individual units avoiding or having refused combat, murdering their officers and noncommissioned officers, drug-ridden, and dispirited; where not near-mutinous [c]onditions [exist] among American forces in Vietnam that have only been exceeded in this century by . . . the collapse of the Tsarist armies in 1916 and 1917."[112] In fairness, the German Army of World War I also lost more staying power than the American military. Nonetheless, the central point remains: U.S. armed forces behaved like other professional militaries, gradually losing their staying power when the possibility of victory appears remote.

Although this chapter presents evidence in support of cohesion theory, it also sheds light on alternative explanations for U.S. military performance in Vietnam. Recall that three arguments compete with cohesion theory to explain why militaries vary in their capacity to hold together and fight under difficult strategic circumstances: nationalism, democracy, and small-group theory. Of these three arguments, the nationalism argument does not apply to this case. The U.S. government did not advance an ideology demanding unconditional loyalty from its citizens, a typical feature of strong forms of nationalist ideologies. Similarly, U.S. forces did not fight to defend the American homeland, a common element of nationalist motivation in militaries.

This case, however, does provide mixed evidence for the democracy argument. True to form, this democratic military lost its staying power the longer the war continued. This outcome fits the prediction that Dan Reiter and Allan Stam present in their book, *Democracies at War*.[113] Cohesion theory, however, provides greater insight into the American experience in Vietnam for three reasons. First, cohesion theory explains the conditions under which staying power deteriorates: not simply over time, but when victory looks unlikely. Second, cohesion theory offers a more comprehensive description of what happens when staying power decays: U.S. units still fought hard on the battlefield but gradually lost their discipline, succumbing to disorder and combat refusals. Again, the U.S. military experience mirrors, to a lesser degree, the loss of staying power seen in the World War I cases detailed in earlier chapters. Third, cohesion theory can explain the American case in Vietnam as well as the aforementioned World War I cases of other professional militaries that lost their staying power gradually. This conclusion means that cohesion theory can account for the behavior of armies created and sustained by democracies as well as nondemocracies.

The most compelling alternative argument to cohesion theory is small-group theory. According to small-group theory, military cohesion comes from how armed forces train and replace personnel within units. In broad terms, the U.S. armed forces followed an individual replacement system: the military did not train individuals in the same units in which they would later fight in combat. Recall that small-group theory argues unit replacement systems provide military organizations with greater cohesion. Individuals should train and fight with the same units. Under these conditions, bonds develop among small groups of personnel that share common experiences with one another. Because they become "bands of brothers," they fight hard on the battlefield. There are two difficulties with this explanation. First, even though the U.S. military used individual replacements, it nonetheless managed to fight with determination and flexibility on the battlefield, even in the later years of the war. Small-group theory would predict otherwise. Finally, small-group theory cannot account for the gradual decay in staying power. It does not provide a reason for the slow deterioration in discipline and dissension. Only cohesion theory can shed light on why professional militaries suffer this fate, in the Vietnam War case as well as in other cases.

8 Conclusion

THIS BOOK HAS EXPLORED why some national armed forces differ in their cohesion, or will to fight. Specifically, cohesion is the capacity of a country's military to fight with determination and flexibility on the battlefield while enduring difficult strategic circumstances. To explain why some national armed forces fight harder than others, I presented *cohesion theory*. According to the theory, military cohesion depends on two key variables: the degree of regime control and the degree of autonomy for military organizations to train and to operate in battle. When combined, these ties produce four types of armed forces, each capable of fighting with differing degrees of military cohesion. Most of the book focused on testing the plausibility of cohesion theory through seven comparative case studies in five chapters.

In this last chapter, I perform three tasks. First, I compare the findings of the historical case studies. These cases show the explanatory power of cohesion theory across several wars. They also show the significant limits to the alternative arguments, especially the claim that democracies create armies with unrivaled staying power. Second, the chapter discusses what these results mean for the field of international security. Third, I provide some policy implications of cohesion theory.

EVALUATING THE PLAUSIBILITY OF COHESION THEORY

To test cohesion theory as a plausible explanation for a military's will to fight, this study examined cases from Europe during both world wars and from the Vietnam War. Chapter 3 focused on German performance on the Western Front from 1944 to 1945, where the democratic forces of Britain and the United

States led the charge against Hitler's military. In contrast to France, Nazi Germany fielded *messianic* armed forces. Created from a high degree of regime control and organizational autonomy, the Wehrmacht fought with strong cohesion. Most units—including reserve formations—fought with determination and flexibility in combat. German forces in the late summer of 1944 recovered enough from devastating defeats in northern France to retreat and to reorganize for a defense of the Siegfried Line. With their backs to the Western Wall, the Wehrmacht tenaciously fought off superior Allied armies, but still mustered the ability to mount an ambitious counterattack in December. Germany's armed forces also displayed a strong staying power. To ensure that the military would not collapse, Hitler imposed brutal coercive measures, including flying courts-martial. A mixture of draconian punishments and ideological zeal maintained the tenacity of the German Army to the very end. Only when the Allies had obtained crushing material superiority and Hitler had taken his own life did the Wehrmacht capitulate.

Chapter 4 assessed the French Army in World War II. In 1940, France fielded an *apathetic* military, held together by a weak degree of regime control and a low degree of organizational autonomy for the armed forces. Domestic divisions undermined any bonds of loyalty and trust from taking root within French military organizations. When the armed forces went to the front, they carried the same divisions that had interfered with its training, created an officer corps hostile to its civilian masters, and fostered an environment of distrust between officers and enlisted personnel.

As a result, French forces displayed a low degree of military cohesion. Without strong organizational bonds, most units fought without determination and flexibility. As cohesion theory predicts, French units did not surrender without a fight, but they gave an uneven performance. Although they inflicted casualties on the German invaders, most French units could not cope with enemy breakthroughs, succumbed to panic, and found it difficult to counterattack at key moments. In addition, the French Army could not display the same staying power for defending the homeland that it had exhibited in World War I and that the Wehrmacht demonstrated in 1945. French military leaders worried more about internal insurrection than a German occupation. They argued early in the conflict for an armistice to avert a domestic uprising as a means to finally end the democratic Third Republic.

The French and German cases also provided an opportunity to test cohesion theory's ability to account for changes in military cohesion from one war to the next. In World War I, both countries fielded *professional* armed forces. Motivated by strong internal bonds that grew out of organizational autonomy to train, the French military could endure terrible casualties as long as an Allied victory appeared plausible. Most units fought with determination and flexibility, especially at crucial battles like the Marne in 1914. More than two decades later, the French military went to war with apathetic armed forces. Domestic turmoil undermined the military's autonomy to train for war.

Similarly, most German units in World War I fought with determination and flexibility. However, because the regime possessed low regime control, it could not fight with the same staying power it would show in World War II. Defeats on the battlefield, especially after the failure of the Spring Offensives of 1918, eventually eroded military support for the war. When victory appeared out of Germany's grasp, the armed forces lost its will to continue. The declining support at home also drained the military's staying power. Nationalists and military officers would later claim that civilian politicians stabbed the German Army in the back, preventing it from winning on the battlefield. The bitter historical reality, however, was the opposite: the German military plainly lost key battles in 1918, and then the home front concluded Germany could no longer win the war.[1]

Chapter 5 explained why the *authoritarian* military of the Soviet Union could withstand the near knockout blow of Operation Barbarossa in 1941. Because of a high degree of regime control, Stalin and Soviet leaders could motivate their armed forces to fight with determination on the battlefield. Most units would defend and counterattack relentlessly. Soviet forces displayed an enormous capacity for suffering large casualties to push back the German invaders. A high degree of regime control also meant that the armed forces possessed the staying power to endure the terrible strategic setbacks of that summer. True, the home front rallied to resist a barbaric invader, but Stalin's police state served as the backbone of the Soviet war effort. One demonstrable flaw of Soviet armed forces was its inability to fight with flexibility. Units frequently proved inflexible and would succumb to panic. Stalin's coercive regime prevented the military from training with autonomy. The state's political interference in training and operations disrupted the formation of the strong internal bonds of trust among military personnel, which help units fight with flexibility on the battlefield. As the war progressed, however, Stalin recognized this problem and gave the armed forces the autonomy to build those bonds. By the end of the war, the Red Army would rival the Wehrmacht's military cohesion.[2]

The last two chapters focused on the Vietnam War. Chapter 6 described the cohesion of the North Vietnamese Army (NVA), also a *messianic* military. Although it endured horrendous casualties, the NVA remained a highly determined military throughout the war. In contrast, a *professional* American military began to slowly lose its staying power to fight the Vietnam War after the Tet Offensives of 1968. Over time, opposition on the home front contributed to the decay in the staying power of American forces. The public concluded that the war was no longer winnable, and new recruits brought this dissatisfaction to the front lines. Consequently, public opinion turned against the conflict and caused dissension within the military, forcing the country's leaders to reduce their military commitment in Vietnam. During the painfully slow drawdown of American forces, the military fell victim to discipline problems in the form of increasing desertions, insubordination, and attacks on officers by enlisted men.

Assessing the Evidence from the Case Studies

The case studies offer evidence in support of cohesion theory's ability to explain the sources of a military's staying power and determination in combat. In this initial test of the theory's plausibility, two broad trends emerge from the historical evidence. First, a high degree of regime control creates militaries capable of fighting with strong staying power. In the German and North Vietnamese cases, their messianic militaries could endure enormous casualties, in what looked like a lost cause, without internal pressure to collapse. Stalin's authoritarian armed forces showed an equal amount of staying power under terrible strategic circumstances.

Without a high degree of regime control, however, pressure for military collapse will emerge as battlefield events call into question the probability of victory. The rate of disintegration from within depends on the degree of military autonomy. When military organizations possess autonomy to train without domestic interference, they can build bonds of loyalty and trust among their personnel. These connections offset pressure to collapse and bolster the armed forces when it looks like they might lose the war they are fighting. In the professional armed forces of World War I Germany, and the United States during the Vietnam War, resistance to the war emerged only after the possibility of achieving the country's war aims looked impossible. Within the German Army, it took four years and a few crucial defeats to convince personnel at the front and the population at home to turn against the war. U.S. forces began to lose their staying power after the Tet Offensive, but went on to draw out their involvement in Vietnam for several more years.

These episodes contrast starkly with France's apathetic armed forces in 1940. Early defeats quickly spread panic within the leadership of the armed forces. Panic rapidly became sharp opposition to continuing the war. Preoccupied with the potential for a communist revolt at home, Generals Weygand and Petain pressed the country's civilian leaders for an armistice. They also did little to hide their belief that President Reynaud and his political predecessors deserved the blame for France's defeat. The German victory gave them a chance to finally do away with their primary enemy: the Third Republic.

Second, a high degree of organizational autonomy allows a country's armed forces to fight with determination and flexibility on the battlefield. Nazi Germany's messianic military could recover from defeats and counterattack. It could reform new units from shattered ones. Panic seldom hindered its ability to fight tenaciously. U.S. forces in Vietnam showed similar abilities in combat against the North Vietnamese and the Viet Cong. Only the best units in France's apathetic armed forces of 1940 could fight with determination, while most showed less resolve. In the Soviet case, a high degree of regime control motivated Stalin's authoritarian army to fight with determination. Most units, however, lacked flexibility. Stalin's regime did not permit the armed forces the autonomy required to produce the internal bonds of loyalty and trust that enable initiative and risk-taking on the battlefield.

EVALUATING ALTERNATIVE ARGUMENTS

What do the results of these historical cases suggest about existing alternative arguments to cohesion theory? A popular argument in the field of international relations contends that democratic armies wage war with unmatched determination because their soldiers fight for a popular government.[3] Moreover, the liberal ideology of democratic states inspires members of the military to risk their lives for their country. The individualism found in democratic countries also encourages members of the armed forces to show initiative on the battlefield. In terms of case selection, this book examined cases that would clearly test the democracy argument. This book analyzed cases where a democracy possessed formidable military capabilities (the United States and France, World War I and II). It also examined examples of a democratic state defending its homeland (France in World Wars I and II).

For these cases, cohesion theory does a better job of explaining a military's will to fight than regime type. Cohesion theory possesses greater explanatory range than the democracy argument. The book's case studies come from some of the most important wars of the past century. While cohesion theory can shed light on all of these cases, the democracy argument can explain only a few. In many of the cases, the unconditional loyalty obtained by regimes and military organizations represents a stronger source of motivation to inspire soldiers to risk their lives than the individualism found in most democracies. A high degree of regime control in illiberal nations, like Nazi Germany, North Vietnam, and the Soviet Union, created militaries that fought with more zeal than their respective democratic opponents, the United States and France. As these cases show, democracies are not the only form of government capable of motivating their soldiers to fight hard.

Nor can the democracy argument explain why military cohesion in a country's armed forces can change from one war to the next. The armies of a democratic France fought with greater tenacity in World War I than they did in World War II. Recall that a high degree of organizational autonomy bolstered its cohesion from 1914 to 1918. Without organizational autonomy during the interwar years, the armed forces could not build the bonds necessary for units to fight with determination and flexibility. Cohesion theory can explain why this change occurred in the French case, as well as why similar changes took place in the German case. Although the democracy argument could account for the battlefield determination and flexibility found in the French military of World War I and the U.S. military during the Vietnam War, it does not lay out the conditions under which these armed forces lose their staying power. Only cohesion theory provides such an explanation. All of these examples point to regime control and organization autonomy, and not a country's form of government, as the crucial determinants of a military's will to fight.

A second alternative to cohesion theory is small-group theory.[4] This argument, embraced by many defense planners, claims that soldiers fight for their friends, those with whom they interact day to day and share experiences. Military cohesion depends on raising units together, training together, and fighting together. From this perspective, unit rotation represents the best strategy for replacing personnel lost in combat. An individual replacement policy cannot forge the same small-unit bonds because new recruits do not train with the personnel in the units they will eventually join. Proponents of this view point to the U.S. performance in Vietnam as evidence for why an individual replacement system undermines military cohesion.

In most of this study's cases, the rate of casualties prevents the formation of primary group ties. For convenience, most of the militaries observed here raised units regionally and trained new recruits together. Nevertheless, the high turnover of personnel caused too much turbulence for these small group bonds to form. Even in the case this argument claims to explain best, the logic and evidence fall short. Although the U.S. military gradually lost its staying power, most units still fought with determination and flexibility in combat. This performance differs from the expectations of small-group theory. In comparison, cohesion theory argues that the U.S. military in Vietnam behaved like the other professional militaries in this study.

The final explanation for cohesion is the claim that nationalism accounts for a military's will to fight. Because the logic of this argument remains vague, it is difficult to evaluate. All countries promote some form of nationalism. Some governments foster feelings of patriotism, either along civic or ethnic lines. Most military organizations follow suit, inculcating love of country during training. Nationalism, according to this view's proponents, is strongest when militaries defend their own homeland.[5]

While the nationalism argument holds considerable intuitive appeal, it offers less explanatory power than cohesion theory. If all countries promote nationalism, then this motivating factor does not vary from one country to another. This is not helpful since militaries differ in their cohesion. If defending the homeland is the key to nationalism, then this study's cases suggest wide variation in military cohesion, as the two French cases show.

IMPLICATIONS FOR THE STUDY OF INTERNATIONAL SECURITY

In addition to offering a new theory of military cohesion, this book contributes to the study of international security in three ways. First, it provides a broad and systematic definition of military cohesion that encompasses not only battlefield performance but also the ability of political-military leaders to hold the armed forces together when victory seems unlikely. Existing concepts allude to battlefield performance and staying power, but scholars until now have not linked them.[6] Assessing military cohesion from this vantage point also follows a growing trend in the field of international security that explores the domestic and societal sources of military effectiveness.[7] This definition, more importantly, helps scholars to better conceptualize all possible variations in staying power. On the one hand, national militaries rarely fight to the last soldier. Even Nazi Germany's armed forces eventually surrendered. On the other hand, national militaries rarely collapse without a fight. Even the French Army in 1940 both suffered and inflicted casualties. This study, then, helps to explain the range of possible outcomes in cohesion between these two extremes.

Second, the book demonstrates the importance of incorporating nonmaterial factors into the field of international security's conceptions of military power. Most theories of international relations conceive of military power in terms of material capabilities. They see military potential in terms of manpower and armaments.[8] However, putting these instruments to use involves a human element, since people, soldiers more precisely, of varying quality and motivation must employ the military technology their government gives them to fight wars. These considerations have increasingly drawn scholars to the subject of combat effectiveness.[9] Not only do militaries differ in their skill, but they can also differ in their will to fight.

Nonmaterial factors like cohesion can also improve net assessments, or how planners gauge prospects for defeating potential opponents on the battlefield.[10] Militaries appearing weak after a net assessment of their material capabilities may actually perform better than estimates suggest, because of their staying power. Examples from the book's case studies illustrate this tendency. American analysts thought that the German Army stood on the verge of collapse at the end of 1944, and they were shocked when the Wehrmacht went on the offensive during the Battle of the Bulge.[11] Similarly, U.S. planners misjudged the determination of the North Vietnamese to withstand the onslaught of American firepower. In the end, the NVA took more casualties than the U.S. military could stand to inflict.[12]

Cohesion theory offers one way to gauge a country's actual military power by understanding the forces that motivate a population and its armed forces. For students of military effectiveness, the theory can help address one of the field's unanswered questions: which national armed forces can fight with enough cohesion to adopt the doctrine and tactics required for success on the contemporary battlefield, what Stephen Biddle calls the "modern system"?[13] As the case studies demonstrate, cohesion represents a crucial determinant of who wins and loses in combat. Contemporary militaries might understand what the "modern system" requires for victory but nevertheless lack the determination and discipline to employ it. For international relations scholars more broadly, the book's findings suggest nonmaterial factors, such as cohesion, deserve a larger role in definitions of military power.

Finally, scholars can employ cohesion theory to explain the performance and staying power of groups other than national militaries. According to the theory, the degree of group cohesion depends on the ability of leaders to promote and to enforce norms of unconditional loyalty. These norms establish the requirements of group membership. Cohesion is strong when a group demands unconditional loyalty from its members and can enforce these demands. Cohesion theory bridges two literatures on collective action: rationalist accounts that emphasize the use of incentives to motivate individuals in a group, and constructivist arguments that focus on the ability of norms to shape how strongly members identify with a group's interests.[14]

Cohesion theory, thus, can apply to several types of groups. Researchers could, for example, use cohesion theory to assess the determination and staying power of nonstate actors such as insurgents, terrorists, or peaceful social movements.[15] Much of the terrorism research today examines the capabilities of these groups to explain their behavior.[16] There is likely wide variation among terrorist groups in the capacity of members to carry out missions with determination and flexibility. Similarly, groups will probably vary in their staying power, or willingness to persevere, even when it looks as though they may not achieve their strategic objectives. For these reasons, it seems reasonable that no two terrorist organizations are alike in their will to fight. Surely al Qaeda, a group responsible for thousands of deaths and on the run from the international community, ranks higher in tenacity than the followers of Bhagwan Shree Rajneesh, whose most infamous act came in 1984 when they contaminated the salsa bar of a Taco Time restaurant with salmonella.[17] Cohesion theory offers a way to differentiate terrorist groups between those that are highly motivated and those that lack the endurance to conduct missions under difficult circumstances.

THE IMPLICATIONS OF COHESION THEORY FOR U.S. NATIONAL SECURITY POLICY

Understanding the sources of military cohesion holds practical benefits for policy-makers. U.S. defense planners will want to know how hard potential adversaries, like North Korea and Iran, might fight on the battlefield. These same officials will also need a framework to assess the military potential of U.S. allies. In its efforts to build an Afghan National Army, for instance, policy-makers should focus on the ingredients necessary to create a cohesive military organization. Cohesion theory can help with both of these tasks.

Evaluating Potential U.S. Adversaries

Estimates of staying power could help planners anticipate whether enemy forces could fight a losing war. In Operation Iraqi Freedom, as U.S. planners anticipated, Saddam Hussein's armed forces proved vulnerable to blitzkrieg operations. However, the fanatical resistance of the Baathist regime's most ardent supporters, the Fedayeen Saddam, surprised coalition forces. These loyalists later participated in the initial insurgency against the U.S. occupation.[18] Whether major combat operations against other adversaries will occur in a similar fashion depends on many factors, not least of which are the sources of an adversary's military cohesion.

Cohesion theory offers another way to assess an adversary's military power. By looking at the degree of regime control and the degree of organizational autonomy, analysts stand a better chance of determining which armies will fold under pressure and which will fight to the death. This can help countries accurately understand the military realities of a potential war. Knowing if a potential enemy bears more resemblance to the messianic armed forces of North Vietnam than to the apathetic military of France should help decision-makers avoid costly mistakes.

Specifically, the book's theory and case studies suggest some broad lessons about defeating these different types of militaries. When fighting messianic militaries, for instance, planners need to understand that victory will likely require the destruction of an enemy's armed forces, and nothing less. Luckily, states rarely create messianic armed forces. The combination of both a high degree of regime control and organizational autonomy will prove too difficult for most states to maintain. The majority of national governments cannot establish a high degree of regime control because their citizens would reject them. Regimes that manage to secure a high degree of regime control, then, will understandably guard their power fiercely, viewing the armed forces as a potential competitor. Under these circumstances, military autonomy, and strong bonds inside the armed forces, could threaten the regime's grip on state power. The armed forces of Nazi Germany, North Vietnam, and Imperial Japan represent three, and perhaps the only, examples of messianic armed forces.

Authoritarian militaries will display a similar tendency to fight to the bitter end, but most of their units will lack flexibility on the battlefield. They operate best with centralized command and control. Severing these links could undermine their combat performance. More important, opponents can use a combination of maneuver and firepower to shock enemy units, but must bring sufficient follow-on forces to prevent the adversary from recovering.[19] During Operation Barbarossa, for example, German armored forces tore gaping holes into Soviet front lines. Without flexibility and training, most Soviet units reacted improperly to these circumstances. German units would then race to the rear to complete the encirclement of disorganized Soviet defenders. With follow-on forces lagging behind the Wehrmacht's armored spearheads, the Germans had trouble solidifying their gains. Put simply, units in authoritarian militaries might prove vulnerable to shock, but they will still fight with determination.

Professional militaries, in contrast, lack the staying power for fighting long wars where the probability of victory is low. The book's case study of the U.S. experience in the Vietnam War provides an important example. Professional militaries fight with determination and flexibility on the battlefield, but internal opposition in the armed forces will slowly emerge the more victory looks impossible. To defeat these militaries, opponents cannot count on the shock of blitzkrieg operations. Instead, they must demonstrate to the enemy that continuing the war will only result in a costly stalemate followed by defeat. These conditions will undermine their staying power, especially if an opponent perceives they have little at stake in a conflict.[20]

Finally, apathetic armed forces will fight with a low degree of cohesion. Most units will lack determination and flexibility, and will struggle against blitzkrieg operations. A few decisive defeats will likely produce panic among some units and paralysis among others. Military leaders will quickly face domestic pressure to end a conflict, either because no one is confident in the armed forces' ability to fight, or because the regime worries about domestic opponents more than foreign ones.

What does cohesion theory suggest about the staying power and battlefield performance of the following three possible U.S. opponents: North Korea, Iran, and China? Below, I use cohesion theory to offer some preliminary observations. The purpose of this limited discussion is not to provide definitive answers about these countries' armed forces. Instead, I suggest what kinds of evidence would help render a judgment about their military cohesion.

Out of the pool of potential U.S. adversaries, North Korean and Iran clearly possess regimes with a high degree of control over their societies. With its highly repressive regime, the North Korean armed forces will either fight as a messianic or an authoritarian military. The key variable to determining which type of military it could send into battle is the degree of military autonomy. Given the government's penchant for internal security, however, the regime seems unlikely to provide the armed forces much independence. After all, the military represents the clearest threat to the regime's survival.[21] The North Korean armed forces, then, seem most likely to fight as an authoritarian military, capable of showing great staying power, battlefield determination, but lacking flexibility.

The Iranian case seems less clear. Classifying the Iranian armed forces depends on the level of regime control and military autonomy. How much control the regime can exert over Iranian society remains in question after the election protests of 2009 and 2010. Although it is hard to imagine these same sustained outbursts occurring in North Korea, the regime in Tehran still retains power and holds a variety of tools at its disposal to suppress dissent. More important, this need to retain control suggests that the regime will also keep a tight hold over the armed forces. The presence of parallel military organizations, such as the Iranian Revolutionary Guard and the regular armed forces, a commitment to indoctrinating service members, and the promotion of officers loyal to the regime all indicate a low degree of military autonomy. Such evidence suggests that Iran would fight as an authoritarian military.[22]

As with the Iranian armed forces, assessing the People Liberation Army of China also requires weighing different pieces of evidence about the nature of regime control and military autonomy. While the Communist Party rulers of China seem to lack the same level of control over Chinese society as their predecessors, the regime in Beijing still retains a dominant hold over the country's politics. Opposition to the regime's single-party rule appears minimal.[23] Only the military possesses the latent capability to overthrow the regime.

The central question, then, for assessing military cohesion of the Chinese armed forces centers around how much freedom from outside interference they enjoy and if they use this independence to focus on training for warfighting. By some accounts, the Chinese military has begun to focus on becoming smaller, more skilled, and largely, more professional.[24] To succeed, this transition will require a government commitment to give the armed forces freedom from political control. If it does succeed, without the regime relinquishing control over the other aspects of Chinese society, the armed forces could fight as a messianic military. If it does not succeed, then China would field an authoritarian military, as North Korea and Iran do today. Another possibility sees the regime loosening its control over China while the military gains autonomy. Under these conditions, China would create a professional military.

Implications for U.S. operations in Afghanistan

What does cohesion theory say about a professional U.S. military fighting currently in Afghanistan? Arguably, a volunteer military represents a key difference between U.S. operations today and those during the Vietnam era. A conscript military draws from a wider cross-section of the country, increasing the burden of service among the entire population. Domestic opposition under these circumstances is more likely. In addition, the level of casualties the United States has suffered in these two modern conflicts is far lower than in Vietnam. Nevertheless, as cohesion theory predicts, the U.S. public has not shown unlimited patience with the two current wars. The popular "surge" strategy used in Iraq, and later employed in Afghanistan, increased U.S. commitments, albeit for a few years.[25] In both cases, the surge preceded a drawdown and eventual withdrawal despite many of the surge supporters believing stability in these countries too fragile for a U.S. departure. Put another way, the United States appears on the verge of leaving Afghanistan because it has tired of the conflict.

Success now depends on the United States and its coalition partners creating a cohesive national military in Afghanistan. Coalition planners can take several approaches to raise and to sustain cohesive Afghan armed forces: they can promote a high degree of regime control, a high degree of military autonomy, or both. With the U.S. commitment to fostering liberal democracies, establishing a regime with a high degree of control over Afghanistan seems out of the question. As such, the United States and its coalition allies must focus on fostering organizational autonomy for Afghanistan's newly raised army, in the hope that it develops strong bonds of loyalty through training. The case studies here would point to the United States in Vietnam, and France in World War I, as relevant examples. In those armed forces, military leaders possessed the autonomy to demand and to enforce norms of unconditional loyalty and trust. They promoted these norms through training. Shielded from outside interference, these militaries could ensure that officers and enlisted personnel embraced these norms. Hence, successfully establishing strong organizational bonds requires the freedom from outside interference to conduct training.

Given these necessary conditions, the Afghan experiment to forge a national army out of whole cloth faces enormous hurdles. Afghanistan suffers from sectarian and ethnic divides. These cleavages will likely undermine the autonomy of the armed forces. The recent civil war has likely reinforced sectarian loyalties at a time when the armed forces strive to foster national loyalties.[26] Furthermore, nascent military bonds, only recently formed, will compete with ethnic and tribal bonds that have existed for decades, if not centuries.

The historical record suggests, moreover, that it takes several years to overcome ethnic divides in national militaries.[27] There are success stories, such as modern India and the Austro-Hungarian Empire.[28] In those cases, military organizations found ways to recognize ethnic cleavages while instilling norms of loyalty for the armed forces at the same time. An officer corps that believed its primary allegiance was to the armed forces, and that its primary duty was to defend the state, provided the key ingredient for strong organizational bonds within the Indian and Habsburg militaries.

Finally, the United States and its coalition partners have few historical models to guide their efforts. The number of cases of foreign powers creating new militaries in countries they occupy is small.[29] Perhaps the closest example comes from the British colonial experience in India. In that case, British officers would command units until Indian officers were capable of leading. Over time, the Indian Army possessed strong organizational bonds.[30] Of course, this process took several decades, time the United States might not want to commit to such an endeavor.

Notes

CHAPTER 1

1. Carl von Clausewitz, *On War*, ed. and trans. Michael Howard and Peter Paret (Princeton: Princeton University Press, 1984), 185.

2. For good accounts, see Ian Kershaw, *The End: The Defiance and Destruction of Hitler's Germany, 1944–1945* (London: Penguin Press, 2012); Max Hasting, *Armageddon: The Battle for Germany, 1944–1945* (New York: Knopf, 2004); and Anthony J. Beevor, *The Fall of Berlin 1945* (London: Penguin, 2002).

3. Julian Jackson, *The Fall of France: The Nazi Invasion of 1940* (Oxford: Oxford University Press, 2003); Karl-Heinz Frieser, *The Blitzkrieg Legend: The 1940 Campaign in the West* (Annapolis, MD: Naval Institute Press, 2005); Ernest May, *Strange Victory: Hitler's Conquest of France* (New York: Hill and Wang, 2000); and John Paul Pallud, *Blitzkrieg in the West* (London: Then and Now, 1991).

4. See Marilyn B. Young, *The Vietnam Wars, 1945–1990* (New York: Harper Collins, 1991); and George C. Herring, *America's Longest War: The United States and Vietnam, 1950–1975*, 3rd ed. (New York: McGraw-Hill, 1996).

5. Douglas Pike, *PAVN: Peoples Army of Vietnam* (New York: Presidio Press, 1986).

6. Rod Nordland, "Troop 'Surge' in Afghanistan Ends with Mixed Results," *New York Times*, September 21, 2012, A4; and Michael O'Hanlon and Hassina Sherjan, *Toughing It Out in Afghanistan* (Washington, DC: Brookings Institution, 2010).

7. Sven Gunner Simonsen, "Leaving Security in Safe Hands: Identity, Legitimacy and Cohesion in the New Afghan and Iraqi Armies," *Third World Quarterly* 30, no. 8 (2009): 1–19.

8. Edward A. Shils and Morris Janowitz, "Cohesion and Disintegration in the Wehrmacht in World War II," *Public Opinion Quarterly* 12, no. 2 (Summer 1948): 280–315. A popular version of this argument can be found in Stephen Ambrose, *Band of Brothers: E Company, 506th Regiment, 101st Airborne from Normandy to Hitler's Eagle's Nest* (New York: Simon and Schuster, 2001).

9. Representative works on ideology include John A. Lynn, *The Bayonets of the Republic: Motivation and Tactics in the Army of Revolutionary France, 1791–1794*, 2nd ed. (Boulder, CO: Westview Press, 1996); and Barry R. Posen, "Nationalism, the Mass Army, and Military Power," *International Security* 18, no. 2 (Fall 1993): 80–124. On the military effectiveness of democracies, see Dan Reiter and Allan Stam, *Democracies at War* (Princeton: Princeton University Press, 2002).

10. My theory builds in particular on the arguments found in Michael Hechter, *Principles of Group Solidarity* (Berkeley: University of California Press, 1987); and Alexander Wendt, *Social Theory of International Politics* (Cambridge: Cambridge University Press, 1999).

11. On terrorist motivation, see Martha Crenshaw, "The Causes of Terrorism," *Comparative Politics* 13, no. 4 (July 1981): 379–99; and Robert Pape, *Dying to Win: The Strategic Logic of Suicide Terrorism* (New York: Random House, 2005). On the ability of groups to recover from defeats, see Jenna Jordan, "When Heads Roll: Assessing the Effectiveness of Leadership Decapitation," *Security Studies* 18, no. 4 (2009): 719–55.

12. How to perform net assessments remains controversial. During the Cold War, analysts recognized the importance of cohesion but seldom made it an explicit or central part of their arguments. See, for example, U.S. Army, Concepts Analysis Agency, *Weapons Effectiveness Indices/Weighted Unit Values (WEI/WUV):* Study Report CAA-SR- 73–18 (Bethesda, MD, 1974); John J. Mearsheimer, *Conventional Deterrence* (Ithaca, NY: Cornell University Press, 1983), 165–88; Barry R. Posen, "Measuring the European Conventional Balance: Coping with Complexity in Threat Assessment," *International Security* 9, no. 3 (Winter 1984/85): 47–88; Eliot A. Cohen, "Toward Better Net Assessment: Rethinking the European Conventional Balance," *International Security* 13, no. 1 (Summer 1988): 50–89; Stephen Peter Rosen, "Net Assessment as an Analytical Concept," in *On Not Confusing Ourselves: Essays on National Security Strategy in Honor of Albert and Roberta Wohlstetter*, ed. Andrew W. Marshall, J. J. Martin, and Henry S. Rowen (Boulder, CO: Westview Press, 1991), 283–301; and Paul Bracken, "Net Assessment: A Practical Guide," *Parameters* 36, no. 1 (Spring 2006): 90–100. For historical examples of planners applying net assessment, see Williamson Murray and Allan Millett, eds., *Calculations: Net Assessments and the Coming of World War II* (New York: Free Press, 1992). Finally, Charles Glaser and Chaim Kaufman incorporate net assessments into international relations theory in their "What Is the Offense-Defense Balance and How Can We Measure It?" *International Security* 22, no. 4 (Spring 1998): 44–82.

13. See Gil Merom, *How Democracies Lose Small Wars: State, Society, and the Failures of France in Algeria, Israel in Lebanon, and the United States in Vietnam* (Cambridge: Cambridge University Press, 2003).

14. Discussing the effects of ethnic divisions in armed forces is Alon Peled, *A Question of Loyalty: Military Manpower Policy in Multiethnic States* (Ithaca, NY: Cornell University Press, 1998); and Timothy D. Hoyt, "Social Structure, Ethnicity, and Military

Effectiveness: Iraq, 1980–2004," in *Creating Military Power: The Sources of Military Effectiveness*, ed. Risa A. Brooks and Elizabeth A. Stanley (Stanford: Stanford University Press, 2007), 55–79. For historical perspective, see Istvan Deak, *Beyond Nationalism: A Social and Political History of the Habsburg Officer Corps, 1848–1918* (Oxford: Oxford University Press, 1990).

15. Michael R. Gordon and Bernard E. Trainor, *Cobra II: The Inside Story of the Invasion and Occupation of Iraq* (New York: Random House, 2006).

16. On the nature of the contemporary battlefield, see Stephen Biddle, *Military Power: Explaining Victory and Defeat in Modern Battle* (Princeton: Princeton University Press, 2004). For alternative views on what it takes to win on the modern battlefield, see Thomas G. Mahnken, *Technology and the American Way of War since 1945* (New York: Columbia University Press, 2008); and Keith Shimko, *The Iraq Wars and America's Military Revolution* (Cambridge: Cambridge University Press, 2010).

17. The collective action problem has a long tradition in political science. See Mancur Olson, Jr., *The Logic of Collective Action* (Cambridge: Harvard University Press, 1965). Treatments that are more recent include Hechter, *Principles of Group Solidarity*; and Russell Hardin, *One for All: The Logic of Group Conflict* (Princeton: Princeton University Press, 1995). The sociological perspective is probably best represented by Charles Tilly, *From Mobilization to Revolution* (Reading: Addison-Wesley, 1978). For a summary of the social psychological literature, see Michael A. Hogg and Dominic Abrams, *Social Identifications: A Social Psychology of Inter-group Relations and Group Processes* (London: Routledge, 1998).

18. Wendt, *Social Theory of International Politics*, articulates this perspective in the field of international relations. Wendt contends that states can form group identities, and thus overcome the obstacles preventing them from cooperating in international politics.

19. For the different roles scholars believe that norms play in international relations theory, see ibid.; and the essays in Peter J. Katzenstein, *The Culture of National Security: Norms and Identity in World Politics* (New York: Columbia University Press, 1996).

20. James Fearon and Alexander Wendt, "Rationalism vs. Constructivism: A Skeptical View," in *Handbook of International Relations*, ed. Walter Carlsnaes, Thomas Risse, and Beth A. Simmons (London: Sage Publications, 2002), 52–72. See also Theo Farrell, "Constructivist Security Studies: Portrait of a Research Program," *International Studies Review* 4, no. 1 (Spring 2002): 49–74.

21. In addition to cohesion, analysts also attempt to determine how much skill armies might possess. For discussions on skill, see the essays in Allan R. Millett and Williamson Murray, eds., *Military Effectiveness*, Vols. 1–3 (Boston: Allen and Unwin, 1988). The role of skill versus technology as a contributor to American victory in the Persian Gulf War is debated by Stephen Biddle, "Victory Misunderstood: What the Gulf War Tells Us about the Future of Conflict," *International Security* 21, no. 2 (Fall 1996): 139–79;

Stephen Biddle, "The Gulf War Debate Redux: Why Skill and Technology Are the Right Answer," *International Security* 22, no. 2 (Fall 1997): 163–74; Daryl G. Press, "Lessons from Ground Combat in the Gulf: The Impact of Training and Technology," *International Security* 22, no. 2 (Fall 1997): 137–46; and Thomas G. Mahnken and Barry D. Watts, "What the Gulf War Can (and Cannot) Tell Us about the Future of Warfare," *International Security* 22, no. 2 (Fall 1997): 151–62.

22. Biddle, *Military Power*, 28–51.

23. Stephen Biddle and Robert Zirkle, "Technology, Civil-Military Relations, and Warfare in the Developing World," *Journal of Strategic Studies* 19, no. 2 (June 1996): 171—212.

24. Risa A. Brooks, *Shaping Strategy: The Civil-Military Politics of Strategic Assessment* (Princeton: Princeton University Press, 2008).

25. James T. Quinlivan, "Coup-proofing: Its Practice and Consequence in the Middle East," *International Security* 24, no. 2 (Fall 1999).

26. Stephen Peter Rosen, *Societies and Military Power: India and Its Armies* (Ithaca, NY: Cornell University Press, 1996); and Stephen Peter Rosen, "Military Effectiveness: Why Society Matters," *International Security* 19, no. 4 (Spring 1995): 5–31. See also the essays in Risa A. Brooks and Elizabeth A. Stanley, eds., *Creating Military Power: The Sources of Military Effectiveness* (Stanford: Stanford University Press, 2007).

27. Elizabeth Kier, *Imagining War: French and British Military Doctrine between the Wars* (Princeton: Princeton University Press, 1997).

28. For an example of a cultural explanation, see Kenneth M. Pollack, "The Influence of Culture on Arab Military Effectiveness," Ph.D. dissertation, Massachusetts Institute of Technology, 1996; and Kenneth Pollack, *Arabs at War* (Lincoln: University of Nebraska Press, 2002). See also Alexander Watson, "Culture and Combat in the Western World, 1900–1945," *Historical Journal* 51, no. 2 (2008): 529–46.

29. Paul C. Stern, "Why Do People Sacrifice for Their Nations?" *Political Psychology* 16, no. 2 (1995): 217–35.

30. See Sun Tzu, *The Art of War*, trans. Samuel B Griffith (Oxford: Oxford University Press, 1963); and von Clausewitz, *On War*.

31. Bruce Allen Watson, *When Soldiers Quit* (New York: Praeger Books, 1997).

32. General overviews of this subject include John Keegan, *The Face of Battle* (New York: Viking, 1986); and Richard Holmes, *Acts of War: The Behavior of Men in Battle* (New York: Free Press, 1985).

33. For a good review of the modern literature, see Hew Strachan, "Training, Morale and Modern War," *Journal of Contemporary History* 41, no. 2 (2006): 211–27; and Nora Kinzer Stewart, *Mates and Muchachos: Unit Cohesion in the Falklands/Malvinas War* (Washington, DC: Brassey's, 1991), 11–30.

34. Deborah Avant, "From Mercenary to Citizen Armies: Explaining Change in the Practice of War," *International Organization* 54, no. 1 (Winter 2000): 41–72.

35. Three recent works emphasize these needs. See Alexander Watson, *Enduring the Great War: Combat, Morale, and Collapse in the German and British Armies, 1914–1918* (Cambridge: Cambridge University Press, 2008); David Fennell, *Combat and Morale in the North African Campaign: The Eighth Army and the Path to El Alamein* (Cambridge: Cambridge University Press, 2010); and David Stephenson, *With Our Backs to the Wall: Victory and Defeat in 1918* (Cambridge: Belknap Press of Harvard University Press, 2011), 244–310.

36. Strachan, "Training, Morale, and Modern War," 214–16; David Englander, "Mutinies and Morale," in *The Oxford Illustrated History of the First World War*, ed. Hew Strachan (Oxford: Oxford University Press): 191–203; and Hew Strachan, "The Soldier's Experience in Two World Wars: Some Historiographical Comparisons," in *Time to Kill: The Soldier's Experience of War in the West, 1939–1945*, ed. Paul Addison and Angus Calder (London: Pimlico, 1997): 369–78.

37. An interesting discussion of the importance of drill in combat is William H. McNeil, *Keeping Together in Time: Dance and Drill in Human History* (Cambridge: Harvard University Press, 1995).

38. Ardant du Picq, *Battle Studies: Ancient and Modern Battle* (Harrisburg, PA: Military Service Publishing Company, 1947).

39. Edward A. Stouffer et al., *The American Soldier: Adjustment during Army Life* (Princeton: Princeton University Press, 1949): 430–85.

40. S. L. A. Marshall, *Men against Fire* (Gloucester, MA: Peter Smith, 1978). Questioning Marshall's findings is Fredric Smoler, "The Secret of the Soldiers Who Didn't Shoot," *American Heritage* 40 (March 1989): 37–45. For a recent defense of Marshall's findings, see Lieutenant-Colonel Dave Grossman, *On Killing: The Psychological Cost of Learning to Kill in War and Society* (Boston: Bay Back Books, 1995), 5–37.

41. Shils and Janowitz, "Cohesion and Disintegration in the Wehrmacht in World War II," 280–315. See also M. I. Gurfein and Morris Janowitz, "Trends in Wehrmacht Morale," *Public Opinion Quarterly* 10, no. 1 (Spring 1946): 78–84.

42. Roger W. Little and Morris Janowitz, "Buddy Relations and Combat Performance," in *The New Military: Changing Patterns of Organization*, ed. Morris Janowitz (New York: Russell Sage Foundation, 1964).

43. Charles C. Moskos, Jr., *The American Enlisted Man: The Rank and File in Today's Military* (New York: Russell Sage Foundation, 1970). For a discussion of unit replacement systems and their effect on military performance, see Christopher C. Straub, *The Unit First: Keeping the Promise of Cohesion* (Washington, DC: National Defense University Press, 1988).

44. Robert Sterling Rush, *Hell in the Hürtgen Forest: The Ordeal and Triumph of an American Infantry Regiment* (Lawrence: University of Kansas Press, 2001).

45. This was the case with the German Army at the end of World War II. See Martin van Creveld, *Fighting Power: German and U.S. Army Performance, 1939–1945* (Westport,

CT: Greenwood Press, 1982): 76; and Omer Bartov, *Hitler's Army: Soldiers, Nazis, and War in the Third Reich* (Oxford: Oxford University Press, 1992): 29–58.

46. The best depiction of Iranian performance during this war comes from Steven R. Ward, *Immortal: A Military History of Iran and Its Armed Forces* (Washington, DC: Georgetown University Press, 2009), 242–98.

47. Bartov, *Hitler's Army*; and Omer Bartov, *The Eastern Front, 1941–1945: German Troops and the Barbarization of Warfare*, 2nd ed. (New York: Palgrave, 2001).

48. Stephen G. Fritz, *Frontsoldaten: The German Soldier in World War II* (Lexington: University of Kentucky Press, 1995); and Stephen Fritz, "'We Are Trying . . . to Change the Face of the World' Ideology and Motivation in the Wehrmacht on the Eastern Front: The View from Below," *Journal of Military History* 60, no. 4 (October 1996): 700–724.

49. Moskos, *The American Enlisted Man*, 146–76, argues that at the very least soldiers need some justification for fighting, what he calls latent ideology.

50. For an interesting account of ideology's role in the Civil War, see James McPherson, *What They Fought For, 1861–1865* (New York: Anchor Books, 1994); Reid Mitchell, *Civil War Soldiers* (London: Penguin Books, 1988); and the essays in Gabor S. Boritt, ed., *Why the Confederacy Lost* (Oxford: Oxford University Press, 1992).

51. See Lynn, *The Bayonets of the Republic.*

52. Dan Reiter, "Nationalism and Military Effectiveness: Post-Meiji Japan," in *Creating Military Power: The Sources of Military Effectiveness*, ed. Risa A. Brooks and Elizabeth A. Stanley (Stanford: Stanford University Press, 2007), 27–54.

53. Posen, "Nationalism, the Mass Army, and Military Power," 80–124.

54. For a detailed description of the toll World War I took on France, see Ian Ousby, *The Road to Verdun: World War I's Most Momentous Battle and the Folly of Nationalism* (New York: Doubleday, 2002), 1–10.

55. Michael Clodfelter, *Warfare and Small Armed Conflicts: A Statistical Reference*, Vol. II: *1900–1991* (Jefferson, NC: McFarland and Company, 1992), 581.

56. Jürgen E. Förster, "Ludendorff and Hitler in Perspective: The Battle for the German Soldier's Mind, 1917–1944," *War in History* 10, no. 3 (2003): 321–34; and Stern, "Why Do People Sacrifice for Their Nations?"

57. Scott Sigmund Gartner, "Opening up the Black Box of War," *Journal of Conflict Resolution* 42, no. 3 (June 1998): 252–58.

58. This summary relies primarily on chapter 3 of Reiter and Stam, *Democracies at War*, 258–83; and Dan Reiter and Allan Stam, "Democracy and Battlefield Military Effectiveness," *Journal of Conflict Resolution* 42, no. 3 (June 1998): 377–89. See also David A. Lake, "Powerful Pacifists: Democratic States and War," *American Political Science Review* 80, no. 1 (March 1992): 24. For other ways democracy contributes to military effectiveness, see Stephen Biddle and Stephen Long, "Democracy and Military Effectiveness: A Closer Look," *Journal of Conflict Resolution* 48, no. 4 (August 2004): 525–46. For alternative views, see Michael C. Desch, *Power and Military Effectiveness: The Fallacy of*

Democratic Triumphalism (Baltimore, MD: Johns Hopkins University Press, 2008); and Alexander Downes, "How Smart and Tough Are Democracies? Reassessing Theories of Democratic Victory in War," *International Security* 33, no. 4 (Spring 2009).

59. For an excellent depiction of Japan's motivation in World War II, see Richard B. Frank, *Downfall: The End of the Japanese Empire* (London: Penguin Books, 1999).

60. Richard Overy, *Russia's War: A History of the Soviet War Effort, 1941–1945* (London: Penguin Books, 1997), esp. 154–63.

61. Andrew J. R. Mack, "Why Big Nations Lose Small Wars: The Politics of Asymmetric Conflict," in *Power, Strategy, and Security*, ed. Klaus Knorr (Princeton: Princeton University Press, 1983), 126–51.

62. Stephen Van Evera discusses this blended approach in his *Guide to Methods for Students of Political Science* (Ithaca, NY: Cornell University Press, 1997), 89–90. Studies in this vein include Barry R. Posen, *The Sources of Military Doctrine: France, Britain, and Germany between the Wars* (Ithaca, NY: Cornell University Press, 1984); Kier, *Imagining War*; Rosen, *Societies and Military Power*; Stephen M. Walt, *Revolution and War* (Ithaca, NY: Cornell University Press, 1996); and Christopher Layne, *The Peace of Illusions: American Grand Strategy from 1940 to the Present* (Ithaca, NY: Cornell University Press, 2006).

63. Foundational works on case studies include Alexander L. George and Timothy J. McKewon, "Case Studies and Theory Development: The Method of Structured Focused Comparison," in *Diplomacy: New Approaches in Theory, History, and Policy*, ed. Gordon Lauren (New York: Free Press, 1979); Harry H. Eckstein, "Case Study and Theory in Political Science," in *Handbook of Political Science*, Vol. 7: *Strategies of Inquiry*, ed. Fred J. Greestein and Nelson W. Polsby (Reading, MA: Addison-Wesley, 1975); and Arend Lijphart, "Comparative Politics and the Comparative Method," *American Political Science Review* 65, no. 4 (September 1971): 682–93.

64. Alexander L. George and Timothy J. McKewon, "Case Studies and Theories of Organizational Decision Making," *Advances in Information Processing in Organizations*, eds. Robert Coulam and Richard Smith (Greenwich, CT: JAI Press, 1985, pp. 21–58.; Van Evera, *Guide to Methods for Students of Political Science*, 64–67.

65. These difficult-to-measure concepts have called into question the reliability of some existing datasets on battlefield effectiveness, especially the HERO/CAA data collection. Raising concern about the reliability of such measures, such as leadership, initiative, and morale, are Desch, *Power and Military Effectiveness*, 59–63; Risa A. Brooks, "Making Military Might: Why Do States Fail and Succeed? A Review Essay," *International Security* 28, no. 2 (Fall 2003): 181–92; John J. Mearsheimer, "Assessing the Conventional Balance: The 3:1 Rule and Its Critics," *International Security* 13, no. 4 (Spring 1989): 66; and Dan Reiter and Allan Stam use the HERO/CAA dataset and argue that their findings remain valid because any error is not systematic. See Reiter and Stam, *Democracies at War*, 71–72. Stephen Biddle uses these data in his CDB90 dataset, but his analysis

does not rely on the variables some scholars have identified as potentially unreliable. See Biddle, *Military Power*, 152–53.

66. Van Evera, *Guide to Methods for Students of Political Science*, 79; and Eckstein, "Case Study and Theory in Political Science," 119–20.

67. See Robert K. Yin, *Case Study Research: Design and Methods*, 2nd ed. (Thousand Oaks, CA: Sage Publications, California, 1994), 40.

68. On increasing a theory's range by applying it to a wider set of cases, see Adam Przeworkski and Henry Tenue, *The Logic of Comparative Social Inquiry* (New York: John Wiley, 1970), 34–39.

CHAPTER 2

1. On the factors contributing to defeat, see Eliot A. Cohen and John Gooch, *Military Misfortunes: The Anatomy of Failure in War* (New York: Vintage, 1990); and Carl von Clausewitz, *On War*, ed. and trans. Michael Howard and Peter Paret (Princeton: Princeton University Press, 1976, 1984), 86–87.

2. One example is the JANUS model used at the Institute for Defense Analysis. For a description of JANUS and its potential limitations, see Stephen Biddle, *Military Power: Explaining Victory and Defeat in Modern Battle* (Princeton: Princeton University Press, 2004), 181–89.

3. See Stephen Peter Rosen, *Societies and Military Power: India and Its Armies* (Ithaca, NY: Cornell University Press, 1996); the essays in Risa A. Brooks and Elizabeth A. Stanley, eds., *Creating Military Power: The Sources of Military Effectiveness* (Stanford: Stanford University Press, 2007); Allan R. Millet and Williamson Murray, eds., *Military Effectiveness*, Vols. 1–3 (Boston: Allen and Unwin, 1988); and Kenneth M. Pollack, "The Influence of Culture on Arab Military Effectiveness," Ph.D. dissertation, Massachusetts Institute of Technology, 1996.

4. Noting the importance of skill are Biddle, *Military Power*; Stephen Biddle, "Victory Misunderstood: What the Gulf War Tells Us about the Future of Conflict," *International Security* 21, no. 2 (Fall 1996): 139–79; Stephen Biddle, "The Gulf War Debate *Redux*: Why Skill *and* Technology Are the Right Answer," *International Security* 22, no. 2 (Fall 1997): 163–74; Daryl G. Press, "Lessons from Ground Combat in the Gulf: The Impact of Training and Technology," *International Security* 22, no. 2 (Fall 1997): 137–46; and Charles L. Glaser and Chaim Kaufmann, "What Is the Offense-Defense Balance and How Can We Measure It?" *International Security* 22, no. 4 (Spring 1998): 51.

5. For a glimpse at combat from the safe vantage point of a book, see John Keegan, *The Face of Battle* (New York: Viking, 1986); Richard Holmes, *Acts of War: The Behavior of Men in Battle* (New York: Free Press, 1985); and Lieutenant-Colonel Dave Grossman, *On Killing: The Psychological Cost of Learning to Kill in War and Society* (Boston: Bay Back Books, 1995).

6. Quoted in Nora Kinzer Stewart, *Mates and Muchachos: Unit Cohesion in the Falklands/Malvinas War* (Washington, DC: Brassey's, 1991), 12.

7. The quotation comes from A. M. Grey, *Warfighting* (New York: Doubleday, 1994), fn. 103.

8. Ardant du Picq, *Battle Studies: Ancient and Modern Battle* (Harrisburg: Military Service Publishing Company, 1947).

9. See the essays in Trevor N. Dupuy, ed., *Understanding Defeat: How to Recover from Loss in Battle to Gain Victory in War* (New York: Paragon House, 1990).

10. von Clausewitz, *On War*, 184.

11. For a useful survey of definitions of cohesion, see Arthur A. Stein, "Conflict and Cohesion: A Review of the Literature," *Journal of Conflict Resolution* 20, no. 1 (March 1976): 143–72.

12. For an overview of the cohesion literature used in the context of military effectiveness, I have relied on Elizabeth Kier, "Homosexuals in the U.S. Military: Open Integration and Combat Effectiveness," *International Security* 23, no.2 (Fall 1998): 5–39.

13. Robert J. MacCoun, "Sexual Orientation and Military Cohesion: A Critical Review of the Evidence," in *Out in Force: Sexual Orientation and the Military*, ed. G. Herek, J. Jobe, and R. Carney (Chicago: University of Chicago Press, 1996), 157–76; and Kier, "Homosexuals in the U.S. Military," 17–21.

14. John J. Mearsheimer, *Conventional Deterrence* (Ithaca, NY: Cornell University Press, 1983).

15. On the pressures to collapse, see David Stevenson, *With Our Backs to the Wall* (Cambridge: Belknap Press of Harvard University Press, 2011), 244–310.

16. Albert O. Hirschman, *Exit, Voice, and Loyalty: Responses to Decline in Firms, Organizations, and States* (Cambridge: Harvard University Press, 1970).

17. Bruce Allen Watson, *When Soldiers Quit* (New York: Praeger Books, 1997).

18. This stands in contrast to the literature on small-unit cohesion, which focuses sometimes on platoons but usually on squads.

19. This distinction comes from Edward A. Shils and Morris Janowitz, "Cohesion and Disintegration in the Wehrmacht in World War II," *Public Opinion Quarterly* 12, no. 2 (Summer 1948): 282–86.

20. See the descriptions in Kanan Makiya, *The Republic of Fear: The Politics of Modern Iraq*, 2nd ed. (Berkeley: University of California Press, 1998), 276–84.

21. For an excellent study of how military organizations deal with the need to improvise in war, see Meir Finkel, *On Flexibility: Recovery from Technological and Doctrinal Surprise on the Battlefield*, trans. Moshe Tiamin (Stanford: Stanford University Press, 2011).

22. Mancur Olson, *The Logic of Collective Action: Public Goods and the Theory of Groups* (Cambridge: Harvard University Press, 1965); and Russell Hardin, *Collective Action* (Washington, DC: Johns Hopkins University Press/Resources for the Future, 1982).

23. This is a key puzzle explored in S. L. A. Marshall's, *Men against Fire* (Gloucester, MA: Peter Smith, 1978).

24. David M. Glantz, *The Failures of Historiography: Forgotten Battles of the German-*

Soviet War (1941–1945) (Fort Leavenworth, KS: U.S. Army, Foreign Military Studies Office, 1997).

25. Exploring how self-interest can persuade individuals to participate in collective action is Russell Hardin, *One for All: The Logic of Group Conflict* (Princeton: Princeton University Press, 1995).

26. Michael Howard, *War in European History* (Oxford: Oxford University Press, 1976), 20–37.

27. Jonathan Bender and Dilip Mookherjee, "Institutional Structure and the Logic of Ongoing Collective Action," *American Political Science Review* 81 (1987): 129–54.

28. Michael Hechter, *Principles of Group Solidarity* (Berkeley: University of California Press, 1987).

29. Hardin, *One for All.*

30. The criticism comes from Alexander Wendt, *Social Theory of International Politics* (Cambridge: Cambridge University Press, 1999), 316–17.

31. Margaret Levi, *Consent, Dissent, and Patriotism* (Cambridge: Cambridge University Press, 1997).

32. David Sloan Wilson and Elliott Sober, "Reintroducing Group Selection to the Human Behavioral Sciences," *Behavioral and Brain Sciences* 17 (1994): 585–654.

33. David Sloan Wilson, "Levels of Selection: An Alternative to Individualism in Biology and the Human Sciences," *Social Networks* 11 (1989): 258.

34. Ibid., 261.

35. One approach incorporates the concept of identity. See George A. Akerlof and Rachel E. Kranton, "Economics and Identity," *Quarterly Journal of Economics* 115, no. 3 (August 2000): 715–53.

36. See Milton Friedman, *Capitalism and Freedom* (Chicago: University of Chicago Press, 1962).

37. See James Coleman, *Foundations of Social Theory* (Cambridge: Harvard University Press, 1990); and, more recently, Dora L. Costa and Matthew E. Kahn, "Civic Engagement and Community Heterogeneity: An Economist's Perspective," *Perspectives on Politics* 1, no. 1 (March 2003): 103–11.

38. The suggestion is from Robert Sugden, "On the Economics of Philanthropy," *Economic Journal* 92, no. 366 (June 1982): 349–50.

39. Making arguments in favor of team preferences is Robert Sugden, "Thinking as a Team: Towards an Explanation of Non-Selfish Behavior," *Social Philosophy and Policy* 10, no. 1 (Winter 1993): 69–89.

40. Henri Tajfel, *Human Groups and Social Categories* (Cambridge: Cambridge University Press, 1981), 36.

41. See Michael A. Hogg and Dominic Abrams, *Social Identifications: A Social Psychology of Inter-group Relations and Group Processes* (London: Routledge, 1998), 31–63.

42. The term "hierarchy of needs" belongs to A. H. Maslow, *Motivation and Personality* (New York: Harper, 1954).

43. The list of group needs comes from Wendt, *Social Theory of International Politics*, 131–32.

44. Ibid., 224–33.

45. Robert Sugden, "Team Preferences," *Economics and Philosophy* 16 (2000): 175–204.

46. Shils and Janowitz, "Cohesion and Disintegration in the Wehrmacht in World War II," 280–315; M. I. Gurfein and Morris Janowitz, "Trends in Wehrmacht Morale," *Public Opinion Quarterly* 10, no. 1 (Spring 1946): 78–84; Roger W. Little and Morris Janowitz, "Buddy Relations and Combat Performance," in *The New Military: Changing Patterns of Organization*, ed. Morris Janowitz (New York: Russell Sage Foundation, 1964); and Christopher C. Straub, *The Unit First: Keeping the Promise of Cohesion* (Washington, DC: National Defense University Press, 1988).

47. James McPherson, *What They Fought For, 1861–1865* (New York: Anchor Books, 1994); Reid Mitchell, *Civil War Soldiers* (London: Penguin Books, 1988); and the essays in Gabor S. Boritt, ed., *Why the Confederacy Lost* (Oxford: Oxford University Press, 1992).

48. On the role of societal stability, see Rosen, *Societies and Military Power*. On the influence of debate about the role of the military on combat effectiveness, see Elizabeth Kier, *Imaging War: French and British Military Doctrine between the Wars* (Princeton: Princeton University Press, 1997).

49. Barry R. Posen, "Nationalism, the Mass Army, and Military Power," *International Security* 18, no. 2 (Fall 1993): 80–124; and Michael C. Desch, "Democracy and Victory," *International Security* 27, no. 2 (Fall 2002): 39–40.

50. The definition is from Stephen M. Walt, *Revolution and War* (Ithaca, NY: Cornell University Press, 1996), 25. On the role of ideology as a mobilizing force, see Goran Therborn, *The Power of Ideology and the Ideology of Power* (London: Verso, 1980).

51. Douglas C. North, *Structure and Change in Economic History* (New York: W. W. Norton and Company, 1981), 53–54.

52. See Harry C. Triandis, *Individualism and Collectivism* (Boulder, CO: Westview Press, 1995), 2 and 21–33.

53. On the appeal of nationalism as a political ideology, see Ronald Grigor Suny, *Revenge of the Past: Nationalism, Revolution, and the Collapse of the Soviet Union* (Stanford: Stanford University Press, 1993), 1–19.

54. My definition of nationalism borrows from Ernest Gellner, *Nations and Nationalism* (Ithaca, NY: Cornell University Press, 1983); Anthony Marx, *Faith in Nation: Exclusionary Origins of Nationalism* (Oxford: Oxford University Press, 2003); and Jack Snyder, *From Voting to Violence: Democratization and Nationalist Conflict* (New York: W. W. Norton and Company, 2000).

55. Michael C. Desch, *Power and Military Effectiveness: The Fallacy of Democratic Triumphalism* (Baltimore, MD: Johns Hopkins University Press, 2008); Posen, "Nationalism, the Mass Army, and Military Power"; John A. Lynn, *The Bayonets of the Republic: Motivation and Tactics in the Army of Revolutionary France, 1791–1794*, 2nd ed. (Boulder,

CO: Westview Press, 1996); Paul C. Stern, "Why Do People Sacrifice for Their Nations?" *Political Psychology* 16, no. 2 (1995): 217–35; and Howard, *War in European History*, 109–15.

56. John Gray, "Totalitarianism, Reform, and Civil Society," in *Totalitarianism at the Crossroads*, ed. Ellen Frankel Paul (New Brunswick, NJ: Transaction Books, 1990), 97–142. On the importance of civil society, see also Joshua Cohen and Joel Rogers, "Secondary Associations in Democratic Governance," *Politics and Society* 20 (1992): 393–472.

57. See Hechter, *Principles of Group Solidarity*.

58. MacGregor Knox, "Conquest, Foreign and Domestic, in Fascist Italy and Nazi Germany," *Journal of Modern History* 56, no. 1 (March 1984): 1–57.

59. On the differences separating the military from society, see Samuel Huntington, *The Soldier and the State: The Theory and Politics of Civil-Military Relations* (Cambridge: Belknap Press of Harvard University Press, 1957), 58–62 and 79.

60. Anthony King, "The Word of Command: Communication in the Military," *Armed Forces and Society* 32, no. 4 (July 2006): 493–512. For an example of a military organization trying to promote these principles during training, see Thomas E. Ricks, *Making the Corps* (New York: Scribner, 1997).

61. Scholars sometimes refer to these norms as organizational culture. See Kier, *Imaging War*, 28. Other key works on organizational culture include Jeffrey W. Legro, "Military Culture and Inadvertent Escalation in World War II," *International Security* 18, no. 4 (Spring 1994): 108–42; and Jeffrey Legro, "Culture and Preferences in the International Cooperation Two-Step," *American Political Science Review* 90, no. 1 (March 1996): 118–37.

62. I am borrowing here from the master variables Alexander Wendt describes as key to form strong group identities. See his *Social Theory of International Politics*, 344–56.

63. On the possibility of developing loyalties outside of the primary group, see Edward Shils, "Primordial, Personal, Sacred and Civil Ties," *British Journal of Sociology* 8, no. 2 (1957): 130–45.

64. The ability of strong bonds forming inside military organizations and motivating personnel is one of the central arguments made in Robert Sterling Rush, *Hell in the Hürtgen Forest: The Ordeal and Triumph of an American Infantry Regiment* (Lawrence: University of Kansas Press, 2001). On this possibility, see Stewart, *Mates and Muchachos*, 11–30.

65. David Segal and Meyer Kestenbaum, "Professional Closure in the Military Labor Market: A Critique of Pure Cohesion," in *The Future of the Army Profession*, ed. Don M. Snider and Gayle L. Watkins (New York: McGraw-Hill, 2002), 453.

66. Hew Strachan, "Training, Morale and Modern War," *Journal of Contemporary History* 41, no. 2 (2006): 216.

67. Uzi Ben-Shalom, Zeev Lehrer, and Eyal Ben-Ari, "Cohesion during Military Operations: A Field Study on Combat Units in the Al-Aqsa Intifada," *Armed Forces and Society* 32, no. 1 (October 2005): 63–79.

68. The perils of political interference are widely recognized. See James T. Quinlivan, "Coup-proofing: Its Practice and Consequence in the Middle East," *International Security* 24, no. 2 (Fall 1999): 131–65; Risa A. Brooks, "Civil-Military Relations and Military Effectiveness: Egypt in the 1967 and 1973 Wars," and Timothy D. Hoyt, "Social Structure, Ethnicity, and Military Effectiveness: Iraq, 1980–2004," in *Creating Military Power: The Sources of Military Effectiveness*, ed. Risa A. Brooks and Elizabeth A. Stanley (Stanford: Stanford University Press, 2007), 106–35 and 55–79, respectively. In the Soviet Army of World War II, see Norman Davies, *No Simple Victory: World War II in Europe, 1939–1945* (London: Viking, 2006), 232.

69. Kier, *Imaging War*, 29–30; and Terrence Lee, "Military Cohesion and Regime Maintenance: Explaining the Role of the Military in 1989," *Armed Forces & Society* 32, no. 1 (2005): 80–104.

70. See Rosen, *Societies and Military Power*, 1–32.

71. See Vladimir Tismaneanu, *Fantasies of Salvation: Democracy, Nationalism and Myth in Post-Communist Europe* (Princeton: Princeton University Press, 1998), 21–31; and Jacob L. Talman, *Myth of the Nation and Vision of the Revolution: Ideological Polarization in the Twentieth Century* (New Brunswick, NJ: Transaction Publishers, 1991).

72. The preceding builds on Daniel Chirot, *Modern Tyrants: The Power and Prevalence of Evil in Our Age* (New York: Free Press, 1994). For the role of resentment in nationalism, see Liah Greenfeld, *Nationalism: Five Roads to Modernity* (Cambridge: Harvard University Press, 1992).

73. For the distinction between causal versus constitutive mechanisms, see Wendt, *Social Theory of International Politics*, 139–92; and Alexander Wendt, "On Constitution and Causation in International Relations," *Review of International Studies* 24 (1998): 101–17.

74. Hirschman, *Exit, Voice, and Loyalty*.

75. For discussions regarding the use of domestic coercion during war, see Allan C. Stam III, *Win, Lose, or Draw: Domestic Politics and the Crucible of War* (Ann Arbor: University of Michigan Press, 1996); and Hein Goemans, *War and Punishment: The Causes of War Termination and the First World War* (Princeton: Princeton University Press, 2000).

76. International relations scholars commonly use the term "authoritarian" to describe a type of government. Here I use it to describe a specific type of military that can exist with many different types of regime as long as they exercise a high degree of control that also undermines their autonomy to train.

77. See Quinlivan, "Coup-proofing," 131–65.

78. Of course, no two authoritarian militaries fight alike. They will vary in their overall military effectiveness for various reasons. The argument I present here applies only to their cohesion, one aspect of their combat performance. Sometimes, for example, civil-military relations will affect their ability to craft military strategy. On other oc-

casions, the degree of external threat neighboring countries pose could also inspire even the most repressive regime to undertake reforms that give the military some independence as a means to improve combat effectiveness. On the strategy, see Risa A. Brooks, *Shaping Strategy: The Civil-Military Politics of Strategic Assessment* (Princeton: Princeton University Press, 2008). On the role of threat, see Caitlin Talmadge, "The Puzzle of Personalist Performance: Iraqi Battlefield Effectiveness in the Iran-Iraq War," *Security Studies* 22, no. 2 (Spring 2013):180–221.

79. On conceptions of military professionalism, see Huntington, *The Soldier and the State*; Morris Janowitz, *The Professional Soldier, a Social and Political Portrait* (New York: Free Press, 1960); and Stanislav Andreski, *Military Organization and Society*, 2nd ed. (Berkeley: University of California Press, 1968).

80. This prediction is similar to the one made about the willingness of democratic regimes to endure long wars. Finding that support for war in democracies decreases over time and when casualties mount are D. Scott Bennett and Allan C. Stam III, "The Duration of Interstate Wars 1816–1985," *American Political Science Review* 90, no. 2 (June 1996): 239–57; D. Scott Bennett and Allan C. Stam, "The Declining Advantages of Democracy: A Combined Model of War Outcomes and Duration," *Journal of Conflict Resolution* 42, no. 3 (June 1998): 344–66; and Branislav L. Slantchev, "How Initiators End Their Wars: The Duration of Warfare and the Terms of Peace," *American Journal of Political Science* 48, no. 4 (October 2004): 813–29. See also Adam Berinky, "Assuming the Costs of War: Events, Elites, and American Public Support for Military Conflict," *Journal of Politics* 69, no. 4 (November 2007): 975–97.

81. Charles C. Moskos, Jr., *The American Enlisted Man* (New York: Russell Sage Foundation, 1970): 146–48.

82. See the discussion in Levi, *Consent, Dissent, and Patriotism*; Kier, *Imagining War*; and Eliot A. Cohen, *Citizens and Soldiers: The Dilemmas of Military Service* (Ithaca, NY: Cornell University Press, 1985).

83. Little and Janowitz, "Buddy Relations and Combat Performance."

84. Shils and Janowitz, "Cohesion and Disintegration in the Wehrmacht in World War II," 280–315.

85. Martin van Creveld, *Fighting Power: German and U.S. Army Performance, 1939–1945* (Westport, CT: Greenwood Press, 1982).

86. Moskos, *The American Enlisted Man*; William Daryl Henderson, *Cohesion: The Human Element in Combat* (Washington, DC: National Defense University, 1985); and Straub, *The Unit First*.

87. Lynn, *The Bayonets of the Republic*.

88. Desch, *Power and Military Effectiveness*, 60–61.

89. Dan Reiter, "Nationalism and Military Effectiveness: Post-Meiji Japan," in *Creating Military Power: The Sources of Military Effectiveness*, ed. Risa A. Brooks and Elizabeth A. Stanley (Stanford: Stanford University Press, 2007), 27–54.

90. Posen, "Nationalism, the Mass Army, and Military Power."

91. Stern, "Why Do People Sacrifice for Their Nations?" 217–35.

92. Triandis, *Individualism and Collectivism*, 2 and 19–23.

93. This section relies on chapter 3 in Dan Reiter and Allan Stam, *Democracies at War* (Princeton: Princeton University Press, 2002), 258–83; and Reiter and Stam, "Democracy and Battlefield Military Effectiveness," *Journal of Conflict Resolution* 42, no. 3 (June 1998): 377–89.

94. David A. Lake, "Powerful Pacifists: Democratic States and War," *American Political Science Review* 80, no. 1 (March 1992): 24.

95. Reiter and Stam, *Democracies at War*, 61.

96. Ibid.

97. Ibid., 64, 164, and 192; see also Bennett and Stam, "The Declining Advantages of Democracy," 344–66; and Scott Sigmund Gartner and Randolph M. Sieverson, "War Expansion and War Outcome," *Journal of Conflict Resolution* 40 (March 1996): 4–15.

CHAPTER 3

1. Martin van Creveld, *Fighting Power: German and U.S. Army Performance, 1939–1945* (Westport, CT: Greenwood Press, 1982), 163; and Norman Davies, *No Simple Victory: World War II in Europe*, 1939–1945 (London: Viking, 2006).

2. Omer Bartov, "The Missing Years: German Workers, German Soldiers," in *Nazism and German Society, 1939–1945*, ed. David F. Crew (London: Routledge, 1994), 46.

3. The murderous record of the Wehrmacht has been well documented. For a recent overview of its war crimes, see Wolfram Wette, *Wehrmacht: History, Myth, Reality*, trans. Deborah Lucas Schneider (Cambridge: Harvard University Press, 2007).

4. Richard Overy, *The Road to War: The Origins of World War II* (London: Macmillan; 2nd ed., New York: Random House, 1989), 25.

5. Excellent overviews on this period include Ian Kershaw, *The End: The Defiance and Destruction of Hitler's Germany, 1944–1945* (London: Penguin Press, 2012); Richard J. Evans, *The Third Reich in Power* (London: Penguin, 2005); Ian Kershaw, *Hitler: Hubris, 1889–1936* (New York: W. W. Norton and Company, 2000); and Klaus P. Fischer, *Nazi Germany: A New History* (New York: Continuum, 1995).

6. Richard Grunberger, *The 12-Year Reich: A Social History of Nazi Germany, 1933–1945* (New York: Holt, Reinhardt, and Winston, 1971), 1–17.

7. Quoted in P. H. H. Bell, *The Origins of the Second World War in Europe*, 2nd ed. (London: Longman, 1997), 79.

8. Fischer, *Nazi Germany*, 267–71.

9. Alan Bullock, *A Study in Tyranny*, rev. ed. (New York: Harper and Row, 1964), 309–11.

10. Gordon R. Craig, *The Politics of the Prussian Army: 1640–1945* (New York: Clarendon Press; rev. ed., Oxford: Oxford University Press, 1955), 468–81.

11. Bell, *The Origins of the Second World War in Europe*, 78–79.

12. Ian Kershaw, *Hitler: Profile in Power* (London: Longman, 1991), 8.

13. Karl Dietrich Bracher, *The German Dictatorship: The Origins, Structure, and Effects of National Socialism*, trans. Jean Steinberg (New York: Praeger, 1970), 22.

14. Norman Rich, *Hitler's War Aims: Ideology, the Nazi State, and the Course of Expansion* (New York: W. W. Norton and Company, 1973), 14.

15. Bracher, *The German Dictatorship*, 251.

16. Bell, *The Origins of the Second World War in Europe*, 89.

17. MacGregor Knox, *Common Destiny: Dictatorship, Foreign Policy, and War in Fascist Italy and Nazi Germany* (Cambridge: Cambridge University Press, 2000), 232.

18. David Schoenbaum, *Hitler's Social Revolution: Class and Status in Nazi Germany, 1933–1939*, 2nd ed. (New York: W. W. Norton, 1980), 55.

19. See the discussion in Fritz Stern, *Dream and Delusions: The Drama of German History* (New York: Alfred A. Knopf, 1987; rev. ed., New Haven: Yale University Press, 1999), 25–50.

20. John Luckas, *The History of Hitler* (New York: Vintage Books, 1997), 176–96.

21. Schoenbaum, *Hitler's Social Revolution*, 35 and 244.

22. Michael Burleigh, *The Third Reich: A New History* (New York: Hill and Wang, 2000), 12.

23. MacGregor Knox, "Conquest, Foreign and Domestic, in Fascist Italy and Nazi Germany," *Journal of Modern History* 56, no. 1 (March 1984): 5, 11, and 20.

24. Schoenbaum, *Hitler's Social Revolution*, 272.

25. Burleigh, *The Third Reich*, 243.

26. Knox, "Conquest, Foreign and Domestic, in Fascist Italy and Nazi Germany," 12.

27. Bracher, *The German Dictatorship*, 251.

28. Knox, "Conquest, Foreign and Domestic, in Fascist Italy and Nazi Germany," 29.

29. Quoted in Randall Schweller, *Deadly Imbalances: Tripolarity and Hitler's Strategy of World Conquest* (New York: Columbia University Press, 1998), 10.

30. Bullock, *A Study in Tyranny*, 368–71.

31. Quoted in Schweller, *Deadly Imbalances*, 109.

32. Quoted in Bracher, *The German Dictatorship*, 288.

33. Overy, *The Road to War*, 38.

34. Schoenbaum, *Hitler's Social Revolution*, 278. On the widespread support of Hitler's regime, see Eric A. Johnson, *Nazi Terror: The Gestapo, Jews and Ordinary Germans* (New York: Basic Books, 2000), 375.

35. Robert Gellately, *Backing Hitler: Consent and Coercion in Nazi Germany* (Oxford: Oxford University Press, 2001), 259; and Bracher, *The German Dictatorship*, 259.

36. Schoenbaum, *Hitler's Social Revolution*, 47–53.

37. Knox, "Conquest, Foreign and Domestic, in Fascist Italy and Nazi Germany," 32.

38. Ibid., 30.

39. Ibid., 31–32.

40. Bartov, "The Missing Years," 47–52.

41. This argument is made in Timothy Mason, *Social Policy in the Third Reich: The Working Class and the "National Community," 1918–1939* (Oxford: Berg Publishers, 1993).

42. David F. Crew, "General Introduction," in *Nazism and German Society, 1939–1945*, ed. David F. Crew (London: Routledge, 1994), 4.

43. Burleigh, *The Third Reich*, 222.

44. Gellately, *Backing Hitler*, 256.

45. Bracher, *The German Dictatorship*, 249.

46. See Knox, "Conquest, Foreign and Domestic, in Fascist Italy and Nazi Germany," 37–38; Knox, *Common Destiny*, 234; and Rich, *Hitler's War Aims*, 12.

47. Omer Bartov, "The Conduct of War: Soldiers and Barbarization of Warfare," *Journal of Modern History* 64, suppl. (December 1992): S41; and Burleigh, *The Third Reich*, 234–38.

48. See George H. Stein, *The Waffen SS: Hitler's Elite Guard, 1939–1945* (Ithaca, NY: Cornell University Press, 1966).

49. Gellately, *Backing Hitler*, 260–61.

50. Ibid., 253 and 256.

51. Christoph Rass, "The Social Profile of the German Army's Combat Units, 1939–1945," in *German Wartime Society, 1939–1945: Politicization, Disintegration, and the Struggle for Survival* (Vol. IX/I of *Germany and the Second World War*), ed. Jorg Echternkamp, trans. Derry Cook-Radmore (Oxford: Clarendon Press, 2008), 617–770.

52. Stephen G. Fritz, *Ostkrieg: Hitler's War of Extermination in the East* (Lexington: University of Kentucky Press, 2011), 374–77.

53. Stephen G. Fritz, "'We Are Trying . . . to Change the Face of the World' Ideology and Motivation in the Wehrmacht on the Eastern Front: The View from Below," *Journal of Military History* 60, no. 4 (October 1996): 687.

54. Omer Bartov, "Soldiers, Nazis, and War in the Third Reich," *Journal of Modern History* 63 (March 1991): 50.

55. Gellately, *Backing Hitler*, 224.

56. The quotations are from Stephen G. Fritz, *Frontsoldaten: The German Soldier in World War II* (Lexington: University of Kentucky Press, 1995), 174 and 175, respectively. Some might worry that the letters German soldiers wrote home might suffer from bias because censors would permit only favorable messages to reach the home front. However, as Stephen Fritz notes, "Still, the flood of letters to and from the front (estimated at 40 to 50 billion total, and in some individual months as many as 500 million) meant that many passed through censorship unopened; and the longer the war continued, the less seriously many *Landers* regarded the censor. As two of the leading authorities on German *Feldpostbriefe* (letters from the field) concluded after studying thousands of missives, '[T]he mass of soldiers expressed their opinions and views in a surprisingly open

and uninhibited fashion.' So despite the problems, much can be gained from a study of letters and diaries, especially if the historian relates these necessarily individual and narrow documents to a wider context." Ibid., 9. Still further, it seems reasonable to assume that if German soldiers did not support the National Socialist cause, historians would find very few letters imbued with ideological language. The number of letters examined by historians like Fritz and Bartov, however, suggests that this was far from the case.

57. This was true in World War I. See B. H. Liddell Hart, *The Real War: 1914–1918* (Boston: Little, Brown and Company; 2nd ed., 1964), 37–38.

58. Quoted in Fritz, *Frontsoldaten*, 15.

59. Ibid., 14.

60. Ibid., 18.

61. Van Creveld, *Fighting Power*, 129.

62. Fritz, *Frontsoldaten*, 235.

63. Ibid., 25.

64. S. Hart, R. Hart, and M. Hughes, *The German Soldier in World War II* (Osceola, WI: Amber Books, 2000), 9; and Van Creveld, *Fighting Power*, 35–37.

65. Hart, Hart, and Hughes, *The German Soldier in World War II*, 8.

66. See Fritz, *Frontsoldaten*, 13, 19, and 22–23.

67. Van Creveld, *Fighting Power*, 121.

68. Bartov, "Soldiers, Nazis, and War in the Third Reich," 46.

69. Fritz, "We Are trying . . . to Change the Face of the World," 697.

70. Richard Overy, *Russia's War: A History of the Soviet War Effort, 1941–1945* (London: Penguin Books, 1997), 187–89.

71. Hart, Hart, and Hughes, *The German Soldier in World War II*, 12.

72. Robert J. O'Neill, *The Germany Army and the Nazi Party, 1933–1939* (New York: James H. Heineman, 1966), 83; and Rich, *Hitler's War Aims*, 51.

73. Jurgen Forester, "The Dynamics of Volksgemeinschaft: The Effectiveness of the German Military Establishment in the Second World War," in *Military Effectiveness 3: World War II*, ed. Allan R. Millet and Williamson Murray (Boston: Allen and Unwin, 1988), 208.

74. Knox, "Conquest, Foreign and Domestic, in Fascist Italy and Nazi Germany," 50–51.

75. See Bartov, "Soldiers, Nazis, and War in the Third Reich," 47.

76. Forester, "The Dynamics of Volksgemeinschaft," 208.

77. Quoted in O'Neill, *The Germany Army and the Nazi Party*, 66.

78. Ibid., 71.

79. Ibid., 73.

80. Bartov, "Soldiers, Nazis, and War in the Third Reich," 54.

81. See Omer Bartov, *Hitler's Army: Soldiers, Nazis, and War in the Third Reich* (Oxford: Oxford University Press, 1992), 29–58, esp. 32–33.

82. Bartov, "Soldiers, Nazis, and War in the Third Reich," 49–50.

83. Documenting the numerical advantage of Allied manpower in Normandy is Niklas Zetterling, *Normandy 1944: German Military Organization, Combat Power and Organizational Effectiveness* (Winnipeg, Manitoba: J. J. Fedorowicz Publishing, 2000), 27–36.

84. These figures come from Overy, *Russia's War*, 240.

85. Charles MacDonald, *The Mighty Endeavor: American Armed Forces in the European Theater in World War II* (Oxford: Oxford University Press, 1969), 314–19.

86. Williamson Murray and Allan R. Millett, *A War to Be Won: Fighting the Second World War* (Cambridge: Belknap Press of Harvard University Press, 2000), 432.

87. On the Allied advance in France, see Martin Blumenson, *Breakout and Pursuit: The United States Army in World War II, European Theater of Operations* (Washington, DC: Office of the Chief of Military History, 1961); and Martin Blumenson, *The Duel for France, 1944* (New York: Da Capo Press, 1963). From the German perspective, see Samuel W. Mitchum, Jr., *Retreat to the Reich: The German Defeat in France, 1944* (Westport, CT: Praeger, 2000), esp. 69–102.

88. Several authors offer differing accounts of the balance of forces. For instance, H. Essame, *The Battle for Germany* (New York: Ace Books, 1969), 12, contends that the Allies fielded 56 full-strength divisions in the West. Still further, Robin Cross, *Fallen Eagle: The Last Days of the Third Reich* (New York: John Wiley and Sons, 1995), 111, contends that the Allies deployed 6,000 medium and 1,700 light tanks. I have decided to rely on the more recent and conservative estimates from Mitchum, *Retreat to the* Reich, 221–22.

89. There is some confusion regarding the commanders of Army Group G. General Blaskowitz originally commanded the army group at the beginning of September, but Hitler replaced him with General Balck in the fall. See Essame, *The Battle for Germany*, 27 and 76; and Cross, *Fallen Eagle*, 127.

90. On the German recovery, see Cross, *Fallen* Eagle, 118–19; and Murray and Millett, *A War to Be* Won, 464–65.

91. Essame, *The Battle for Germany*, 72–73.

92. Charles Whiting, *Siegfried: The Nazis' Last Stand* (New York: Jove Books, 1983), 39.

93. Ibid., 25–27.

94. Ibid., 24.

95. Ibid., 28–29.

96. For a good depiction of the early American assault on Aachen, see Peter R. Mansoor, *The GI Offensive in Europe: The Triumph of American Infantry Divisions, 1941–1945* (Lawrence: University of Kansas Press, 1999), 185–87. Along with his summary of events at Aachen, Mansoor argues that American forces were superior in combat effectiveness to German forces. For a better assessment of German performance, see Zetterling, *Normandy 1944*, 429–38.

97. The account is from Whiting, *Siegfried*, 36–37.

98. For details of Operation Market Garden, see Cornelius Ryan, *A Bridge Too Far* (New York: Simon and Schuster, 1974).

99. Cross, *Fallen Eagle*, 120–21.

100. For a good summary of the final assault on Aachen, see David T. Zabecki, ed., *World War II in Europe: An Encyclopedia* (New York: Garland Publishing, 1999), 1349–50.

101. Ted Ballard, *Rhineland: The U.S. Army Campaigns of World War II* (Washington, DC: U.S. Government Printing Office, 1995), 14.

102. Roger Cirrilo, *Ardennes-Alsace: The U.S Army Campaigns of World War II* (Washington, DC: U.S. Government Printing Office, 1994), 3.

103. John Keegan, *The Second World War* (New York: Random House),, 441.

104. Quoted in B. H. Liddell Hart, *History of the Second World War* (New York: G. P. Putnam and Sons, 1971), 647.

105. Ibid.

106. Ibid., 639.

107. Keegan, *The Second World War*, 444–45.

108. Michael D. Doubler, *Closing with the Enemy: How GIs Fought the War in Europe, 1944–45* (Lawrence, KS: University of Kansas Press, 1994), 207–14.

109. MacDonald, *The Mighty Endeavor*, 388.

110. Calvocoressi, Wint, and Pritchard, *The Penguin History of World War II*, (London: Penguin, 1999), 550.

111. Liddell Hart, *History of the Second World War*, 667–68.

112. Ian C.B. Dear and Rosemary R. D. Foot, *The Oxford Companion to World War II* (Oxford: Oxford University Press, 1995), 39–40

113. Albert Seaton, *The German Army, 1933–1945* (New York: Meridian, 1982), 242.

114. Calvocoressi, Wint, and Pritchard, *The Penguin History of World War II*, 552.

115. Murray and Millett, *A War to Be Won*, 471.

116. Seaton, *The German Army*, 232–42.

117. Quoted in Matthew Cooper, *The German Army: Its Political and Military Failures* (New York: Stein and Day, 1978), 540.

118. Bracher, The *German Dictatorship*, 462.

119. Quoted in Cooper, *The German Army*, 542.

120. Bracher, *The German Dictatorship*, 463.

121. Fritz, *Ostkrieg*, 454. This number does not include the approximately ten thousand German citizens shot for similar crimes.

122. Ibid., 224.

123. Knox, *Common Destiny*, 237–38.

124. Bartov, "The Conduct of War," S45.

125. Van Creveld, *Fighting Power*, 113–14.

126. Fritz, *Frontsoldaten*, 90 and 159; and Knox, *Common Destiny*, 235.

127. Bartov, *Hitler's Army*, 59–61.

128. Bartov, "Soldiers, Nazis, and War in the Third Reich," 51; and Bartov, "The Conduct of War," S33.

129. Cross, *Fallen Eagle*, 159. This was the first time missiles were used as battlefield weapons. The Karl Moser broke down after firing four shells.

130. Liddell Hart, *History of the Second World War*, 678.

131. Russell F. Weigley, *Eisenhower's Lieutenants: The Campaign of France and Germany, 1944–1945* (Bloomington: Indiana University Press, 1981), 639.

132. Again, there are conflicting accounts as to the size of the Allied forces on the German border. Ballard, *Rhineland*, 32–33, offers different figures, claiming that they numbered ninety divisions and 4.5 million men. Because their estimates are more consistent with other approximations in the literature, I have relied on the descriptions given by Weigley, *Eisenhower's Lieutenants*, 667–68; and Gerhard Weinberg, *A World at Arms: A Global History of World War II* (Cambridge: Cambridge University Press, 1994), 811.

133. Mansoor, *The GI Offensive in Europe*, 235.

134. Cooper, *The German Army*, 527.

135. *The Oxford Companion to World War II*, 381.

136. Cross, *Fallen Eagle*, 168–71.

137. On Eisenhower's decision-making, see Donald E. Shepardson, "The Fall of Berlin and the Rise of a Myth," *Journal of Military History* 62 (January 1998): 135–54.

138. Murray and Millett, *A War to Be Won*, 480.

139. Cross, *Fallen Eagle*, 169.

140. Knox, *Common Destiny*, 238.

(150.Norman Rich, *Great Power Diplomacy: 1814–1914* (New York: McGraw Hill, 1992).

151. For an introduction to Germany's domestic problems before World War I, see David Calleo, *The German Problem Reconsidered: Germany and the World Order, 1870 to the Present* (Cambridge: Cambridge University Press, 1978), 123–60.

152. Roger Chickering, *Imperial Germany and the Great War: 1914–1918* (Cambridge: Cambridge University Press, 1998), 4.

153. See David Blackbourn and Geoff Eley, *The Peculiarities of German History: Bourgeois Society and Politics in Nineteenth-Century Germany* (Oxford: Oxford University Press, 1984).

154. Chickering, *Imperial Germany and the Great War*, 1.

155. Ibid., 7.

156. See Jack Snyder, *Myths of Empire: Domestic Politics and International Ambition* (Ithaca, NY: Cornell University Press, 1991), 66–111.

157. An excellent summary of this literature is David E. Kaiser, "Germany and the Origins of the First World War," *Journal of Modern History* 55 (1983): 442–74.

158. Geoff Eley, "The Wilhelmine Right," in *Society and Politics in Wilhelmine Germany*, ed. Richard J. Evans (London: Croom Helm Ltd., 1978), 125–28; and Roger Chickering, *We Men Who Feel Most German: A Cultural Study of the Pan-German League, 1886–1914* (Boston: Unwin Hyman, 1984), 60–63, 66, and 213.

159. On the comparison between nationalism in Nazi and Wilhelmine Germany, see Juan Linz, "Political Space and Fascism as a Late-Comer," in *Who Were the Fascists?* ed. Stein Larsen et al. (Oslo-Bergen: Universitetsforlaget, 1980), 153–91; and Ian Kershaw, *The Nazi Dictatorship* (London, 1985), chs. 3 and 4.

160. Quoted in Hew Strachan, *The First World War*, Vol. I: *To Arms* (Oxford: Oxford University Press, 2001), 121–22.

161. Chickering, *Imperial Germany and the Great War*, 15–17.

162. Martin Kitchen, *The German Officer Corps, 1890–1914* (Oxford: Oxford University Press, 1968), 143–86.

163. See the discussion in Eric Dorn Brose, *The Kaiser's Army: The Politics of Military Technology in Germany during the Machine Age, 1870–1918* (Oxford: Oxford University Press, 2001), 7–25.

164. On training during this period, see Matthias Strohn, *The German Army and the Defense of the Reich: Military Doctrine and the Conduct of the Defensive Battle, 1918–1939* (Cambridge: Cambridge University Press, 2011); Bruce I. Gudmundsson, *Stormtroop Tactics: Innovation in the German Army, 1914–1918* (London: Praeger, 1989); and Martin Samuels, *Doctrine and Dogma: German and British Infantry Tactics in the First World War* (New York: Greenwood Press, 1992).

165. Karl Demeter, *The German Officer-Corps in Society and State, 1650–1945* (New York: Praeger, 1965), 246–47.

166. Craig, *The Politics of the Prussian Army: 1640–1945*, 217–19.

167. David G. Herrmann, *The Arming of Europe and the Making of the First World War* (Princeton: Princeton University Press, 1996), 220–25.

168. Martin Samuels, *Command or Control? Command, Training and Tactics in the British and German Armies, 1888–1918* (London: Frank Cass, 1995), 77 and 81.

169. Demeter, *The German Officer-Corps in Society and State*, 185, 233–34, and 246–47.

170. Samuels, *Command or Control?* 77–78.

171. Technically, there was no "German Army," since several of the former and larger German states still exercised administrative control over their units. In practice, however, all units came under the control of the former Prussian Army in war. See Kitchen, *The German Officer Corps*, 19.

172. For the first years on the Western Front, see John Keegan, *The First World War* (New York: Vintage Books, 1998), 71–137.

173. Ian Ousby, *The Road to Verdun: World War I's Most Momentous Battle and the Folly of Nationalism* (New York: Doubleday, 2002), 6–8.

174. John Moiser, *The Myth of the Great War: How the Germans Won the Battles and How the Americans Saved the Allies* (New York: Harper Collins, 2001), 225.

175. David Eggenberger, *An Encyclopedia of Battles*, 2nd ed. (New York: Dover Publications, 1985), 84.

176. Ibid., 9.

177. Rod Paschall, *The Defeat of Imperial Germany, 1917–1918* (Chapel Hill, NC: Da Capo Press, 1994), 79.

178. My summary of Cambrai relies on Liddell Hart, *The Real War*, 344–56; and Keegan, *The First World War*, 369–71.

179. Paschall, *The Defeat of Imperial Germany*, 115.

181. Chickering, *Imperial Germany and the Great War*, 140–44.

182. Ibid., 103–14.

183. Ibid., 148–49 and 156–57.

184. Ibid., 165.

185. Ibid., 160.

186. Paschall, *The Defeat of Imperial Germany*, 157–62.

187. H. E. Goemans, *War and Punishment: The Causes of War Termination and the First World War* (Princeton: Princeton University Press, 2000), 264.

188. Ibid., 271.

189. Chickering, *Imperial Germany and the Great War*, 187–89.

190. Goemans, *War and Punishment*, 278.

191. Wilhelm Deist, "The Military Collapse of the German Empire: The Reality behind the Stab-in-the-Back Myth," *War in History* 3, no. 2 (April 1996): 186–207.

192. These calculations come from Tim Travers, "The Allied Victories, 1918," in *The Oxford Illustrated History of the First World War*, ed. Hew Strachan (Oxford: Oxford University Press, 1998), 281–85.

193. Spencer C. Tucker, *The Great War: 1914–1918* (Bloomington: Indiana University Press, 1998), 171.

194. My discussion of Germany's deteriorating staying power relies on David Stevenson, *With Our Backs to the Wall* (Cambridge: Belknap Press of Harvard University Press, 2011); and Alexander Watson, *Enduring the Great War: Combat, Morale, and Collapse in the German and British Armies, 1914–1918* (Cambridge: Cambridge University Press, 2008).

195. See Liddell Hart, *The Real War*, 429–38.

196. Eggenberger, *An Encyclopedia of Battles*, 18; and Peter Simkins, *The First World War*, Vol. 3: *The Western Front 1917–1918* (Oxford: Osprey Publishing, 2002), 62–65.

197. Scott Stephenson, *The Final Battle: Soldiers of the Western Front and the German Revolution of 1918* (Cambridge: Cambridge University Press, 2009).

198. Hew Strachan, "Training, Morale and Modern War," *Journal of Contemporary History* 41, no. 2 (2006): 211–14; and Bartov, *Hitler's Army*, 32–33.

199. It is interesting to note that Woodrow Wilson considered Wilhemine Germany democratic before World War I. See Ido Oren, "The Subjectivity of the 'Democratic' Peace: Changing U.S. Perceptions of Imperial Germany," *International Security* 20, no. 2 (Fall 1995): 147–87.

CHAPTER 4

1. Good overviews of the literature are provided by Julian Jackson, *The Fall of France: The Nazi Invasion of 1940* (Oxford: Oxford University Press: 2003); Martin S. Alexander, "The Fall of France, 1940," *Journal of Strategic Studies* 13, no. 10 (March 1990): 10–44; John C. Cairns, "Some Recent Historians and the 'Strange Defeat' of 1940," *Journal of Modern History* 46, no. 1 (March 1974): 60–85; and Peter Jackson, "Recent Journeys along the Road back to France, 1940," *Historical Journal* 39, no. 2 (1996): 497–510.

2. Traditional accounts that focus on domestic politics and organizational flaws in the military include Marc Bloch, *Strange Defeat: A Statement of Evidence Written in 1940*, trans. Gerald Hopkins (New York: W. W. Norton and Company, 1968); and William L. Shirer, *The Collapse of the Third Republic: An Inquiry into the Fall of France in 1940* (New York: Simon and Schuster, 1969). A controversial history from a Vichy sympathizer comes from Jacques Benoist-Mechin, *Sixty Days That Shook the World: The Fall of France, 1940*, trans. Peter Wiles (New York: G. P. Putnam and Sons, 1963).

3. The most recent work in this vein is Karl-Heinz Frieser, *The Blitzkrieg Legend: The 1940 Campaign in the West* (Annapolis, MD: Naval Institute Press, 2005). See also Robert Doughty, *The Seeds of Disaster: The Development of French Army Doctrine, 1919–1939* (Hamden, CT: Archon, 1985); Barry Posen, *The Sources of Military Doctrine: France, Britain, and Germany between the World Wars* (Ithaca, NY: Cornell University Press, 1984), 86–94; and Elizabeth Kier, *Imaging War: French and British Military Doctrine between the Wars* (Princeton: Princeton University Press, 1997). Emphasizing problems in French command and control is Martin S. Alexander, T*he Republic in Danger: General Maurice Gamelin and the Politics of French Defense, 1933–1940* (Cambridge: Cambridge University Press, 1992); and Martin Alexander, "After Dunkirk: The French Army's Performance against 'Case Red,' 25 May to 25 June 1940," *War in History* 14, no. 2 (2007): 219–64.

4. On the flaws in French security policy, see Robert J. Young, *In Command of France: French Foreign Policy and Military Planning, 1933–1940* (Cambridge: Harvard University Press, 1978). For an examination of Allied planning, see Jeffrey Gunsberg, *Divided and Conquered: The French High Command and the Defeat of the West, 1940* (Westport, CT: Greenwood Press, 1980).

5. Eugenia C. Kiesling, *Arming against Hitler: France and the Limits of Military Planning* (Lawrence: University of Kansas Press, 1996).

6. Ernest May, *Strange Victory: Hitler's Conquest of France* (New York: Hill and Wang, 2000).

7. The key article remains Don W. Alexander, "Repercussions of the Breda Variant,"

French Historical Studies 8, no. 3 (Spring 1974): 459–88. See also Jackson, *The Fall of France*; Douglas Porch, "Military Culture and the Fall of France: A Review Essay," *International Security* 24, no. 4 (Spring 2000): 157–80; and Michael C. Desch, "Planning War in Peacetime," *Joint Forces Quarterly*, 30 (Spring 2002): 94–104.

8. See Hew Strachan, *The First World War*, Vol. I: *To Arms* (Oxford: Oxford University Press, 2001), 124–31; and Jean-Jacques Becker, *The Great War and the French People*, trans. Arnold Pomerans (New York: St. Martin's Press, 1986). On the national revival prior to the war, see David Ralston, *The Army of the Republic: The Place of the Military in the Political Evolution of France, 1871–1914* (Cambridge, MA: MIT Press, 1967).

9. My account draws upon Eugen Weber, *The Hollow Years: France in the 1930s* (New York: W. W. Norton and Company, 1994); Philip Charles Farwell Bankwitz, *Maxime Weygand and Civil-Military Relations in Modern France* (Cambridge: Harvard University Press, 1967); and Jean-Baptiste Duroselle, *France and the Nazi Threat: The Collapse of French Diplomacy 1932–1939* (New York: Enigma Books, 2004).

10. See Strachan, *The First World War*, Vol. I, 124–31.

11. Shirer, *The Collapse of the Third Republic*, 137–54.

12. Alistair Horne, *The French Army and Politics, 1870–1970* (New York: Peter Bedrick Books, 1984), 50–51.

13. See the discussion in Kier, *Imagining War*, 56–88.

14. Doughty, *The Seeds of Disaster*, 14–22.

15. Marie de La Gorce, *The French Army: A Military-Political History*, trans. Kenneth Douglas (New York: George Braziller, 1963), 221.

16. The quotations are from Shirer, *The Collapse of the Third Republic*, 200–201.

17. La Gorce, *The French Army*, 237.

18. Andre Maurois, *Why France Fell*, trans. Denver Lindsey (London: Bodley Head, 1941), 155–56.

19. Horne, *The French Army and Politics*, 45.

20. May, *Strange Victory*, 455.

21. The journal entry is quoted in Horne, *The French Army and Politics*, 54.

22. Ronald Chalmers Hood III, "Bitter Victory: French Military Effectiveness during the Second World War," in *Military Effectiveness*, Vol. 3, ed. Allan R. Millet and Williamson Murray (Boston: Allen and Unwin, 1988), 223.

23. Jolyon Howorth and Patricia Chilton, eds., *Defense and Dissent in Contemporary France* (New York: Croom Helm, 1984), 2.

24. Kier, *Imagining War*, 58; and Shelby Cullom Davis, *The French War Machine* (London: Allen and Unwin, 1937), 88–91.

25. Horne, *The French Army and Politics*, 48–49.

26. La Gorce, *The French Army*, 220 and 225–226.

27. Ibid., 196.

28. Bankwitz, *Maxime Weygand and Civil-Military Relations in Modern France*, 151.

29. Kiesling, *Arming against Hitler*, chs.3 and 4; Porch, "Military Culture and the Fall of France in 1940," 170; and Horne, *The French Army and Politics*, 53.

30. Kier, *Imagining War*, 84–85.

31. Saul K. Padover, "France in Defeat: Causes and Consequences," *World Politics* 2, no. 3 (April 1950): 305–7.

32. On how the military disengaged from politics, see Michael Geyer, "The Crisis of Military Leadership in the 1930s," *Journal of Strategic Studies* 14, no. 4 (December 1991): 459; and George A. Kelly, "The French Army Re-Enters Politics: 1940–1945," *Political Science Quarterly* 76, no. 3 (September 1961): 370–71.

33. La Gorce, *The French Army*, 233.

34. Ibid., 250.

35. Doughty, *The Seeds of Disaster*, 12.

36. Ibid., 30 and 33–34.

37. Colonel A. Goutard, *The Battle of France: 1940*, trans. Capt. A. P. R. Burgess (New York: Ives Washburn, 1959), 15; and Bloch, *Strange Defeat*, 48.

38. Noting, in particular, the disdain generals Corap and Huntzinger had for reservists is May, *Strange Victory*, 389.

39. Bloch, *Strange Defeat*, 123–24.

40. Ibid., 107.

41. La Gorce, *The French Army*, 245 and 303–4.

42. Ibid., 244.

43. Horne, *The French Army and Politics*, 51.

44. Bloch, *Strange Defeat*, 122.

45. La Gorce, *The French Army*, 246–51.

46. Weber, *The Hollow Years*, 249–50.

47. Porch, "Military Culture and the Fall of France in 1940," 169.

48. Bloch, *Strange Defeat*, 90, 139–43, 163, and 168. These same struggles hampered arms production in France. See Maurois, *Why France Fell*, 53.

49. Bloch, *Strange Defeat*, 139–42.

50. Horne, *The French Army and Politics*, 56.

51. Alistair Horne, *To Lose a Battle: France 1940*, 2nd ed. (New York: Penguin Books, 1979), 104.

52. Goutard, *The Battle of France*, 82.

53. Ibid., 150.

54. Jackson, *The Fall of France*, 153.

55. Bloch, *Strange Defeat*, 94 and 124.

56. Dorothy S. White, *Seeds of Discord: De Gaulle, Free France and the Allies* (Syracuse: Syracuse University Press, 1964), 39.

57. Weber, *The Hollow Years*, 249.

58. The battles of May and June 1940 remain controversial. I have relied on well-

known works such as Horne, *To Lose a Battle*; and Goutard, *The Battle of France*; as well as more recent treatments such as May, *Strange Victory*; Jackson, *The Fall of France*; and the very detailed discussion of the battles of Sedan in Frieser, *The Blitzkrieg Legend*. Another excellent narrative of events is Robert Allan Doughty, *The Breaking Point: Sedan and the Fall of France, 1940* (Hamden, CT: Archon Books, 1990). See also John Paul Pallud, *Blitzkrieg in the West* (London: Then and Now, 1991).

59. My description of German planning comes from Frieser, *The Blitzkrieg Legend*, 12–98; and John J. Mearsheimer, *Conventional Deterrence* (Ithaca, NY: Cornell University Press, 1983), 112–29.

60. See Posen, *The Sources of Military Doctrine*, 86–94.

61. For a description of the evolution of the Dyle Plan and its Breda variant, see May, *Strange Victory*, 304–22.

62. Len Deighton, *Blitzkrieg From the Rise of Hitler to the Fall of Dunkirk* (New York: Ballantine Books, 1979), 170.

63. My description of the balance of forces reflects the figures given in R. Ernest Dupuy and Trevor N. Dupuy, *The Encyclopedia of Military History* (New York: Harper and Row, 1977), 1057–58; Goutard, *The Battle of France*, 21–31, 37, and 39–40; and Posen, *The Sources of Military Doctrine*, 83.

64. Deighton, *Blitzkrieg*, 190–94.

65. For details of this encounter, see Jeffrey A. Gunsburg, "The Battle of the Belgian Plain, 12–14 May 1940: The First Great Tank Battle," *Journal of Military History* 56 (April 1992): 207–44.

66. Frieser, *The Blitzkrieg Legend*, 100–144; Doughty, *The Breaking Point*, 33–41.

67. On XV Corps's advance to the Meuse, see Kenneth Macksey, *Rommel: Battles and Campaigns* (New York: Da Capo Press, 1997), 28–33.

68. Horne, *To Lose a Battle*, 262.

69. May, *Strange Victory*, 321.

70. Ibid., 424.

71. Ibid., 412.

72. Horne, *To Lose a Battle*, 318.

73. Ibid., 315–19.

74. Ibid.

75. Deighton, *Blitzkrieg*, 214.

76. Some scholars have incorrectly argued that all the units available to contain the German breakout from the Meuse were of poor quality. This overstates the case and gives too much credit to German strategy for the French defeat, a fact that even German officers would later dispute when they described the breakout as nearly unsuccessful. For arguments asserting that the French units were of poor quality, see Desch, "Planning War in Peacetime," 94–104. A more accurate statement is that the units along the Meuse varied in quality, but they remained constant in their lack

of enthusiasm to counterattack. From this point forward, I indicate the type of unit (Regular, A Reserve, or B Reserve) the German forces faced. My coding reflects Horne, *To Lose a Battle*, passim.

77. Horne, *To Lose a Battle*, 320.
78. May, *Strange Victory*, 408.
79. Horne, *To Lose a Battle*, 365.
80. May, *Strange Victory*, 435.
81. Ibid., 413.
82. Horne, *To Lose a Battle*, 365.
83. Frieser, *The Blitzkrieg Legend*, 145–97; and May, *Strange Victory*, 429.
84. Goutard, *The Battle of France*, 131.
85. Horne, *To Lose a Battle*, 286.
86. May, *Strange Victory*, 431.
87. Horne, *To Lose a Battle*, 338–43.
88. Ibid., 344–45.
89. Horne, *To Lose a Battle*, p. 307.
90. Goutard, *The Battle of France*, 129.
91. Horne, *To Lose a Battle*, 338–43.
92. Doughty, *The Breaking Point*, p. 4.
93. Horne, *To Lose a Battle*, 348.
94. Ibid., 369–74.
95. May, *Strange Victory*, 410.
96. Goutard, *The Battle of France*, 128–29.
97. Horne, *To Lose a Battle*, 353.
98. Goutard, *The Battle of France*, 137.
99. May, *Strange Victory*, 432.
100. Williamson Murray and Allan R. Millett, *A War to Be Won: Fighting the Second World War* (Cambridge: Belknap Press, 2000), 74–75; and Goutard, *The Battle of France*, 143–46.
101. Quoted in Horne, *To Lose a Battle*, 411–12.
102. May, *Strange Victory*, 433.
103. Horne, *To Lose a Battle*, 415–17.
104. Murray and Millett, *A War to Be Won*, 76.
105. Frieser, *Blitzkrieg Legend*, 252–64.
106. The quotation is from Jackson, *The Fall of France*, 57–58.
107. Quoted in ibid., 86–87.
108. Goutard, *The Battle of France*, 216.
109. B. H. Liddell Hart, *History of the Second World War* (New York: G. P. Putnam and Sons, 1971), 75–77.
110. These estimates come from Michael Clodfelter, *Warfare and Small Armed*

Conflicts: A Statistical Reference II: *1900–1991* (Jefferson, NC: McFarland and Company, 1992), 796–97.

111. For an argument that Weygand chose this strategy in order to precipitate a French defeat, see Philip C. F. Bankwitz, "Maxime Weygand and the Fall of France: A Study in Civil-Military Relations," *Journal of Modern History* 31, no. 3 (September 1959): 231–32.

112. Alexander, "After Dunkirk," 219–64.

113. Weber, *The Hollow Years*, 273; Jackson, *The Fall of France*, 103–6; Horne, *To Lose a Battle*, 566–78; and Horne, *The French Army and Politics*, 62–67.

114. These concerns began to trouble Reynaud as early as May 24. See Jackson, *Fall of France*, 134–35.

115. Quoted in Liddell Hart, *History of the Second World War*, 85.

116. Eliot A. Cohen and John Gooch, *Military Misfortunes: The Anatomy of Failure in War* (New York: Vintage Books, 1990), 225–26.

117. The story is recounted in Padover, "France in Defeat," 324.

118. Jackson, *The Fall of France*, 134.

119. Ibid., 135.

120. Bloch, *Strange Defeat*, 170.

121. Bankwitz, "Maxime Weygand and the Fall of France," 237–38.

122. Ibid., 236.

123. Liddell Hart, *History of the Second World War*, 85.

124. Bloch, *Strange Defeat*, 134.

125. Jackson, *Fall of France*, 180; and Bloch, *Strange Defeat*, 111.

126. Padover, "France in Defeat," 312; and Horne, *The French Army and Politics*, 63–65.

127. One study concludes that an indicator of the French Army's weakness was the few number of counterattacks it mounted. See Cohen and Gooch, *Military Misfortunes*, 224.

128. Bloch, *Strange Defeat*, 39.

129. Padover, "France in Defeat," 322.

131. "Per Cap Loss Infliction" is a measure of the casualties inflicted on an opponent controlling for force size. Attacker Per Cap Loss Infliction, then, is defender casualties/attacker force size. "Loss-Infliction Ratios" compare the per capita magnitude of casualties one side inflicted on the other. Attacker's Loss Infliction Ratio, then, is attacker per capita loss infliction/defender per capita loss infliction. The casualty data for the table comes from Clodfelter, *Warfare and Small Armed Conflicts*, 489; Jackson, *The Fall of France*, 179–80; Frieser, *The Blitzkrieg Legend*, 317–19; and Alexander, "After Dunkirk," 219–64.

132. Edmund Taylor, "Democracy Demoralized: The French Collapse," *Public Opinion Quarterly* 4, no. 4 (December 1940): 637.

133. Cohen and Gooch, *Military Misfortunes*, 226.

134. Quoted in Jackson, *The Fall of France*, 135. See also Padover, "France in Defeat," 312; and Horne, *The French Army and Politics*, 63–65.

135. Jackson, *The Fall of France*, 142.

136. Maurois, *Why France Fell*, 151–52.

137. Good summaries of the early Third Republic include A. J. Mayeur and M. Rebérioux, *The Third Republic from Its Origins to the Great War, 1871–1914* (Cambridge: Cambridge University Press, 1982); and Eugen Weber, *Peasants into Frenchmen: The Modernization of Rural France, 1870–1914* (Stanford: Stanford University Press, 1977).

138. Roger Price, *A Concise History of France* (Cambridge: Cambridge University Press, 1993), 195–97.

139. Horne, *The French Army and Politics*, 22.

140. Jean-Jacques Becker, *The Great War and the French People* (Lexington Spa: Berg, 1983), 3–6 and 324–25.

141. Strachan, *The First World War*, 129–30.

142. These problems are detailed in Douglas Porch, *The March to the Marne: The French Army 1871–1914* (Cambridge: Cambridge University Press, 1981), esp. 59–61.

143. La Gorce, *The French Army*, 21.

144. Price, *A Concise History of France*, 205–6; and Eugen Weber, *The Nationalist Revival in France, 1905–1914* (Berkeley: University of California Press, 1959), 94–128.

145. La Gorce, *The French Army*, 62 and 81.

146. David G. Herrmann, *The Arming of Europe and the Making of the First World War* (Princeton: Princeton University Press, 1996), 223.

147. Ibid., 175.

148. Strachan, *The First World War*, 226–27; and Herrmann, *The Arming of Europe and the Making of the First World War*, 223.

149. Ralston, *The Army of the Republic*, 373.

150. Gerhard Ritter, *The Schlieffen Plan: Critique of a Myth* (Westport, CT: Greenwood Press, 1979).

151. My summary of the Battle of the Marne comes from Spencer C. Tucker, *The Great War: 1914–1918* (Bloomington: Indiana University Press, 1998), 30–32.

152. Trevor Dupuy and Gay Hammerman, *Stalemate in the Trenches: November, 1914–March, 1918* (New York: Franklin Watts, 1977).

153. These figures come from Ian Ousby, *The Road to Verdun: World War I's Most Momentous Battle and the Folly of Nationalism* (New York: Doubleday, 2002), 6–8.

154. Clodfelter, *Warfare and Small Armed Conflicts*, 783.

155. Jean-Jacques Becker, *The Great War and the French People* (Lexington Spa: Berg, 1983), 3.

156. John Keegan, *The First World War* (New York: Vintage Books, 1998), 329–32.

157. Price, *A Concise History of France*, 213–17.

158. Clodfelter, *Warfare and Small Armed Conflicts*, 727.

159. See Bruce Allen Watson, *When Soldiers Quit: Studies in Military Disintegration* (Westport, CT: Praeger, 1997), 44–66.

160. Clodfelter, *Warfare and Small Armed Conflicts*, 727.

161. Leonard V. Smith, *Between Mutiny and Obedience: The Case of the French Fifth Infantry Division during World War I* (Princeton: Princeton University Press, 1994), 228–29; John Moiser, *The Myth of the Great War: How the Germans Won the Battles and How the Americans Saved the Allies* (New York: Harper Collins, 2001), 305–6; Rod Paschall, *The Defeat of Imperial Germany, 1917–1918* (Chapel Hill, NC: Da Capo Press, 1994), 51; and Tucker, *The Great War*, 131.

162. Clodfelter, *Warfare and Small Armed Conflicts*, 783.

163. My account of the Second Battle of the Marne reflects Peter Simkins, *The First World War*, Vol. 3: *The Western Front 1917–1918* (Oxford: Osprey Publishing, 2002), 62.

164. Porch, *The March to the Marne*, 254.

165. This is the argument made in Kiesling, *Arming against Hitler*, 174–75; and Kier, *Imagining War*, 39–88.

166. Jeffery A. Gunsberg, "The Battle of Gembloux, 14–15 May 1940: The 'Blitzkrieg' Checked," *Journal of Military History* 64, no. 1 (January 2000): 97–140.

167. See Porch, "The Fall of France," 179–80; and Norman Davies, *No Simple Victory: World War II in Europe, 1939–1945* (London: Viking, 2006), 256–57.

168. Kiesling, *Arming against Hitler*, 174.

169. See Porch, "The Marne and After," 363–85; and Desch, "Planning War in Peacetime," 94–104.

170. See Kiesling, *Arming against Hitler*, 10 and 171.

171. Ibid., 62–84.

CHAPTER 5

1. Casualty figures are from David M. Glantz, *Before Stalingrad: Barbarossa, Hitler's Invasion of Russia, 1941*, 2nd ed. (Stroud, Gloucestershire: Tempus Publishing, 2003), 200.

2. The amount of autonomy given the armed forces varied during the history of the Soviet Union. See Dale R. Herspring, *Russian Civil-Military Relations* (Bloomington: University of Indiana Press, 1996), 3–55 and 181–92.

3. Richard Overy, *The Dictators: Hitler's Germany and Stalin's Russia* (New York: W. W. Norton, 2004); Geoffrey Roberts, *Stalin's War: From the World War to the Cold War, 1939–1953* (New Haven: Yale University Press, 2006); Alan Bullock, *Hitler and Stalin: Parallel Lives* (New York: Harper Collins, 1991); and Simon Sebag-Montefiore, *Stalin: The Court of the Red Tsar* (London: Phoenix Press, 2004).

4. On Soviet ideology during the period, see Robert Gellately, *Lenin, Stalin, Hitler: The Age of Social Catastrophe* (New York: Vintage, 2008); and Overy, *The Dictators*, 54–97.

5. See Jeffrey Brook, *Thank You, Comrade Stalin! Soviet Public Culture from Revolu-*

tion to Cold War (Princeton: Princeton University Press, 2000); and Robert Conquest, *Reflections of a Ravaged Century* (New York: W. W. Norton and Company, 2001), 34–56.

6. Geoffrey Hosking, "The Second World War and Russian National Consciousness," *Past and Present* 175, no. 1 (May 2002): 162–87; and Overy, *The Dictators*, 98–131.

7. Catherine Merridale, *Ivan's War: Life and Death in the Red Army, 1939–1945* (New York: Picador, 2006), 30.

8. Alexander Werth, *Russia at War, 1941–1945* (New York: E. P. Dutton and Company, 1964), 249.

9. Richard Overy, *Russia's War: A History of the Soviet War Effort, 1941–1945* (London: Penguin Books, 1997), 21–35.

10. Shelia Kirkpatrick, *Stalin's Peasants: Resistance and Survival in the Russian Village after Collectivization* (Oxford: Oxford University Press, 1994), 80–82; and Overy, *Russia's War*, 23.

11. Robert Conquest, *The Great Terror: A Reassessment* (Oxford: Oxford University Press, 1990), 224.

12. Norman Davies, *No Simple Victory: World War II in Europe, 1939–1945* (New York: Viking, 2006), 176.

13. For an excellent overview of the Soviet military before World War II, see David M. Glantz, *Stumbling Colossus: The Red Army on the Eve of World War* (Lawrence: University of Kansas Press, 2011).

14. Roger R. Reese, *Stalin's Reluctant Soldiers: A Social History of the Red Army, 1925–1941* (Lawrence: University of Kansas Press, 1996), 8.

15. Roger R. Reese, *Red Commanders: A Social History of the Soviet Army Officer Corps, 1918–1991* (Lawrence: University of Kansas Press, 2005), 128.

16. Reese, *Stalin's Reluctant Soldiers*, 206.

17. Mark von Hagen, *Soldiers in the Proletarian Dictatorship: The Red Army and the Soviet Socialist State, 1917–1930* (Ithaca, NY: Cornell University Press, 1993), 82 and 275–90.

18. Boris Gorbachevsky, *Through the Maelstrom: A Red Army Soldier's War on the Eastern Front, 1942–1945*, trans. Stuart Britton (Lawrence: University of Kansas Press, 2008), 13.

19. Hagen, *Soldiers in the Proletarian Dictatorship*, 278.

20. Merridale, *Ivan's War*, 23–48.

21. John Erickson, *The Soviet High Command: A Military-Political History, 1918–1941*, 3rd ed. (London: Frank Cass, 2001), 479.

22. Reese, *Red Commanders*, 114.

23. David M. Glantz, *Colossus Reborn: The Red Army at War, 1941–1943* (Lawrence: University of Kansas Press, 2005), 565.

24. Davies, *No Simple Victory*, 232.

25. Overy, *Russia's War*, 213.

26. Davies, *No Simple Victory*, 232.

27. Reese, *Red Commanders*, 98.

28. Bernd Bonwetsch, "Stalin, the Red Army, and the 'Great Patriotic War,'" in *Nazism and Stalinism: Dictatorships in Comparison*, ed. Ian Kershaw and Moshe Lewin (Cambridge: Cambridge University Press, 1997), 193.

29. Reese, *Stalin's Reluctant Soldiers*, 108.

30. Erickson, *The Soviet High Command*, 554.

31. Ibid.

32. Ibid., 555.

33. Ibid., 555, 557–58, and 567.

34. Reese, *Stalin's Reluctant Soldiers*, 163–64.

35. Merridale, *Ivan's War*, 69.

36. David Stahel, *Kiev 1941: Hitler's Battle for Supremacy in the East* (Cambridge: Cambridge University Press, 2012), 28.

37. Reese, *Red Commanders*, 150.

38. On prewar problems in training and shortages, see Glantz, *Colossus Reborn*, 110–18; and Stahel, *Kiev 1941*, 28–29.

39. Glantz, *Colossus Reborn*, 135.

40. Michael Clodfelter, *Warfare and Armed Conflicts: A Statistical Encyclopedia of Casualty and Other Figures, 1494–2007*, 3rd ed. (Jefferson, NC: McFarland and Company, 2008), 499.

41. David M. Glantz and Jonathan House, *When Titans Clashed: How the Red Army Stopped Hitler* (Lawrence: University of Kansas Press, 1995), 49.

42. Ibid., 52.

43. Merridale, *Ivan's War*, 112.

44. Glantz, *Before Stalingrad*, 35.

45. Christopher Bellamy, *Absolute War: Soviet Russia in the Second World War* (New York: Vintage Books, 2007), 193.

46. Stahel, *Kiev 1941*, 32.

47. Glantz and House, *When Titans Clashed*, 80.

48. Ibid., 58.

49. My overview of events at Smolensk rely on Werth, *Russia at War*, 168–74; Glantz, *Before Stalingrad*, 298–99; Keith E. Bonn, ed., *Slaughterhouse: The Handbook of the Eastern Front* (Bedford, PA: Aberjona Press, 2005), 398; and Glantz and House, *When Titans Clashed*, 59–63, 74, and 293.

50. Glantz and House, *When Titans Clashed*, 59.

51. Ibid.

52. David Glantz, *Barbarossa Derailed: The Battle for Smolensk, 10 July–10 September 1941*, Vol.1: *The German Advance, The Encirclement Battle, and the First and Second Soviet Counteroffensives, 10 July–24 August 1941* (London: Helion and Company, 2010).

53. Clodfelter, *Warfare and Armed Conflicts*, 500.

54. Glantz, *Before Stalingrad*, 92; and Glantz and House, *When Titans Clashed*, 60.

55. Glantz and House, *When Titans Clashed*, 61.

56. Ibid., 74.

57. Casualties for the battle come from Bryan Perrett, *The Battle Book: Crucial Conflicts in History from 1469 BC to the Present* (London: Arms and Armour, 1992), 273; and Glantz and House, *When Titans Clashed*, 293.

58. Quoted in Max Hastings, *Inferno: The World at War, 1939–1945* (New York: Alfred Knopf, 2011), 154.

59. Glantz and House, *When Titans Clashed*, 60.

60. Overy, *Russia's War*, 87.

61. Ibid.

62. Ibid., 61.

63. Overy, *Russia's War*, 90–91.

64. Perrett, *The Battle Book*, 157.

65. Clodfelter, *Warfare and Armed Conflicts*, 501; and Perrett, *The Battle Book*, 157.

66. Overy, *Russia's War*, 90–91. See also Keegan, *The Second World War* (New York: Random House), 193.

67. Glantz and House, *When Titans Clashed*, 75.

68. Ibid., 121.

69. Glantz, *Before Stalingrad*, 119–30.

70. Quoted in Stahel, *Kiev 1941*, 126.

71. Werth, *Russia at War*, 205.

72. Quoted in Stahel, *Kiev 1941*, 224.

73. Glantz, *Before Stalingrad*, 127.

74. Merridale, *Ivan's War*, 118.

75. Reese, *Stalin's Reluctant Soldiers*, 243.

76. Overy, *Russia's War*, 117.

77. Clodfelter, *Warfare and Armed Conflicts*, 482.

78. Ibid.

79. Glantz and House, *When Titans Clashed*, 83.

80. Clodfelter, *Warfare and Armed Conflicts*, 482.

81. Glantz, *Before Stalingrad*, 191.

82. Glantz and House, *When Titans Clashed*, 83.

83. Ibid., 85.

84. Alan Clark, *Barbarossa: The Russian-German Conflict, 1941–1945* (London: Phoenix, 1965), 163–65; and Glantz and House, *When Titans Clashed*, 85.

85. Werth, *Russia at War*, 254–55; and Glantz and House, *When Titans Clashed*, 85.

86. Overy, *Russia's War*, 119.

87. On the weather and logistical problems, see Richard J. Evans, *The Third Reich at War* (London: Penguin, 2008), 206; On wild German optimism, see Werth, *Russia at War*, 245.

88. Evans, *The Third Reich at War*, 207.

89. Overy, *Russia's War*, 119.

90. Evans, *The Third Reich at War*, 211; and Clark, *Barbarossa*, 182.

91. Glantz, *Before Stalingrad*, 191–92.

92. Hastings, *Inferno*, 155.

93. See John Barber, "Popular Reactions in Moscow in the German Invasion of June 22, 1941," in *Operation Barbarossa: The German Attack on the Soviet Union, June 22, 1941*, ed. Joseph L. Wieczynski (Salt Lake City, UT: Charles Schlacks, 1993), 10–11; and Gellately, *Lenin, Stalin, Hitler*, 473.

94. Reese, *Stalin's Reluctant Soldiers*, 308.

95. Glantz, *Colossus Reborn*, 135; and Gellately, *Lenin, Stalin, Hitler*, 475.

96. Reese, *Stalin's Reluctant Soldiers*, 18.

97. Merridale, *Ivan's War*, 91.

98. Ibid., 140–41.

99. Davies, *No Simple Victory*, 238.

100. Glantz, *Colossus Reborn*, 124; Gorbachevsky, *Through the Maelstrom*, 178–79; and Merridale, *Ivan's War*, 81.

101. Merridale, *Ivan's War*, 8l.

102. John Erickson, "Red Army Battlefield Performance 1941–45: The System and the Soldier," in *Time to Kill: The Soldier's Experience of War in the West, 1939–1945*, ed. Paul Addison and Angus Calder (London: Pimlico, 1997).

103. Davies, *No Simple Victory*, 99; and Glantz, *Colossus Reborn*, 124.

104. Gorbachevsky, *Through the Maelstrom*, 98.

105. Merridale, *Ivan's War*, 120–21.

106. Robert W. Thurston, *Life and Terror in Stalin's Russia, 1934–1941* (New Haven: Yale University Press, 1998), 199–226.

107. Glantz, *Colossus Reborn*, 124.

108. Merridale, *Ivan's War*, 103.

109. Glantz, *Colossus Reborn*, 123.

110. On training, see Glantz, *Colossus Reborn*, 14; and Reese, *Stalin's Reluctant Soldiers*, 71.

111. Quoted in Max Hastings, *Armageddon: The Battle for Germany, 1944–1945* (New York: Alfred A. Knopf, 2004), 125.

112. Quoted in Hastings, *Armageddon*, 112.

113. See Reese, *Stalin's Reluctant Soldiers*, 218–19; Merridale, *Ivan's War*, 16 and 134.

CHAPTER 6

1. Although my argument may also apply to the Viet Cong insurgents, my focus in this chapter is on the North Vietnamese Army.

2. Michael Lee Lanning and Dan Cragg, *Inside the VC and the NVA* (New York: Ivy Books, 1992), 45; and Brian Bond, *The Pursuit of Victory: From Napoleon to Saddam Hussein*, 2nd ed. (Oxford: Oxford University Press, 1998), 189.

3. These figures come from the *Statistical Abstract of the United States* (Washington, DC: Government Printing Office, 1994), 297 and 330–33. For details of the American bombing campaigns, see Mark Clodfelter, *The Limits of Air Power: The American Bombing of North Vietnam* (New York: Free Press, 1989); and Robert A. Pape, *Bombing to Win: Air Power and Coercion in War* (Ithaca, NY: Cornell University Press, 1996), 174–210.

4. See Huynh Kim Khanh, *Vietnamese Communism, 1925–1945* (Ithaca, NY: Cornell University Press, 1982).

5. Alexander B. Woodside, *Community and Revolution in Modern Vietnam* (Boston: Houghton Mifflin, 1976).

6. Summarizing the nationalist movements and their success against the French is Stanley Karnow, *Vietnam: A History* (New York: Viking Press, 1983; rev. ed., London: Penguin Books, 1997), 101–38.

7. Lanning and Cragg, *Inside the VC and the NVA*, 23.

8. William Duiker, *Sacred War: Nationalism and Revolution in a Divided Vietnam* (New York: McGraw Hill, 1995), 252.

9. On the strategies used to win support of the population, see David Chanoff and Doan Van Toai, *Portrait of the Enemy* (New York: Random House, 1986).

10. Ibid., 255.

11. On the educational system in North Vietnam, see Harrison Salisbury, *Behind the Lines—Hanoi* (New York: Harper and Row, 1967), 114–20.

12. Konrad Kellen, *Conversations with Enemy Soldiers in Late 1968/Early 1969: A Study of Motivation and Morale*, RM-6131–1-ISA/ARPA (Santa Monica, CA: Rand Corporation, 1970), 1, 8–9, and 39.

13. Konrad Kellen, *A View of the VC: Elements of Cohesion in the Enemy Camp, 1966–1967*, RM-5462–1-ISA-ARPA (Santa Monica, CA: Rand Corporation, 1969), 76.

14. Quoted in Karnow, *Vietnam*, 20.

15. Brian Michael Jenkins, *Why the North Vietnamese Will Keep Fighting*, P-4395–1 (Santa Monica, CA: Rand Corporation, 1972), 1.

16. Ronald H. Spector, *After Tet: The Bloodiest Year in Vietnam* (New York: Vintage Books, 1993), 87–88.

17. The last two quotations come from Lanning and Cragg, *Inside the VC and the NVA*, 2 and 10–11, respectively.

18. Spector, *After Tet*, 9 and 25.

19. Douglas Pike, *PAVN: People's Army of Vietnam* (New York: Da Capo Press, 1991), 146–57.

20. Ibid., 147, 152, and 158.

21. Konrad Kellen, *A Profile of the PAVN Soldier in South Vietnam*, RM-5013–1-ISA/ARPA (Santa Monica, CA: RAND Corporation, June 1966), 17.

22. Quoted in Spector, *After Tet*, 88.

23. Lanning and Cragg, *Inside the VC and the NVA*, 46–49.

24. William Darryl Henderson, *Why the Vietcong Fought: A Study in Motivation and Control in a Modern Army in Combat* (Westport, CT: Greenwood Press, 1979), 40–47.

25. Lanning and Cragg, *Inside the VC and the NVA*, 50.

26. Andres Sweetland, *Rallying Potential among the North Vietnamese Armed Forces*, RM-6375–1-ARPA (Santa Monica, CA: Rand Corporation, 1966), 2.

27. Si Dunn, *The History of the 1st Cavalry Division* (Austin, TX: 1st Cavalry Division Association, 1984), 138–39.

28. My description of the battle is largely from the perspective of American forces because most of the best available sources about the war concern the experiences of the U.S. military.

29. Mark Woodruff, *The Defeat of the Viet Cong and the North Vietnamese Army: 1961–1973* (Arlington, VA: Vandamere Press, 1999), 68–69.

30. Ibid., 70–71.

31. Ibid.

32. Ibid., 71–72.

33. My description of events at Landing Zone X-Ray relies heavily on Lieutenant-General Harold G. Moore, USA (Ret.) and Joseph L. Galloway, *We Were Soldiers Once . . . and Young: Ia Drang—The Battle That Changed the War in Vietnam*, 2nd ed. (New York: Harper Collins, 1993); and John A. Cash, "Fight at Ia Drang, 14–16 November 1965," in *Seven Firefights in Vietnam* (Washington, DC: Office of the Chief of Military History of the United States Army, 1985).

34. Woodruff, *The Defeat of the Viet Cong and the North Vietnamese Army*, 71.

35. Moore and Galloway, *We Were Soldiers Once*, 68.

36. Ibid., 69.

37. Ibid., 71.

38. Woodruff, *The Defeat of the Viet Cong and the North Vietnamese Army*, 74.

39. The quotation and description of Alpha Company's engagement come from Moore and Galloway, *We Were Soldiers Once*, 79.

40. Ibid., 81.

41. Woodruff, *The Defeat of the Viet Cong and the North Vietnamese Army*, 74–75.

42. Ibid., 77.

43. Ibid.

44. Woodruff, *The Defeat of the Viet Cong and the North Vietnamese Army*, 79; and Moore and Galloway, *We Were Soldiers Once*, 7.

45. Woodruff, *The Defeat of the Viet Cong and the North Vietnamese Army*, 79.

46. Ibid.

47. Ibid.

48. Ibid.

49. Moore and Galloway, *We Were Soldiers Once*, 188.

50. Ibid., 189.

51. Woodruff, *The Defeat of the Viet Cong and the North Vietnamese Army*, 79.

52. Ibid., 81.

53. The figures come from Spencer C. Tucker, ed., *The Encyclopedia of the Vietnam War* (Oxford: Oxford University Press, 2000), 187. Andrew F. Krepinevich, Jr., *The Army and Vietnam*, 3rd ed. (Baltimore, MD: Johns Hopkins University Press, 1990), 168, offers smaller overall numbers because he does not include the engagements before and after the American attack at Landing Zone X-Ray.

54. Russell W. Glenn, *Reading Athena's Dance Card: Men against Fire in Vietnam* (Annapolis, MD: U.S. Naval Institute Press, 2000).

55. See Tucker, *The Encyclopedia of the Vietnam War*, 396.

56. Quoted in Duiker, *Sacred War*, 219.

57. Anders Sweetland, "Rallying Potential among the North Vietnamese Armed Forces," RAND Corporation, RM-6375-1-ARPA, December 1970, 25.

58. Jon M. Van Dyke, *North Vietnam's Strategy for Survival* (Palo Alto, CA: Pacific Books, 1972), 25–40.

59. Duiker, *Sacred War*, 199–202.

60. Quoted in Karnow, *Vietnam*, 22.

61. Quoted in Spector, *After Tet*, 91.

62. Kellen, *Conversations with Enemy Soldiers in Late 1968/Early 1969*, 102.

63. My discussion draws heavily from Woodruff, *The Defeat of the Viet Cong and the North Vietnamese Army*, 148–59.

64. Lewis Sorley, *A Better War: The Unexamined Victories and Final Tragedy of America's Last Years in Vietnam* (New York: Harcourt, 1999), 138–39; and Shelby L. Stanton, *The Rise and Fall of an American Army: U.S. Ground Forces in Vietnam, 1965–1973* (Novato, CA: Presidio Press, 1985), 295.

(66. Lanning and Cragg, *Inside the VC and the NVA*, 22.

67. Quoted in Karnow, *Vietnam*, 20–21.

68. Ibid., 23.

CHAPTER 7

1. Harry G. Summers, Jr., *The Vietnam War Almanac*, 2nd ed. (Novato, CA: Presidio Press, 1999), 29–59.

2. On the coercive air campaigns, see Mark Clodfelter, *The Limits of Air Power: The American Bombing of North Vietnam* (New York: Free Press, 1989).

3. George C. Herring, *America's Longest War: The United States and Vietnam, 1950–1975*, 3rd ed. (New York: McGraw Hill, 1996), 159.

4. My description of U.S. performance during this period draws from Gregory A. Daddis, *No Sure Victory: Measuring U.S. Army Effectiveness and Progress in the Vietnam War* (Oxford: Oxford University Press, 2011).

5. See Seymour Martin Lipset and Gary Marks, *It Didn't Happen Here: Why Socialism Failed in the United States* (New York: W. W. Norton and Company, 2000), 266–68.

6. Eliot A. Cohen, *Citizens and Soldiers: The Dilemmas of Military Service* (Ithaca, NY: Cornell University Press, 1985), 136.

7. Michael Walzer, *Obligations: Essays on Disobedience, War and Citizenship* (New York: Simon and Schuster, 1970), 89.

8. Allan R. Millett and Peter Maslowski, *For the Common Defense: A Military History of the United States of America* (New York: Free Press, 1984), 229. In contrast, 620,000 American soldiers died in the Civil War.

9. Mark A. Lorell and Charles Kelley, Jr., *Casualties, Public Opinion and Presidential Policy During the Vietnam War*, R-3060-AF (Santa Monica, CA: Rand Corporation, 1985), 16–30.

10. Paul Johnson, *A History of the American People*, 2nd ed. (New York: Harper Books, 1999), 879.

11. Ibid., 880.

12. Herring, *America's Longest War*, 126.

13. Johnson, *A History of the American People*, 881.

14. Ronald H. Spector, *After Tet: The Bloodiest Year in Vietnam* (New York: Vintage Books, 1993), 28.

15. Herring, *America's Longest War*, 154–55.

16. David Halberstam, *The Best and the Brightest* (New York: Random House, 1972; rev. ed., London: Penguin Books, 1983), 720–21; and Herring, *America's Longest War*, 154–55.

17. James Gavin, *Crisis Now* (New York: Random House, 1968), 48.

18. Spector, *After Tet*, 27.

19. Jonathan Shay, *Achilles in Vietnam: Combat Trauma and the Undoing of Character* (New York: Touchstone, 1995), 6.

20. Morris Janowitz, *The Professional Soldier: A Social and Political Portrait* (New York: Free Press, 1971), 42–43.

21. Richard Holmes, *Acts of War: The Behavior of Men in Battle* (New York: Free Press, 1985), 53.

22. Charles C. Moskos, Jr., *The American Enlisted Man* (New York: Russell Sage Foundation, 1970), 18.

23. Janowitz, *The Professional Soldier*, 43–44.

24. Ibid., 47.

25. Holmes, *Acts of War*, 46.

26. Moskos, *The American Enlisted Man*, 156.

27. Janowitz, *The Professional Soldier*, 61.

28. S. L. A. Marshall, *Men against Fire: The Problem of Battle Command* (New York: William Morrow and Company, 1947; rev. ed., Norman: University of Oklahoma Press, 2000), 22.

29. My description of events during the Tet Offensive relies heavily on Don Oberdorfer, *Tet! The Turning Point in the Vietnam War,* 2nd ed. (New York: Doubleday and Co., 1971; rev. ed., Baltimore, MD: Johns Hopkins University Press, 2001).

30. For a summary of North Vietnamese thinking, see Mark Woodruff, *The Defeat of the Viet Cong and the North Vietnamese Army: 1961–1973* (Arlington, VA: Vandamere Press, 1999), 29–31.

31. Spencer C. Tucker, ed., *The Encyclopedia of the Vietnam War* (Oxford: Oxford University Press, 2000), 396–97.

32. Ibid., 397.

33. On communist coordination problems, see Woodruff, *The Defeat of the Viet Cong and the North Vietnamese Army*, 32–33.

34. John Whiteclay Chambers II, ed., *The Oxford Companion to American Military History* (Oxford: Oxford University Press, 1999), 716–18.

35. For good accounts, see Keith William Nolan, *Battle for Hue: Tet 1968*, rev. ed. (Novato, CA: Presidio, 1996); and Nicholas Warr, *Phase Line Green: The Battle for Hue, 1968* (Annapolis, MD: Naval Institute Press, 1997).

36. Tucker, *The Encyclopedia of the Vietnam War*, 396–97.

37. Oberdorfer, *Tet!*, 197–236.

38. Spector, *After Tet*, 1–4.

39. For a good summary of events at Khe Sanh, see Moyers S. Shore, *The Battle for Khe Sanh* (Washington, DC: History and Museums Division, Headquarters, U.S. Marine Corps, 1969).

40. Tucker, *The Encyclopedia of the Vietnam War*, 205.

41. Ibid., 206.

42. Woodruff, *The Defeat of the Viet Cong and the North Vietnamese Army*, 103–6.

43. Tucker, *The Encyclopedia of the Vietnam War*, 206.

44. Woodruff, *The Defeat of the Viet Cong and the North Vietnamese Army*, 108–9.

45. Chambers, *The Oxford Companion to American Military History*, 365.

46. Quoted in Herring, *America's Longest War*, 197–98.

47. Quoted in B. H. Liddell Hart, *The Decisive Wars of History: A Study of Strategy* (London: G. Bell and Sons, 1929), 131.

48. Ibid., 147.

49. Moskos, *The American Enlisted Man*, 3.

50. Herbert Y. Schandler, *The Unmaking of a President: Lyndon Johnson and Vietnam* (Princeton: Princeton University Press, 1977), 326.

51. Herring, *America's Longest War*, 159.

52. Ibid., 163–65.

53. Spector, *After Tet*, XVI and 55.

54. David W. Levy, *The Debate over Vietnam*, 2nd ed. (Baltimore, MD: Johns Hopkins University Press, 1995), 132.

55. Lorell and Kelley, *Casualties, Public Opinion and Presidential Policy during the Vietnam War*, 16.

56. John E. Mueller, *War, Presidents, and Public Opinion* (New York: John Wiley and Sons, 1973), 23–61.

57. Spector, *After Tet*, 5.

58. Quoted in Robert S. McNamara with Brian VanDeMark, *In Retrospect: The Tragedy and the Lessons of Vietnam* (New York: Times Books, 1995), 310–11.

59. Marilyn B. Young, *The Vietnam Wars: 1945–1990* (New York: Harper Collins, 1991), 225.

60. Spector, *After Tet*, 11.

61. Ibid., 19.

62. Ibid., 5.

63. Stanley Karnow, *Vietnam: A History* (New York: Viking Press, 1983; rev. ed., London: Penguin Books, 1997), 570; and Herring, *America's Longest War*, 197.

64. Spector, *After Tet*, 7.

65. Herring, *America's Longest War*, 197.

66. Ibid., 199.

67. Young, *The Vietnam Wars*, 214.

68. Ibid., 206.

69. Herring, *America's Longest War*, 200.

70. Ibid., 190; see also Melvin Small, *Johnson, Nixon and the Doves* (New Brunswick, NJ: Rutgers University Press, 1988), 226–34.

71. Young, *The Vietnam Wars*, 297–98.

72. Ibid., 197.

73. Spector, *After Tet*, 5.

74. Herring, *America's Longest War*, 191.

75. Levy, *The Debate over Vietnam*, 156; and Howard Zinn, *The Twentieth Century: A People's History* (New York: Harper and Row, 1980), 189.

76. Zinn, *The Twentieth Century*, 189.

77. Herring, *America's Longest War*, 186.

78. Ibid., 187.

79. Spector, *After Tet*, 33 and 35.

(82. Karnow, *Vietnam*, 581.

83. Johnson, *A History of the American People*, 883–84.

84. Young, *The Vietnam Wars*, 204.

85. Spector, *After Tet*, XVII.

86. Zinn, *The Twentieth Century*, 194–95.

87. Herring, *America's Longest War*, 150.

88. Ibid., 199.

89. Spector, *After Tet*, 11.

90. Ibid., 155.

91. Young, *The Vietnam Wars*, 238; and Zinn, *The Twentieth Century*, 192.

92. Levy, *The Debate over Vietnam*, 167.

94. My description relies on Samuel Zaffiri, *Hamburger Hill, May 11–20, 1969* (Novato, CA: Presidio Press, 1988); and Tucker, *The Encyclopedia of the Vietnam War*, 159–60.

95. Shelby L. Stanton, *The Rise and Fall of an American Army: U.S. Ground Forces in Vietnam, 1965–1973* (Novato, CA: Presidio Press, 1985), 299–301.

96. Guenter Lewy, *American in Vietnam* (Oxford: Oxford University Press, 1978), 144.

97. For an optimistic appraisal of American performance in the closing years of Vietnam, see Lewis Sorley, *A Better War: The Unexamined Victories and Final Tragedy of America's Last Years in Vietnam* (London: Harcourt, 1999).

98. Kyle Longley, *Grunts: The American Combat Soldier in Vietnam* (New York: M. E. Sharpe, 2008), 145.

99. Cited in Guenter Lewy, "The American Experience in Vietnam," in *Combat Effectiveness*, ed. Sam C. Sarkesian (London: Sage Publications, 1980), 95.

100. See Spector, *After Tet*, 242–59.

101. Ibid., 37.

102. Quoted in Daddis, *No Sure Victory*, 187–88.

103. The data for desertions, insubordination, and fragging all come from Lewy, "The American Experience in Vietnam," 94–107.

104. Longley, *Grunts*, 137.

105. Daddis, *No Sure Victory*, 189.

106. Robert D. Schulzinger, *A Time for War: The United States and Vietnam, 1941–1975* (Oxford: Oxford University Press, 1997), 279.

107. Longley, *Grunts*, 139.

108. Sorley, *A Better War*, 289.

109. Daddis, *No Sure Victory*, 198–99.

111. Quoted in Zinn, *The Twentieth Century*, 203.

112. Colonel Robert D. Heinl, Jr., "The Collapse of the Armed Forces," *Armed Forces Journal* (June 7, 1971), reprinted in Marvin Gettleman et al., *Vietnam and America: A Documented History* (New York: Grove Press, 1995), 327.

113. Dan Reiter and Allan Stam, *Democracies at War* (Princeton: Princeton University Press, 2002), 64, 164, and 192; see also D. Scott Bennett and Allan C. Stam, "The

Declining Advantages of Democracy: A Combined Model of War Outcomes and Duration," *Journal of Conflict Resolution* 42, no. 3 (1998): 344–66; and Scott Sigmund Gartner and Randolph M. Sieverson, "War Expansion and War Outcome," *Journal of Conflict Resolution* 40 (March 1996): 4–15.

CHAPTER 8

1. David Stevenson, *With Our Backs to the Wall* (Cambridge: Belknap Press of Harvard University Press, 2011); Alexander Watson, *Enduring the Great War: Combat, Morale, and Collapse in the German and British Armies, 1914–1918* (Cambridge: Cambridge University Press, 2008); and Wilhelm Deist, "The Military Collapse of the German Empire: The Reality behind the Stab-in-the-Back Myth," *War in History* 3 no. 2 (1996): 186–207.

2. David Glantz and Jonathan House, *When Titans Clashed: How the Red Army Stopped Hitler* (Lawrence: University of Kansas Press, 1995), 286–89.

3. Dan Reiter and Allan Stam, *Democracies at War* (Princeton: Princeton University Press, 2002), 258–83; and Reiter and Stam, "Democracy and Battlefield Military Effectiveness," *Journal of Conflict Resolution* 42, no. 3 (June 1998): 377–89. See also David A. Lake, "Powerful Pacifists: Democratic States and War," *American Political Science Review* 80, no. 1 (March 1992): 24. For other ways in which democracy contributes to military effectiveness, see Stephen Biddle and Stephen Long, "Democracy and Military Effectiveness: A Closer Look," *Journal of Conflict Resolution* 48, no. 4 (August 2004): 525–46.

4. Edward A. Shils and Morris Janowitz, "Cohesion and Disintegration in the Wehrmacht in World War II," *Public Opinion Quarterly* 12, no. 2 (1948).

5. Michael C. Desch, *Power and Military Effectiveness: The Fallacy of Democratic Triumphalism* (Baltimore, MD: Johns Hopkins University Press, 2008); Dan Reiter, "Nationalism and Military Effectiveness: Post-Meiji Japan," in *Creating Military Power: The Sources of Military Effectiveness*, ed. Risa A. Brooks and Elizabeth A. Stanley (Stanford: Stanford University Press, 2007); Barry R. Posen, "Nationalism, the Mass Army, and Military Power," *International Security* 18, no. 2 (Fall 1993).

6. Bruce Allen Watson, *When Soldiers Quit* (New York: Praeger Books, 1997); and the essays in Trevor N. Dupuy, ed., *Understanding Defeat: How to Recover from Loss in Battle to Gain Victory in War* (New York: Paragon House, 1990).

7. See Stephen Peter Rosen, *Societies and Military Power: India and Its Armies* (Ithaca, NY: Cornell University Press, 1996); Risa A. Brooks, *Shaping Strategy: The Civil-Military Politics of Strategic Assessment* (Princeton: Princeton University Press, 2008); and the essays in Risa A. Brooks and Elizabeth A. Stanley, eds., *Creating Military Power: The Sources of Military Effectiveness* (Stanford: Stanford University Press, 2007). The foundational study on this subject remains Allan R. Millett and Williamson Murray, eds., *Military Effectiveness*, 3 vols. (Boston: Allen and Unwin, 1988).

8. For examples of theorists who see military power in material terms (especially pig

iron!), see John J. Mearsheimer, *The Tragedy of Great Power Politics* (New York: W. W. Norton, 2001), esp. ch. 4; and Charles Glaser and Chaim Kaufman "What Is the Offense-Defense Balance and How Can We Measure It?" *International Security* 22, no. 4 (Spring 1998): 44–82. All three of these scholars, however, recognize the importance of quality and skill when assessing a country's military power.

9. Stephen Biddle, *Military Power: Explaining Victory and Defeat in Modern Battle* (Princeton: Princeton University Press, 2004); Millett and Murray, eds., *Military Effectiveness.*

10. For examples of net assessments, see Barry R. Posen, "Measuring the European Conventional Balance: Coping with Complexity in Threat Assessment," *International Security* 9, no. 3 (Winter 1984/85): 47–88; Eliot A. Cohen, "Toward Better Net Assessment: Rethinking the European Conventional Balance," *International Security* 13, no. 1 (Summer 1988): 50–89; Stephen Peter Rosen, "Net Assessment as an Analytical Concept," in *On Not Confusing Ourselves: Essays on National Security Strategy in Honor of Albert and Roberta Wohlstetter*, ed. Andrew W. Marshall, J. J. Martin, and Henry S. Rowen (Boulder, CO: Westview Press, 1991), 283–301; and Paul Bracken, "Net Assessment: A Practical Guide," *Parameters* 36, no.1 (Spring 2006): 90–100.

11. Williamson Murray and Allan R. Millett, *A War to Be Won: Fighting the Second World War* (Cambridge: Belknap Press of Harvard University Press, 2000), 434–45.

12. Andrew F. Krepinevich, Jr., *The Army and Vietnam*, 3rd ed. (Baltimore, MD: Johns Hopkins University Press, 1990).

13. On the nature of the contemporary battlefield, see Biddle, *Military Power*, 28–51, and *Explaining Victory and Defeat in Modern Battle* (Princeton: Princeton University Press, 2004). For alternative views on what it takes to win on the modern battlefield, see Thomas G. Mahnken, *Technology and the American Way of War since 1945* (New York: Columbia University Press, 2008); and Keith Shimko, *The Iraq Wars and America's Military Revolution* (Cambridge: Cambridge University Press, 2010).

14. For rationalist accounts, see Mancur Olson, Jr., *The Logic of Collective Action* (Cambridge: Harvard University Press, 1965); Michael Hechter, *Principles of Group Solidarity* (Berkeley: University of California Press, 1987); and Russell Hardin, *One for All: The Logic of Group Conflict* (Princeton: Princeton University Press, 1995). For more constructivist approaches, see Alexander Wendt, *Social Theory of International Politics* (Cambridge: Cambridge University Press, 1999); and Robert Sugden, "Thinking as a Team: Towards an Explanation of Non-Selfish Behavior," *Social Philosophy and Policy* 10, no. 1 (Winter 1993): 69–89. Arguing for combining the approaches are James Fearon and Alexander Wendt, "Rationalism vs. Constructivism: A Skeptical View," in *Handbook of International Relations*, ed. Walter Carlsnaes, Thomas Risse, and Beth A. Simmons (London: Sage Publications, 2002), 52–72.

15. Sidney Tarrow, *Power in Movement: Social Movements and Contentious Politics*, 2nd ed. (Cambridge: Cambridge University Press, 1998).

16. For a recent example, see the debate between Bruce Hoffman, "The Myth of Grass-Roots Terrorism: Why Bin Laden Still Matters," *Foreign Affairs* 87, no. 3 (May/June 2008): 133–38; and Marc Sageman, "Does Osama Still Call the Shots? Debating the Containment of al Qaeda's Leadership," *Foreign Affairs* 87, no. 4 (July/August): 163–67.

17. Seth W. Carus, "The Rajneeshees, 1984," in *Toxic Terror: Assessing Terrorist Use of Chemical and Biological Weapons*, ed. Jonathan B. Tucker (Cambridge: MIT Press, 2000), 115–37.

18. Michael R. Gordon and Bernard E. Trainor, *Cobra II: The Inside Story of the Invasion and Occupation of Iraq* (New York: Random House, 2006).

19. Through armored penetrations to the rear of defending forces, or by encircling them, blitzkrieg operations aim to shock enemy units into collapsing. For overviews on the use of maneuver to create shock, see Robert Leonhard, *The Art of Maneuver: Maneuver-Warfare Theory and AirLand Battle* (Novato, CA: Presidio Press, 1994); Johnathan M. House, *Combined Arms Warfare in the Twentieth Century* (Lawrence: University of Kansas Press, 2001), 105–84; and Robert Michael Citino, *The Path to Blitzkrieg: Doctrine and Training in the German Army, 1920–1939* (Boulder, CO: Lynne Rienner Publishers, 1999). In addition to maneuver, aerial and artillery bombardments can also produce shock.

20. Gil Merom, *How Democracies Lose Small Wars: State, Society, and the Failures of France in Algeria, Israel in Lebanon, and the United States in Vietnam* (Cambridge: Cambridge University Press, 2003).

21. Daniel Byman and Jennifer Lind, "Pyongyang's Survival Strategy: Tools of Authoritarian Control in North Korea," *International Security* 35, no. 1 (Summer 2010): 44–74.

22. See the discussion in Steven R. Ward, *Immortal: A Military History of Iran and Its Armed Forces* (Washington, DC: Georgetown University Press, 2009), esp. 301–8.

23. See Jessica Teets, "Let Many Civil Societies Bloom: Regional Ideational Variation in China," *China Quarterly*, 213 (January 2013): 19–38..

24. "Annual Report to Congress: Military Power of the People's Republic of China, 2009," Washington, DC: Department of Defense, www.defense.gov/pubs/pdfs/China_Military_Power_Report_2013.pdf., accessed May 29, 2013.

25. Thomas E. Ricks, *The Gamble: General Petraeus and the American Military Adventure in Iraq* (New York: Penguin Press HC, 2009); and Michael O'Hanlon and Hassina Sherjan, *Toughing It Out in Afghanistan* (Washington, DC: Brookings Institution, 2010).

26. On the hardening of identities during civil wars, see Chaim Kaufman, "Possible and Impossible Solutions to Ethnic Civil Wars," *International Security* 20, no. 4 (Spring 1996): 136–75.

27. Alon Peled, *A Question of Loyalty: Military Manpower Policy in Multiethnic States* (Ithaca, NY: Cornell University Press, 1998).

28. See Rosen, *Societies and Military Power*; and Istvan Deak, *Beyond Nationalism:*

A Social and Political History of the Habsburg Officer Corps, 1848–1918 (Oxford: Oxford University Press, 1990).

29. On the problems facing foreign occupiers, see David M. Edelstein, *Occupational Hazards: Success and Failure in Military Occupation* (Ithaca, NY: Cornell University Press, 2008).

30. Byron Farwell, *Armies of the Raj: From the Mutiny to Independence, 1858–1947* (New York: W. W. Norton and Company, 1990), 329–47.

Bibliography

Akerlof, George A., and Rachel E. Kranton. "Economics and Identity." *Quarterly Journal of Economics* 115, no. 3 (August 2000).

Alexander, Don W. "Repercussions of the Breda Variant." *French Historical Studies* 8, no. 3 (Spring 1974).

Alexander, Martin S. "After Dunkirk: The French Army's Performance against 'Case Red,' 25 May to 25 June 1940." *War in History* 14, no. 2 (2007).

———. "The Fall of France, 1940." *Journal of Strategic Studies* 13, no. 10 (March 1990).

———. *The Republic in Danger: General Maurice Gamelin and the Politics of French Defense, 1933–1940*. Cambridge: Cambridge University Press, 1992.

Ambrose, Stephen. *Band of Brothers: E Company, 506th Regiment, 101st Airborne from Normandy to Hitler's Eagle's Nest*. New York: Simon and Schuster, 2001.

Andreski, Stanislav. *Military Organization and Society*, 2nd ed. Berkeley: University of California Press, 1968.

"Annual Report to Congress: Military Power of the People's Republic of China, 2009." Washington, DC: Department of Defense, www.defense.gov/pubs/pdfs/China_Military_Power_Report_2013.pdf., accessed May 29, 2013.

Avant, Deborah. "From Mercenary to Citizen Armies: Explaining Change in the Practice of War." *International Organization* 54, no. 1 (Winter 2000).

Ballard, Ted. *Rhineland: The US Army Campaigns of World War II*. Washington, DC: U.S. Government Printing Office, 1995.

Bankwitz, Philip Charles Farwell. *Maxime Weygand and Civil-Military Relations in Modern France*. Cambridge, MA: Harvard University Press, 1967.

———. "Maxime Weygand and the Fall of France: A Study in Civil-Military Relations." *Journal of Modern History* 31, no. 3 (September 1959).

Barber, John. "Popular Reactions in Moscow in the German Invasion of June 22, 1941."

In *Operation Barbarossa: The German Attack on the Soviet Union, June 22, 1941*, ed. Joseph L. Wieczynski. Salt Lake City, UT: Charles Schlacks, 1993.

Bartov, Omer. "The Conduct of War: Soldiers and Barbarization of Warfare." *Journal of Modern History* 64 (December 1992).

———. *The Eastern Front, 1941–1945: German Troops and the Barbarization of Warfare*, 2nd ed. New York: Palgrave, 2001.

———. *Hitler's Army: Soldiers, Nazis, and War in the Third Reich*. Oxford: Oxford University Press, 1992.

———. "The Missing Years: German Workers, German Soldiers." In *Nazism and German Society, 1939–1945*, ed. David F. Crew. London: Routledge, 1994.

———. "Soldiers, Nazis, and War in the Third Reich." *Journal of Modern History* 63 (March 1991).

Becker, Jean-Jacques. *The Great War and the French People*. Translated by Arnold Pomerans. New York: St. Martin's Press, 1986.

Beevor, Anthony J. *The Fall of Berlin 1945*. London: Penguin, 2002.

Bell, P. H. H. *The Origins of the Second World War in Europe*, 2nd ed. London: Longman, 1997.

Bellamy, Christopher. *Absolute War: Soviet Russia in the Second World War*. New York: Vintage Books, 2007.

Ben-Shalom, Uzi, Zeev Lehrer, and Eyal Ben-Ari. "Cohesion during Military Operations: A Field Study on Combat Units in the Al-Aqsa Intifada." *Armed Forces and Society* 32, no. 1 (October 2005).

Bender, Jonathan, and Dilip Mookherjee. "Institutional Structure and the Logic of Ongoing Collective Action." *American Political Science Review* 81 (1987).

Bennett, D. Scott, and Allan C. Stam III. "The Declining Advantages of Democracy: A Combined Model of War Outcomes and Duration." *Journal of Conflict Resolution* 42, no. 3 (June 1998).

———. "The Duration of Interstate Wars 1816–1985." *American Political Science Review* 90, no. 2 (June 1996).

Benoist-Mechin, Jacques. *Sixty Days That Shook the World: The Fall of France, 1940*. Translated by Peter Wiles. New York: G. P. Putnam Sons, 1963.

Berinky, Adam. "Assuming the Costs of War: Events, Elites, and American Public Support for Military Conflict." *Journal of Politics* 69, no. 4 (November 2007).

Biddle, Stephen. "The Gulf War Debate *Redux*: Why Skill *and* Technology Are the Right Answer." *International Security* 22, no. 2 (Fall 1997).

———. *Military Power: Explaining Victory and Defeat in Modern Battle*. Princeton: Princeton University Press, 2004.

———. "Victory Misunderstood: What the Gulf War Tells Us about the Future of Conflict." *International Security* 21, no. 2 (Fall 1996).

Biddle, Stephen, and Stephen Long. "Democracy and Military Effectiveness: A Closer Look." *Journal of Conflict Resolution* 48, no. 4 (August 2004).

Biddle, Stephen, and Robert Zirkle. "Technology, Civil-Military Relations, and Warfare in the Developing World." *Journal of Strategic Studies* 19, no. 2 (June 1996).

Blackbourn, David, and Geoff Eley. *The Peculiarities of German History: Bourgeois Society and Politics in Nineteenth-Century Germany*. Oxford: Oxford University Press, 1984.

Bloch, Marc. *Strange Defeat: A Statement of Evidence Written in 1940*. Translated by Gerald Hopkins. New York: W. W. Norton and Company, 1968.

Blumenson, Martin. *Breakout and Pursuit: The United States Army in World War II, European Theater of Operations*. Washington, DC: Office of the Chief of Military History, 1961.

———. *The Duel for France, 1944*. New York: Da Capo Press, 1963.

Bond, Brian. *The Pursuit of Victory: From Napoleon to Saddam Hussein*, 2nd ed. Oxford: Oxford University Press, 1998.

Bonn, Keith E. *Slaughterhouse: The Handbook of the Eastern Front*. Bedford, PA: Aberjona Press, 2005.

Bonwetsch, Bernd. "Stalin, the Red Army, and the 'Great Patriotic War.'" In *Nazism and Stalinism: Dictatorships in Comparison*, ed. Ian Kershaw and Moshe Lewin. Cambridge: Cambridge University Press, 1997.

Boritt, Gabor S., ed. *Why the Confederacy Lost*. Oxford: Oxford University Press, 1992.

Bracher, Karl Dietrich. *The German Dictatorship: The Origins, Structure, and Effects of National Socialism*. Translated by Jean Steinberg. New York: Praeger, 1970.

Bracken, Paul. "Net Assessment: A Practical Guide." *Parameters* 36, no.1 (Spring 2006).

Brook, Jeffrey. *Thank You, Comrade Stalin! Soviet Public Culture from Revolution to Cold War*. Princeton: Princeton University Press, 2000.

Brooks, Risa A. "Civil-Military Relations and Military Effectiveness: Egypt in the 1967 and 1973 Wars." In *Creating Military Power: The Sources of Military Effectiveness*, ed. Risa A. Brooks and Elizabeth A. Stanley. Stanford: Stanford University Press, 2007.

———. "Making Military Might: Why Do States Fail and Succeed? A Review Essay." *International Security* 28, no. 2 (Fall 2003).

———. *Shaping Strategy: The Civil-Military Politics of Strategic Assessment*. Princeton: Princeton University Press, 2008.

Brooks, Risa A., and Elizabeth A. Stanley. *Creating Military Power: The Sources of Military Effectiveness*. Stanford: Stanford University Press, 2007.

Brose, Eric Dorn. *The Kaiser's Army: The Politics of Military Technology in Germany during the Machine Age, 1870–1918*. Oxford: Oxford University Press, 2001.

Bullock, Alan. *Hitler and Stalin: Parallel Lives*. New York: Harper Collins, 1991.

———. *A Study in Tyranny*. Rev. ed. New York: Harper and Row, 1964.

Burleigh, Michael. *The Third Reich: A New History*. New York: Hill and Wang, 2000.

Byman, Daniel, and Jennifer Lind. "Pyongyang's Survival Strategy: Tools of Authoritarian Control in North Korea." *International Security* 35, no. 1 (Summer 2010).

Cairns, John C. "Some Recent Historians and the 'Strange Defeat' of 1940." *Journal of Modern History* 46, no. 1 (March 1974).

Calleo, David. *The German Problem Reconsidered: Germany and the World Order, 1870 to the Present*. Cambridge: Cambridge University Press, 1978.

Carus, Seth W. "The Rajneeshees, 1984." In *Toxic Terror: Assessing Terrorist Use of Chemical and Biological Weapons*, ed. Jonathan B. Tucker. Cambridge: MIT Press, 2000.

Cash, John A. "Fight at Ia Drang, 14–16 November 1965." In *Seven Firefights in Vietnam*. Washington, DC: Office of the Chief of Military History of the United States Army, 1985.

Chambers II, John Whiteclay, ed. *The Oxford Companion to American Military History*. Oxford: Oxford University Press, 1999.

Chanoff, David, and Doan Van Toai. *Portrait of the Enemy*. New York: Random House, 1986.

Chickering, Roger. *Imperial Germany and the Great War: 1914–1918*. Cambridge: Cambridge University Press, 1998.

———. *We Men Who Feel Most German: A Cultural Study of the Pan-German League, 1886–1914*. Boston: Unwin Hyman, 1984.

Chirot, Daniel. *Modern Tyrants: The Power and Prevalence of Evil in Our Age*. New York: Free Press, 1994.

Cirrilo, Roger. *Ardennes-Alsace: The U.S. Army Campaigns of World War II*. Washington, DC: U.S. Government Printing Office, 1994.

Citino, Robert Michael. *The Path to Blitzkrieg: Doctrine and Training in the German Army, 1920–1939*. Boulder, CO: Lynne Rienner Publishers, 1999.

Clark, Alan. *Barbarossa: The Russian-German Conflict, 1941–1945*. London: Phoenix, 1965.

Clausewitz, Carl von. *On War*. Edited and translated by Michael Howard and Peter Paret. Princeton: Princeton University Press, 1976, 1984.

Clodfelter, Michael. *The Limits of Air Power: The American Bombing of North Vietnam*. New York: Free Press, 1989.

———. *Warfare and Armed Conflicts: A Statistical Encyclopedia of Casualty and Other Figures, 1494–2007*, 3rd ed. Jefferson, NC: McFarland and Company, 2008.

———. *Warfare and Small Armed Conflicts: A Statistical Reference* II: *1900–1991*. Jefferson, NC: McFarland and Company, 1992.

Cohen, Eliot A. *Citizens and Soldiers: The Dilemmas of Military Service*. Ithaca, NY: Cornell University Press, 1985.

———. "Toward Better Net Assessment: Rethinking the European Conventional Balance." *International Security* 13, no. 1 (Summer 1988).

Cohen, Eliot A., and John Gooch. *Military Misfortunes: The Anatomy of Failure in War.* Vintage: New York, 1990.

Cohen, Joshua, and Joel Rogers. "Secondary Associations in Democratic Governance." *Politics and Society* 20 (1992).

Coleman, James. *Foundations of Social Theory.* Cambridge: Harvard University Press, 1990.

Conquest, Robert. *Reflections of a Ravaged Century.* New York: W. W. Norton and Company, 2001.

Costa, Dora L., and Matthew E. Kahn. "Civic Engagement and Community Heterogeneity: An Economist's Perspective." *Perspectives on Politics* 1, no. 1 (March 2003).

Craig, Gordon R. *The Politics of the Prussian Army: 1640–1945.* New York: Clarendon Press. Rev. ed., Oxford: Oxford University Press, 1955.

Crenshaw, Martha. "The Causes of Terrorism." *Comparative Politics* 13, no. 4 (July 1981).

Creveld, Martin van. *Fighting Power: German and U.S. Army Performance, 1939–1945.* Westport, CT: Greenwood Press, 1982.

Crew, David F. "General Introduction." In *Nazism and German Society, 1939–1945*, ed. David F. Crew. London: Routledge, 1994.

Cross, Robin. *Fallen Eagle: The Last Days of the Third Reich.* New York: John Wiley and Sons, 1995.

Daddis, Gregory A. *No Sure Victory: Measuring U.S. Army Effectiveness and Progress in the Vietnam War.* Oxford: Oxford University Press, 2011.

Davies, Norman. *No Simple Victory: World War II in Europe, 1939–1945.* London: Viking, 2006.

Davis, Shelby Cullom. *The French War Machine.* London: Allen and Unwin, 1937.

Deak, Istvan. *Beyond Nationalism: A Social and Political History of the Habsburg Officer Corps, 1848–1918.* Oxford: Oxford University Press, 1990.

Deighton, Len. *Blitzkrieg From the Rise of Hitler to the Fall of* Dunkirk. New York: Ballantine Books, 1979.

Deist, Wilhelm. "The Military Collapse of the German Empire: The Reality behind the Stab-in-the-Back Myth." *War in History* 3, no. 2 (April 1996).

Demeter, Karl. *The German Officer-Corps in Society and State, 1650–1945.* New York: Praeger, 1965.

Desch, Michael C. "Democracy and Victory." *International Security* 27, no. 2 (Fall 2002).

———. "Planning War in Peacetime." *Joint Forces Quarterly* 30 (Spring 2002).

———. *Power and Military Effectiveness: The Fallacy of Democratic Triumphalism.* Baltimore, MD: Johns Hopkins University Press, 2008.

Doughty, Robert Allan. *The Breaking Point: Sedan and the Fall of France, 1940.* Hamden, CT: Archon Books, 1990.

———. *The Seeds of Disaster: The Development of French Army Doctrine, 1919–1939.* Hamden, CT: Archon, 1985.

Downes, Alexander. "How Smart and Tough Are Democracies? Reassessing Theories of Democratic Victory in War." *International Security* 33, no. 4 (Spring 2009).

Duiker, William. *Sacred War: Nationalism and Revolution in a Divided Vietnam*. New York: McGraw Hill, 1995.

Dunn, Si. *The History of the 1st Cavalry Division*. Austin, TX: 1st Cavalry Division Association, 1984.

Dupuy, Trevor N. *Understanding Defeat: How to Recover from Loss in Battle to Gain Victory in War*. New York: Paragon House, 1990.

Dupuy, Trevor N., David Bongard, and Richard Anderson. *Hitler's Last Gamble: The Battle of the Bulge, December 1944–January 1945*. New York: Harper Collins, 1994.

Dupuy, R. Ernest, and Trevor N. Dupuy. *The Encyclopedia of Military History*. New York: Harper and Row, 1977.

Dupuy, Trevor N., and Gay Hammerman. *Stalemate in the Trenches: November, 1914–March, 1918*. New York: Franklin Watts, 1977.

Duroselle, Jean-Baptiste. *France and the Nazi Threat: The Collapse of French Diplomacy 1932–1939*. New York: Enigma Books, 2004.

Dyke, Jon M. Van. *North Vietnam's Strategy for Survival*. Palo Alto, CA: Pacific Books, 1972.

Eckstein, Harry H. "Case Study and Theory in Political Science." In *Handbook of Political Science 7: Strategies of Inquiry*, ed. Fred J. Greestein and Nelson W. Polsby. Reading, MA: Addison-Wesley, 1975.

Edelstein, David M. *Occupational Hazards: Success and Failure in Military Occupation*. Ithaca, NY: Cornell University Press, 2008.

Eggenberger, David. *An Encyclopedia of Battles*, 2nd ed. New York: Dover Publications, 1985.

Eley, Geoff. "The Wilhelmine Right." In *Society and Politics in Wilhelmine Germany*, ed. Richard J. Evans. London: Croom Helm Ltd., 1978.

Ellis, John, and Michael Cox. *The World War I Databook*. London: Aurum Press, 2001.

———. *World War II: A Statistical Survey*. New York: Facts on File, 1995.

Englander, David. "Mutinies and Military Morale." In *The Oxford Illustrated History of the First World War*, ed. Hew Strachan. Oxford: Oxford University Press, 1998.

Erickson, John. "Red Army Battlefield Performance 1941–45: The System and the Soldier." In *Time to Kill: The Soldier's Experience of War in the West, 1939–1945*, ed. Paul Addison and Angus Calder. London: Pimlico, 1997.

———. *The Soviet High Command: A Military-Political History, 1918–1941*, 3rd ed. London: Frank Cass, 2001.

Essame, H. *The Battle for Germany*. New York: Ace Books, 1969.

Evans, Richard J. *The Third Reich at War*. London: Penguin, 2008.

———. *The Third Reich in Power*. London: Penguin, 2005.

Evera, Stephen Van. *Guide to Methods for Students of Political Science*. Ithaca, NY: Cornell University Press, 1997.

Farrell, Theo. "Constructivist Security Studies: Portrait of a Research Program." *International Studies Review* 4, no. 1 (Spring 2002).

Farwell, Byron. *Armies of the Raj: From the Mutiny to Independence, 1858–1947*. New York: W. W. Norton and Company, 1990.

Fearon, James, and Alexander Wendt. "Rationalism vs. Constructivism: A Skeptical View." In *Handbook of International Relations*, ed. Walter Carlsnaes, Thomas Risse, and Beth A. Simmons. London: Sage Publications, 2002.

Fennell, David. *Combat and Morale in the North African Campaign: The Eighth Army and the Path to El Alamein*. Cambridge: Cambridge University Press, 2010.

Finkel, Meir. *On Flexibility: Recovery from Technological and Doctrinal Surprise on the Battlefield*. Translated by Moshe Tiamin. Stanford: Stanford University Press, 2011.

Fischer, Klaus P. *Nazi Germany: A New History*. New York: Continuum, 1995.

Forester, Jurgen. "The Dynamics of Volksgemeinschaft: The Effectiveness of the German Military Establishment in the Second World War." In *Military Effectiveness* 3: *World War II*, ed. Allan R. Millet and Williamson Murray. Boston: Allen and Unwin, 1988.

Förster, Jürgen E. "Ludendorff and Hitler in Perspective: The Battle for the German Soldier's Mind, 1917–1944." *War in History* 10, no. 3 (2003).

Frank, Richard B. *Downfall: The End of the Japanese Empire*. London: Penguin Books, 1999.

Friedman, Milton. *Capitalism and Freedom*. Chicago: University of Chicago Press, 1962.

Frieser, Karl-Heinz. *The Blitzkrieg Legend: The 1940 Campaign in the West*. Annapolis, MD: Naval Institute Press, 2005.

Fritz, Stephen G. *Frontsoldaten: The German Soldier in World War II*. Lexington: University of Kentucky Press, 1995.

———. *Ostkrieg: Hitler's War of Extermination in the East*. Lexington: University of Kentucky Press, 2011.

———. "'We Are Trying . . . to Change the Face of the World'—Ideology and Motivation in the Wehrmacht on the Eastern Front: The View from Below." *Journal of Military History* 60 (October 1996).

Frothingham, Arthur L. *Handbook of War: Facts and Peace Problems*. New York: Committee on Organized Education National Security League, 1919, ch. 9.

Gartner, Scott Sigmund. "Opening up the Black Box of War." *Journal of Conflict Resolution* 42, no. 3 (June 1998).

Gartner, Scott Sigmund, and Randolph M. Sieverson. "War Expansion and War Outcome." *Journal of Conflict Resolution* 40 (March 1996).

Gavin, James. *Crisis Now*. New York: Random House, 1968.

Gellately, Robert. *Backing Hitler: Consent and Coercion in Nazi Germany*. Oxford: Oxford University Press, 2001.

———. *Lenin, Stalin, Hitler: The Age of Social Catastrophe*. New York: Vintage 2008.

Gellner, Ernest. *Nations and Nationalism*. Ithaca, NY: Cornell University Press, 1983.

George, Alexander L., and Timothy J. McKewon. "Case Studies and Theory Development: The Method of Structured Focused Comparison." In *Diplomacy: New Approaches in Theory, History, and Policy*, ed. Gordon Lauren. New York: Free Press, 1979.

Geyer, Michael. "The Crisis of Military Leadership in the 1930s." *Journal of Strategic Studies* 14, no. 4 (December 1991).

Glantz, David M. *Barbarossa Derailed: The Battle for Smolensk, 10 July–10 September 1941, Volume 1: The German Advance, The Encirclement Battle, and the First and Second Soviet Counteroffensives, 10 July–24 August 1941*. London: Helion and Company, 2010.

———. *Before Stalingrad: Barbarossa, Hitler's Invasion of Russia, 1941*, 2nd ed. Stroud, Gloucestershire: Tempus Publishing, 2003.

———. *Colossus Reborn: The Red Army at War, 1941–1943*. Lawrence: University of Kansas Press, 2005.

———. *The Failures of Historiography: Forgotten Battles of the German-Soviet War. 1941–1945*. Fort Leavenworth, KS: U.S. Army, Foreign Military Studies Office, 1997.

———. "The Soviet-German War 1941–1945: Myths and Realities: A Survey Essay." Paper presented as the 20th Anniversary Distinguished Lecture at the Strom Thurmond Institute of Government and Public Affairs, Clemson University, South Carolina, October 11, 2001.

———. *Stumbling Colossus: The Red Army on the Eve of World War*. Lawrence: University of Kansas Press, 2011.

Glantz, David, and Jonathan House. *When Titans Clashed: How the Red Army Stopped Hitler*. Lawrence: University of Kansas Press, 1995.

Glaser, Charles, and Chaim Kaufman. "What Is the Offense-Defense Balance and How Can We Measure It?" *International Security* 22, no. 4 (Spring 1998).

Glenn, Russell W. *Reading Athena's Dance Card: Men against Fire in Vietnam*. Annapolis, MD: U.S. Naval Institute Press, 2000.

Goemans, Hein. *War and Punishment: The Causes of War Termination and the First World War*. Princeton: Princeton University Press, 2000.

Gorbachevsky, Boris. *Through the Maelstrom: A Red Army Soldier's War on the Eastern Front, 1942–1945*. Translated by Stuart Britton. Lawrence: University of Kansas Press, 2008.

Gordon, Michael R., and Bernard E. Trainor. *Cobra II: The Inside Story of the Invasion and Occupation of Iraq*. New York: Random House, 2006.

Goutard, Colonel A. *The Battle of France: 1940*. Translated by Capt. A. P. R. Burgess. New York: Ives Washburn, 1959.

Gray, John. "Totalitarianism, Reform, and Civil Society." In *Totalitarianism at the Crossroads*, ed. Ellen Frankel Paul. New Brunswick, NJ: Transaction Books, 1990.

Gray, Randal, and Christopher Argyle. *Chronicle of the First World War, Volume I: 1914–1916*. New York: Facts on File, 1990.

Greenfeld, Liah. *Nationalism: Five Roads to Modernity*. Cambridge: Harvard University Press, 1992.

Grey, A. M. *Warfighting*. New York: Doubleday, 1994.

Grossman, Lieutenant-Colonel Dave. *On Killing: The Psychological Cost of Learning to Kill in War and Society*. Boston: Bay Back Books, 1995.

Grunberger, Richard. *The 12-Year Reich: A Social History of Nazi Germany, 1933–1945*. New York: Holt, Reinhardt, and Winston, 1971.

Gudmundsson, Bruce I. *Stormtroop Tactics: Innovation in the German Army, 1914–1918*. London: Praeger, 1989.

Gunsberg, Jeffery A. "The Battle of Gembloux, 14–15 May 1940: The 'Blitzkrieg' Checked." *Journal of Military History* 64, no. 1 (January 2000).

———. "The Battle of the Belgian Plain, 12–14 May 1940: The First Great Tank Battle." *Journal of Military History* 56 (April 1992).

———. *Divided and Conquered: The French High Command and the Defeat of the West, 1940*. Westport, CT: Greenwood Press. 1980.

Gurfein, M. I., and Morris Janowitz. "Trends in Wehrmacht Morale." *Public Opinion Quarterly* 10, no. 1 (Spring 1946).

Hagen, Mark von. *Soldiers in the Proletarian Dictatorship: The Red Army and the Soviet Socialist State, 1917–1930*. Ithaca, NY: Cornell University Press, 1993.

Halberstam, David. *The Best and the Brightest*. New York: Random House, 1972. Rev. ed., London: Penguin Books, 1983.

Hanlon, Michael O', and Hassina Sherjan. *Toughing It Out in Afghanistan*. Washington, DC: Brookings Institution, 2010.

Hardin, Russell. *Collective Action*. Washington, DC: Johns Hopkins University Press/Resources for the Future, 1982.

———. *One for All: The Logic of Group Conflict*. Princeton: Princeton University Press, 1995.

Hart, B. H. Liddell. *The Decisive Wars of History: A Study of Strategy*. London: G. Bell and Sons, 1929.

———. *History of the Second World War*. New York: G. P. Putnam and Sons, 1971.

———. *The Real War: 1914–1918*, 2nd ed. Boston: Little, Brown and Company, 1964.

Hart, S., R. Hart, and M. Hughes. *The German Soldier in World War II*. Osceola, WI: Amber Books, 2000.

Hastings, Max. *Armageddon: The Battle for Germany, 1944–1945*. New York: Alfred A. Knopf, 2004.

———. *Inferno: The World at War, 1939–1945*. New York: Alfred Knopf, 2011.

Hechter, Michael. *Principles of Group Solidarity*. Berkeley: University of California Press, 1987.

Heinl, Jr., Colonel Robert D. "The Collapse of the Armed Forces." *Armed Forces Journal* (June 7, 1971), reprinted in Marvin Gettleman et al., *Vietnam and America: A Documented History*. New York: Grove Press, 1995.

Henderson, William Darryl. *Cohesion: The Human Element in Combat*. Washington, DC: National Defense University, 1985.

———. *Why the Vietcong Fought: A Study in Motivation and Control in a Modern Army in Combat*. Westport, CT: Greenwood Press, 1979.

Herring, George C. *American's Longest War: The United States and Vietnam, 1950–1975*, 3rd ed. New York: McGraw Hill, 1996.

Herrmann, David G. *The Arming of Europe and the Making of the First World War*. Princeton: Princeton University Press, 1996.

Herspring, Dale R. *Russian Civil-Military Relations*. Bloomington: University of Indiana Press, 1996.

Hirschman, Albert O. *Exit, Voice, and Loyalty: Responses to Decline in Firms, Organizations, and States*. Cambridge: Harvard University Press, 1970.

Hoffman, Bruce. "The Myth of Grass-Roots Terrorism: Why Bin Laden Still Matters." *Foreign Affairs* 87, no. 3 (May/June 2008).

Hogg, Michael A., and Dominic Abrams. *Social Identifications: A Social Psychology of Inter-group Relations and Group Processes*. London: Routledge, 1998.

Holmes, Richard. *Acts of War: The Behavior of Men in Battle*. New York: Free Press, 1985.

Hood III, Ronald Chalmers. "Bitter Victory: French Military Effectiveness during the Second World War." In *Military Effectiveness, Volume 3*, ed. Allan R. Millet and Williamson Murray. Boston: Allen and Unwin, 1988.

Horne, Alistair. *The French Army and Politics, 1870–1970*. New York: Peter Bedrick Books, 1984.

———. *To Lose a Battle: France 1940*, 2nd ed. New York: Penguin Books, 1979.

Hosking, Geoffrey. "The Second World War and Russian National Consciousness." *Past and Present* 175, no. 1 (May 2002).

House, Jonathan M. *Combined Arms Warfare in the Twentieth Century*. Lawrence: University of Kansas Press, 2001.

Howard, Michael. *War in European History*. Oxford: Oxford University Press, 1976.

Howorth, Jolyon, and Patricia Chilton, eds. *Defense and Dissent in Contemporary France*. New York: Croom Helm, 1984.

Hoyt, Timothy D. "Social Structure, Ethnicity, and Military Effectiveness: Iraq, 1980–2004." In *Creating Military Power: The Sources of Military Effectiveness*, ed. Risa A. Brooks and Elizabeth A. Stanley. Stanford: Stanford University Press, 2007.

Huntington, Samuel P. *The Soldier and the State: The Theory and Politics of Civil-Military Relations*. Cambridge: Belknap Press of Harvard University Press, 1957.

Jackson, Julian. *The Fall of France: The Nazi Invasion of 1940*. Oxford: Oxford University Press, 2003.

Jackson, Peter. "Recent Journeys along the Road Back to France, 1940." *Historical Journal* 39, no. 2 (1996).

Janowitz, Morris. *The Professional Soldier: A Social and Political Portrait*. New York: Free Press, 1971.

Jenkins, Brian Michael. *Why the North Vietnamese Will Keep Fighting*. P-4395–1. Santa Monica, CA: Rand Corporation, 1972.

Johnson, Eric A. *Nazi Terror: The Gestapo, Jews and Ordinary Germans*. New York: Basic Books, 2000.

Jordan, Jenna. "When Heads Roll: Assessing the Effectiveness of Leadership Decapitation." *Security Studies* 18, no. 4 (2009).

Kaiser, David E. "Germany and the Origins of the First World War." *Journal of Modern History* 55 (1983).

Karnow, Stanley. *Vietnam: A History*. New York: Viking Press, 1983. Rev. ed., London: Penguin Books, 1997.

Katzenstein, Peter J. *The Culture of National Security: Norms and Identity in World Politics*. New York: Columbia University Press, 1996.

Kaufmann, Chaim. "Possible and Impossible Solutions to Ethnic Civil Wars." *International Security* 20, no. 4 (Spring 1996).

Keegan, John. *The Face of Battle*. New York: Viking, 1986.

———. *The First World War*. New York: Vintage Books, 1998.

Kellen, Konrad. *Conversations with Enemy Soldiers in Late 1968/Early 1969: A Study of Motivation and Morale*. RM-6131–1-ISA/ARPA. Santa Monica, CA: Rand Corporation, 1970.

———. *A Profile of the PAVN Soldier in South Vietnam*. RM-5013–1-ISA/ARPA. Santa Monica, CA: Rand Corporation, June 1966.

———. *A View of the VC: Elements of Cohesion in the Enemy Camp, 1966–1967*. RM-5462–1-ISA-ARPA. Santa Monica, CA: Rand Corporation, 1969.

Kelly, George A. "The French Army Re-Enters Politics: 1940–1945." *Political Science Quarterly* 76, no. 3 (September 1961).

Kershaw, Ian. *The End: The Defiance and Destruction of Hitler's Germany, 1944–1945*. London: Penguin Press, 2011.

———. *Hitler: Hubris, 1889–1936*. New York: W. W. Norton and Company, 2000.

———. *Hitler: Profile in Power*. London: Longman: 1991.

———. *The Nazi Dictatorship: Problems and Perspectives of Interpretation*. (London: Edward Arnold, 1985.

Kershaw, Robert. *War without Garlands: Operation Barbarossa 1941–1942*. Surrey: Ian Allan Publishing, 2009.

Khanh, Huynh Kim. *Vietnamese Communism, 1925–1945*. Ithaca, NY: Cornell University Press, 1982.

Kier, Elizabeth. "Homosexuals in the U.S. Military: Open Integration and Combat Effectiveness." *International Security* 23 (Fall 1998).

———. *Imagining War: French and British Military Doctrine between the Wars*. Princeton: Princeton University Press, 1997.

Kiesling, Eugenia C. *Arming against Hitler: France and the Limits of Military Planning*. Lawrence: University of Kansas Press, 1996.

King, Anthony. "The Word of Command: Communication in the Military." *Armed Forces and Society* 32, no. 4 (July 2006).

Kirkpatrick, Shelia. *Stalin's Peasants: Resistance and Survival in the Russian Village after Collectivization*. Oxford: Oxford University Press, 1994.

Kitchen, Martin. *The German Officer Corps, 1890–1914*. Oxford: Oxford University Press, 1968.

Knox, MacGregor. *Common Destiny: Dictatorship, Foreign Policy, and War in Fascist Italy and Nazi Germany*. Cambridge: Cambridge University Press, 2000.

———. "Conquest, Foreign and Domestic, in Fascist Italy and Nazi Germany." *Journal of Modern History* 56, no. 1 (March 1984).

Krepinevich, Jr., Andrew F. *The Army and Vietnam*, 3rd ed. Baltimore, MD: Johns Hopkins University Press, 1990.

La Gorce, Marie de. *The French Army: A Military-Political History*. Translated by Kenneth Douglas. New York: George Braziller, 1963.

Lake, David A. "Powerful Pacifists: Democratic States and War." *American Political Science Review* 80, no. 1 (March 1992).

Lanning, Michael Lee, and Dan Cragg. *Inside the VC and the NVA*. New York: Ivy Books, 1992.

Layne, Christopher. *The Peace of Illusions: American Grand Strategy from 1940 to the Present*. Ithaca, NY: Cornell University Press, 2006.

Legro, Jeffrey W. "Culture and Preferences in the International Cooperation Two-Step." *American Political Science Review* 90, no. 1 (March 1996).

———. "Military Culture and Inadvertent Escalation in World War II." *International Security* 18, no. 4 (Spring 1994).

Leonhard, Robert. *The Art of Maneuver: Maneuver-Warfare Theory and AirLand Battle*. Novato, CA: Presidio Press, 1994.

Levi, Margaret. *Consent, Dissent, and Patriotism*. Cambridge: Cambridge University Press, 1997.

Levy, David W. *The Debate over Vietnam*, 2nd ed. Baltimore, MD: Johns Hopkins University Press, 1995.

Lewy, Guenter. "The American Experience in Vietnam." In *Combat Effectiveness*, ed. Sam C. Sarkesian. London: Sage Publications, 1980.

———. *American in Vietnam*. Oxford: Oxford University Press, 1978.

Lijphart, Arend. "Comparative Politics and the Comparative Method." *American Political Science Review* 65, no. 4 (September 1971).

Linz, Juan. "Political Space and Fascism as a Late-Comer." In *Who Were the Fascists?*, ed. Stein Larsen et al. Oslo-Bergen: Universitetsforlaget, 1980.

Lipset, Seymour Martin, and Gary Marks. *It Didn't Happen Here: Why Socialism Failed in the United States*. New York: W. W. Norton and Company, 2000.

Little, Roger W., and Morris Janowitz. "Buddy Relations and Combat Performance." In *The New Military: Changing Patterns of Organization*, ed. Morris Janowitz. New York: Russell Sage Foundation, 1964.

Longley, Kyle. *Grunts: The American Combat Soldier in Vietnam*. New York: M. E. Sharpe, 2008.

Lorell, Mark, and Charles Kelley, Jr. *Casualties, Public Opinion and Presidential Policy during the Vietnam War*. R-3060-AF. Santa Monica, CA: Rand Corporation, 1985.

Luckas, John. *The History of Hitler*. New York: Vintage Books, 1997.

Lynn, John A. *The Bayonets of the Republic: Motivation and Tactics in the Army of Revolutionary France, 1791–1794*, 2nd. ed. Boulder, CO: Westview Press, 1996.

MacCoun, Robert J. "Sexual Orientation and Military Cohesion: A Critical Review of the Evidence." In *Out in Force: Sexual Orientation and the Military*, ed. G. Herek, J. Jobe, and R. Carney. Chicago: University of Chicago Press, 1996.

MacDonald, Charles B. *The Might Endeavor: American Armed Forces in the European Theater in World War II*. Oxford: Oxford University Press, 1969.

———. *The Siegfried Line Campaign*. Washington, DC: Army Center for Military History, 1990.

Mack, Andrew J. R. "Why Big Nations Lose Small Wars: The Politics of Asymmetric Conflict." In *Power, Strategy, and Security*, ed. Klaus Knorr. Princeton: Princeton University Press, 1983.

Macksey, Kenneth. *Rommel: Battles and Campaigns*. New York: Da Capo Press, 1997.

Mahnken, Thomas G. *Technology and the American Way of War since 1945*. New York: Columbia University Press, 2008.

Mahnken, Thomas G., and Barry D. Watts. "What the Gulf War Can (and Cannot) Tell Us about the Future of Warfare." *International Security* 22, no. 2 (Fall 1997).

Makiya, Kanan. *The Republic of Fear: The Politics of Modern Iraq*, 2nd ed. Berkeley: University of California Press, 1998.

Mansoor, Peter R. *The GI Offensive in Europe: The Triumph of American Infantry Divisions, 1941–1945*. Lawrence: University of Kansas Press, 1999.

Marshall, Andrew W., J. J. Martin, and Henry S. Rowen, eds. *On Not Confusing Ourselves: Essays in Honor of Albert and Roberta Wohlstetter*, Boulder, CO: Westview Press, 1991.

Marshall, S. L. A. *Men against Fire: The Problem of Battle Command*. New York: William Morrow and Company, 1947. Rev. ed., Norman: University of Oklahoma Press, 2000.

Marx, Anthony. *Faith in Nation: Exclusionary Origins of Nationalism*. Oxford: Oxford University Press, 2003.

Maslow, A. H. *Motivation and Personality*. New York: Harper, 1954.

Mason, Timothy. *Social Policy in the Third Reich: The Working Class and the "National Community," 1918–1939*. Oxford: Berg Publishers, 1993.

Maurois, Andre. *Why France Fell*. Translated by Denver Lindsey. London: Bodley Head, 1941.

May, Ernest. *Strange Victory: Hitler's Conquest of France*. New York: Hill and Wang, 2000.

Mayeur, A. J., and M. Reberioux. *The Third Republic from Its Origins to the Great War, 1871–1914*. Cambridge: Cambridge University Press, 1982.

McNamara, Robert S., with Brian VanDeMark. *In Retrospect: The Tragedy and the Lessons of Vietnam*. New York: Times Books, 1995.

McNeil, William H. *Keeping Together in Time: Dance and Drill in Human History*. Cambridge: Harvard University Press, 1995.

McPherson, James. *What They Fought For, 1861–1865*. New York: Anchor Books, 1994.

Mearsheimer, John J. "Assessing the Conventional Balance: The 3:1 Rule and Its Critics." *International Security* 13, no. 4 (Spring 1989).

———. *Conventional Deterrence*. Ithaca, NY: Cornell University Press, 1983.

———. *The Tragedy of Great Power Politics*. New York: W. W. Norton, 2001.

Merom, Gil. *How Democracies Lose Small Wars: State, Society, and the Failures of France in Algeria, Israel in Lebanon, and the United States in Vietnam*. Cambridge: Cambridge University Press, 2003.

Merridale, Catherine. *Ivan's War: Life and Death in the Red Army, 1939–1945*. New York: Picador, 2006.

Millett, Allan R., and Peter Maslowski. *For the Common Defense: A Military History of the United States of America*. New York: Free Press, 1984.

Millett, Allan R., and Williamson Murray, eds. *Military Effectiveness, Volumes 1–3*. Boston: Allen and Unwin, 1988.

Mitchell, Reid. *Civil War Soldiers*. London: Penguin Books, 1988.

Mitchum, Jr., Samuel W. *Retreat to the Reich: The German Defeat in France, 1944*. Westport, CT: Praeger, 2000.

Moiser, John. *The Myth of the Great War: How the Germans Won the Battles and How the Americans Saved the Allies*. New York: Harper Collins, 2001.

Moore, Lieutenant-General Harold G., and Joseph L. Galloway. *We Were Soldiers Once . . . and Young: Ia Drang—The Battle That Changed the War in Vietnam*, 2nd ed. New York: Harper Collins, 1993.

Moskos, Jr., Charles C. *The American Enlisted Man: The Rank and File in Today's Military*. New York: Russell Sage Foundation, 1970.

Mueller, John E. *War, Presidents, and Public Opinion*. New York: John Wiley and Sons, 1973.

Murray, Williamson, and Allan Millett. *Calculations: Net Assessments and the Coming of World War II*. New York: Free Press, 1992.

———. *A War to Be Won: Fighting the Second World War*. Cambridge: Belknap Press of Harvard University Press, 2000.

Nolan, Keith William. *Battle for Hue: Tet 1968*. Rev. ed. Novato, CA: Presidio, 1996.

Nordland, Rod. "Troop 'Surge' in Afghanistan Ends with Mixed Results." *New York Times*, September 21, 2012.

North, Douglas C. *Structure and Change in Economic History*. New York: W. W. Norton and Company, 1981.

O'Neill, Robert J. *The Germany Army and the Nazi Party, 1933–1939*. New York: James H. Heineman, 1966.

Oberdorfer, Don. *Tet! The Turning Point in the Vietnam War*. New York: Doubleday and Co., 1971. 2nd. ed., Baltimore, MD: Johns Hopkins University Press, 2001.

Olson, Mancur. *The Logic of Collective Action: Public Goods and the Theory of Groups*. Cambridge: Harvard University Press, 1971.

Oren, Ido. "The Subjectivity of the 'Democratic' Peace: Changing U.S. Perceptions of Imperial Germany." *International Security* 20, no. 2 (Fall 1995).

Ousby, Ian. *The Road to Verdun: World War I's Most Momentous Battle and the Folly of Nationalism*. New York: Doubleday, 2002.

Overy, Richard. *The Dictators: Hitler's Germany and Stalin's Russia*. New York: W. W. Norton, 2004.

———. *The Road to War: The Origins of World War II*. New York: Random House, 1990.

———. *Russia's War: A History of the Soviet War Effort, 1941–1945*. London: Penguin Books, 1997.

Padover, Saul K. "France in Defeat: Causes and Consequences." *World Politics* 2, no. 3 (April 1950).

Pallud, John Paul. *Blitzkrieg in the West*. London: Then and Now, 1991.

Pallud, John Paul. *Blitzkrieg in the West*. London: Then and Now, 1991.

Pape, Robert A. *Bombing to Win: Air Power and Coercion in War*. Ithaca, NY: Cornell University Press, 1996.

———. *Dying to Win: The Strategic Logic of Suicide Terrorism*. New York: Random House, 2005.

Paschall, Rod. *The Defeat of Imperial Germany, 1917–1918*. Chapel Hill, NC: Da Capo Press, 1994.

Peled, Alon. *A Question of Loyalty: Military Manpower Policy in Multiethnic States*. Ithaca, NY: Cornell University Press, 1998.

Perrett, Bryan. *The Battle Book: Crucial Conflicts in History from 1469 BC to the Present*. London: Arms and Armour, 1992.

Picq, Ardant du. *Battle Studies: Ancient and Modern Battle*. Harrisburg: Military Service Publishing Company, 1947.

Pike, Douglas. *PAVN: People's Army of Vietnam*. New York: Da Capo Press, 1991.
Pollack, Kenneth M. *Arabs at War*. Lincoln: University of Nebraska Press, 2002.
———. "The Influence of Culture on Arab Military Effectiveness." Ph.D. dissertation, Massachusetts Institute of Technology, 1996.
Porch, Douglas. *The March to the Marne: The French Army 1871–1914*. Cambridge: Cambridge University Press, 1981.
———. "The Marne and After: A Reappraisal of French Strategy in the First World War." *The Journal of Military History* 53, no. 4 (October 1989), 363–86.
———. "Military Culture and the Fall of France: A Review Essay." *International Security* 24, no. 4 (Spring, 2000).
Posen, Barry R. "Measuring the European Conventional Balance: Coping with Complexity in Threat Assessment." *International Security* 9, no. 3 (Winter 1984/85).
———. "Nationalism, the Mass Army, and Military Power." *International Security* 18, no. 2 (Fall 1993).
———. *The Sources of Military Doctrine: France, Britain, and Germany between the Wars*. Ithaca, NY: Cornell University Press, 1984.
Press, Daryl G. "Lessons from Ground Combat in the Gulf: The Impact of Training and Technology." *International Security* 22, no. 2 (Fall 1997).
Price, Roger. *A Concise History of France*. Cambridge: Cambridge University Press, 1993.
Przeworkski, Adam, and Henry Tenue. *The Logic of Comparative Social Inquiry*. New York: John Wiley, 1970.
Quinlivan, James T. "Coup-proofing: Its Practice and Consequence in the Middle East." *International Security* 24, no. 2 (Fall 1999).
Ralston, David. *The Army of the Republic: The Place of the Military in the Political Evolution of France, 1871–1914*. Cambridge, MA: MIT Press, 1967.
Rass, Christoph. "The Social Profile of the German Army's Combat Units, 1939–1945." In *German Wartime Society, 1939–1945: Politicization, Disintegration, and the Struggle for Survival*. Volume 9/1 of *Germany and the Second World War*. Edited by Jorg Echternkamp; translated by Derry Cook-Radmore. Oxford: Clarendon Press, 2008.
Reese, Roger R. *Red Commanders: A Social History of the Soviet Army Officer Corps, 1918–1991*. Lawrence: University of Kansas Press, 2005.
———. *Stalin's Reluctant Soldiers: A Social History of the Red Army, 1925–1941*. Lawrence: University of Kansas Press, 1996.
Reiter, Dan. "Nationalism and Military Effectiveness: Post-Meiji Japan." In *Creating Military Power: The Sources of Military Effectiveness*, ed. Risa A. Brooks and Elizabeth A. Stanley. Stanford: Stanford University Press, 2007.
Reiter, Dan, and Allan Stam. *Democracies at War*. Princeton: Princeton University Press, 2002.
———. "Democracy and Battlefield Military Effectiveness." *Journal of Conflict Resolution* 42 (June 1998).

Rich, Norman. *Great Power Diplomacy: 1814–1914*. New York: McGraw Hill, 1992.

———. *Hitler's War Aims: Ideology, the Nazi State, and the Course of Expansion*. New York: W. W. Norton and Company, 1973.

Ricks, Thomas E. *The Gamble: General Petraeus and the American Military Adventure in Iraq*. New York: Penguin Press HC, 2009.

———. *Making the Corps*. New York: Scribner, 1997.

Ritter, Gerhard. *The Schlieffen Plan: Critique of a Myth*. Westport, CT: Greenwood Press, 1979.

Roberts, Geoffrey. *Stalin's War: From the World War to the Cold War, 1939–1953*. New Haven: Yale University Press, 2006.

Rosen, Stephen Peter. "Military Effectiveness: Why Society Matters." *International Security* 19, no. 4 (Spring 1995).

———. "Net Assessment as an Analytical Concept." In *On Not Confusing Ourselves: Essays on National Security Strategy in Honor of Albert and Roberta Wohlstetter*, ed. Andrew W. Marshall, J. J. Martin, and Henry S. Rowen. Boulder, CO: Westview Press, 1991.

———. *Societies and Military Power: India and Its Armies*. Ithaca, NY: Cornell University Press, 1996.

Rush, Robert Sterling. *Hell in the Hürtgen Forest: The Ordeal and Triumph of an American Infantry Regiment*. Lawrence: University of Kansas Press, 2001.

Ryan, Cornelius. *A Bridge Too Far*. New York: Simon and Schuster, 1974.

Sageman, Marc. "Does Osama Still Call the Shots? Debating the Containment of al Qaeda's Leadership." *Foreign Affairs* 87, no. 4 (July/August 2008).

Salisbury, Harrison. *Behind the Lines—Hanoi*. New York: Harper and Row, 1967.

Samuels, Martin. *Command or Control? Command, Training and Tactics in the British and German Armies, 1888–1918*. London: Frank Cass, 1995.

———. *Doctrine and Dogma: German and British Infantry Tactics in the First World War*. New York: Greenwood Press, 1992.

Schandler, Herbert Y. *The Unmaking of a President: Lyndon Johnson and Vietnam*. Princeton: Princeton University Press, 1977.

Schoenbaum, David. *Hitler's Social Revolution: Class and Status in Nazi Germany, 1933–1939*, 2nd ed. New York: W. W. Norton, 1980.

Schulzinger, Robert D. *A Time for War: The United States and Vietnam, 1941–1975*. Oxford: Oxford University Press, 1997.

Schweller, Randall. *Deadly Imbalances: Tripolarity and Hitler's Strategy of World Conquest*. New York: Columbia University Press, 1998.

Seaton, Albert. *The German Army, 1933–1945*. New York: Meridian, 1982.

Sebag-Montefiore, Simon. *Stalin: The Court of the Red Tsar*. London: Phoenix Press, 2004.

Segal, David, and Meyer Kestenbaum. "Professional Closure in the Military Labor Market: A Critique of Pure Cohesion." In *The Future of the Army Profession*, ed. Don M. Snider and Gayle L. Watkins. New York: McGraw Hill, 2002.

Shay, Jonathan. *Achilles in Vietnam: Combat Trauma and the Undoing of Character*. New York: Touchstone, 1995.

Shepardson, Donald E. "The Fall of Berlin and the Rise of a Myth." *Journal of Military History* 62 (January 1998).

Shils, Edward. "Primordial, Personal, Sacred and Civil Ties." *British Journal of Sociology* 8, no. 2 (1957).

Shils, Edward A., and Morris Janowitz. "Cohesion and Disintegration in the Wehrmacht in World War II." *Public Opinion Quarterly* 12, no. 2 (1948).

Shimko, Keith. *The Iraq Wars and America's Military Revolution*. Cambridge: Cambridge University Press, 2010.

Shirer, William L. *The Collapse of the Third Republic: An Inquiry into the Fall of France in 1940*. New York: Simon and Schuster, 1969.

Shore, Moyers S. *The Battle for Khe Sanh*. Washington, DC: History and Museums Division, Headquarters, U.S. Marine Corps, 1969.

Simkins, Peter. *The First World War, Volume 3: The Western Front 1917–1918*. Oxford: Osprey Publishing, 2002.

Simonsen, Sven Gunner. "Leaving Security in Safe Hands: Identity, Legitimacy and Cohesion in the New Afghan and Iraqi Armies." *Third World Quarterly* 30, no. 8 (2009).

Slantchev, Branislav L. "How Initiators End Their Wars: The Duration of Warfare and the Terms of Peace." *American Journal of Political Science* 48, no. 4 (October 2004).

Small, Melvin. *Johnson, Nixon and the Doves*. New Brunswick, NJ: Rutgers University Press, 1988.

Smith, Leonard V. *Between Mutiny and Obedience: The Case of the French Fifth Infantry Division during World War I*. Princeton: Princeton University Press, 1994.

Smoler, Fredric. "The Secret of the Soldiers Who Didn't Shoot." *American Heritage* 40 (March 1989).

Snyder, Jack. *From Voting to Violence: Democratization and Nationalist Conflict*. New York: W. W. Norton and Company, 2000.

———. *Myths of Empire: Domestic Politics and International Ambition*. Ithaca, NY: Cornell University Press, 1991.

Sorley, Lewis. *A Better War: The Unexamined Victories and Final Tragedy of America's Last Years in Vietnam*. New York: Harcourt Books, 1999.

Spector, Ronald H. *After Tet: The Bloodiest Year in Vietnam*. New York: Vintage Books, 1993.

Stahel, David. *Kiev 1941: Hitler's Battle for Supremacy in the East*. Cambridge: Cambridge University Press, 2012.

Stam III, Allan C. *Win, Lose, or Draw: Domestic Politics and the Crucible of War*. Ann Arbor: University of Michigan Press, 1996.

Stanton, Shelby L. *The Rise and Fall of an American Army: U.S. Ground Forces in Vietnam, 1965–1973*. Novato, CA: Presidio Press, 1985.

Statistical Abstract of the United States. Washington, DC: Government Printing Office, 1994.

Stein, Arthur A. "Conflict and Cohesion: A Review of the Literature," *Journal of Conflict Resolution* 20, no. 1 (March 1976).

Stein, George H. *The Waffen SS: Hitler's Elite Guard, 1939–1945*. Ithaca, NY: Cornell University Press, 1966.

Stephenson, David. *With Our Backs to the Wall: Victory and Defeat in 1918*. Cambridge: Belknap Press of Harvard University Press, 2011.

Stephenson, Scott. *The Final Battle: Soldiers of the Western Front and the German Revolution of 1918*. Cambridge: Cambridge University Press, 2009.

Stern, Fritz. *Dream and Delusions: The Drama of German History*. New York: Alfred A. Knopf, 1987. Rev. ed., New Haven: Yale University Press, 1999.

Stern, Paul C. "Why Do People Sacrifice for Their Nations?" *Political Psychology* 16, no. 2 (1995).

Stewart, Nora Kinzer. *Mates and Muchachos: Unit Cohesion in the Falklands/Malvinas War*. Washington, DC: Brassey's, 1991.

Stouffer, Edward A., et al. *The American Soldier: Adjustment during Army Life*. Princeton: Princeton University Press, 1949.

Strachan, Hew. *The First World War, Volume I: To Arms*. Oxford: Oxford University Press, 2001.

———. "The Soldier's Experience in Two World Wars: Some Historiographical Comparisons." In *Time to Kill: The Soldier's Experience of War in the West, 1939–1945*, ed. Paul Addison and Angus Calder. London: Pimlico, 1997.

———. "Training, Morale and Modern War." *Journal of Contemporary History* 41, no. 2 (2006).

Straub, Christopher C. *The Unit First: Keeping the Promise of Cohesion*. Washington, DC: National Defense University Press, 1988.

Strohn, Matthias. *The German Army and the Defense of the Reich: Military Doctrine and the Conduct of the Defensive Battle, 1918–1939*. Cambridge: Cambridge University Press, 2011.

Stubbs, Kevin D. *Race to the Front: The Materiel Foundations of Coalition Strategy in the Great War*. Westport, CT: Praeger, 2002.

Sugden, Robert. "On the Economics of Philanthropy." *Economic Journal* 92, no. 366 (June 1982).

———. "Team Preferences." *Economics and Philosophy* 16 (2000).

———. "Thinking as a Team: Towards an Explanation of Non-Selfish Behavior." *Social Philosophy and Policy* 10, no. 1 (Winter 1993).

Summers, Jr., Harry G. *The Vietnam War Almanac*, 2nd ed. Novato, CA: Presidio Press, 1999.

Suny, Ronald Grigor. *Revenge of the Past: Nationalism, Revolution, and the Collapse of the Soviet Union*. Stanford: Stanford University Press, 1993.

Sweetland, Andres. *Rallying Potential among the North Vietnamese Armed Forces*. RM-6375–1-ARPA. Santa Monica, CA: Rand Corporation, 1966.

Tajfel, Henri. *Human Groups and Social Categories*. Cambridge: Cambridge University Press, 1981.

Talmadge, Caitlin. "The Puzzle of Personalist Performance: Iraqi Battlefield Effectiveness in the Iran-Iraq War," *Security Studies* 22, no. 2 (Spring 2013).

Talman, Jacob L. *Myth of the Nation and Vision of the Revolution: Ideological Polarization in the Twentieth Century*. New Brunswick, NJ: Transaction Publishers, 1991.

Tarrow, Sidney. *Power in Movement: Social Movements and Contentious Politics*, 2nd ed. Cambridge: Cambridge University Press, 1998.

Taylor, Edmund. "Democracy Demoralized: The French Collapse." *Public Opinion Quarterly* 4, no. 4 (December 1940).

Teets, Jessica. "Let Many Civil Societies Bloom: Regional Ideational Variation in China." *China Quarterly*, 213 (January 2013).

Therborn, Goran. *The Power of Ideology and the Ideology of Power*. London: Verso, 1980.

Thurston, Robert W. *Life and Terror in Stalin's Russia, 1934–1941*. New Haven: Yale University Press, 1998.

Tilly, Charles. *From Mobilization to Revolution*. Reading, MA: Addison-Wesley, 1978.

Tismaneanu, Vladimir. *Fantasies of Salvation: Democracy, Nationalism and Myth in Post-Communist Europe*. Princeton: Princeton University Press, 1998.

Travers, Tim. "The Allied Victories, 1918." In *The Oxford Illustrated History of the First World War*, ed. Hew Strachan. Oxford: Oxford University Press, 1998.

Triandis, Harry C. *Individualism and Collectivism*. Boulder, CO: Westview Press, 1995.

Tucker, Spencer C. *The Great War: 1914–1918*. Bloomington: Indiana University Press, 1998.

———, ed. *The Encyclopedia of the Vietnam War*. Oxford: Oxford University Press, 2000.

Tzu, Sun. *The Art of War*. Translated by Samuel B. Griffith. Oxford: Oxford University Press, 1963.

U.S. Army, Concepts Analysis Agency. *Weapons Effectiveness Indices/Weighted Unit Values (WEI/WUV)*: Study Report CAA-SR- 73–18. Bethesda, MD, 1974.

Walt, Stephen M. *Revolution and War*. Ithaca, NY: Cornell University Press, 1996.

Walzer, Michael. *Obligations: Essays on Disobedience, War and Citizenship*. New York: Simon and Schuster, 1970.

Ward, Steven R. *Immortal: A Military History of Iran and Its Armed Forces*. Washington, DC: Georgetown University Press, 2009.

Warr, Nicholas. *Phase Line Green: The Battle for Hue, 1968*. Annapolis, MD: Naval Institute Press, 1997.

Watson, Alexander. "Culture and Combat in the Western World, 1900–1945." *Historical Journal* 51, no. 2 (2008).

———. *Enduring the Great War: Combat, Morale, and Collapse in the German and British Armies, 1914–1918*. Cambridge: Cambridge University Press, 2008.

Watson, Bruce Allen. *When Soldiers Quit: Studies in Military Disintegration*. Westport, CT: Praeger, 1997.

Weber, Eugen. *The Hollow Years: France in the 1930s*. New York: W. W. Norton and Company, 1994.

———. *The Nationalist Revival in France, 1905–1914*. Berkeley: University of California Press, 1959.

———. *Peasants into Frenchmen: The Modernization of Rural France, 1870–1914*. Stanford: Stanford University Press, 1977.

Weigley, Russell F. *Eisenhower's Lieutenants: The Campaign of France and Germany, 1944–1945*. Bloomington: Indiana University Press, 1981.

Weinberg, Gerhard. *A World at Arms: A Global History of World War II*. Cambridge: Cambridge University Press, 1994.

Wendt, Alexander. "On Constitution and Causation in International Relations." *Review of International Studies* 24 (1998).

———. *Social Theory of International Politics*. Cambridge: Cambridge University Press, 1999.

Werth, Alexander. *Russia at War, 1941–1945*. New York: E. P. Dutton and Company, 1964.

Wette, Wolfram. *Wehrmacht: History, Myth, Reality*. Translated by Deborah Lucas Schneider. Cambridge: Harvard University Press, 2007.

White, Dorothy S. *Seeds of Discord: De Gaulle, Free France and the Allies*. Syracuse: Syracuse University Press, 1964.

Whiting, Charles. *Siegfried: The Nazis' Last Stand*. New York: Jove Books, 1983.

Wilson, David Sloan. "Levels of Selection: An Alternative to Individualism in Biology and the Human Sciences." *Social Networks* 11 (1989).

Wilson, David Sloan, and Elliott Sober. "Reintroducing Group Selection to the Human Behavioral Sciences." *Behavioral and Brain Sciences* 17 (1994).

Woodruff, Mark. *The Defeat of the Viet Cong and the North Vietnamese Army: 1961–1973*. Arlington, VA: Vandamere Press, 1999.

Woodside, Alexander B. *Community and Revolution in Modern Vietnam*. Boston: Houghton Mifflin, 1976.

Yin, Robert K. *Case Study Research: Design and Methods*, 2nd ed. Thousand Oaks, CA: SAGE Publications, 1994.

Young, Marilyn B. *The Vietnam Wars, 1945–1990*. New York: Harper Collins, 1991.

Young, Robert J. *In Command of France: French Foreign Policy and Military Planning, 1933–1940*. Cambridge: Harvard University Press, 1978.

Zabecki, David T., ed. *World War II in Europe: An Encyclopedia*. New York: Garland Publishing, 1999.

Zaffiri, Samuel. *Hamburger Hill, May 11–20, 1969*. Novato, CA: Presidio Press, 1988.

Zetterling, Niklas. *Normandy 1944: German Military Organization, Combat Power and Organizational Effectiveness*. Winnipeg, Manitoba: J. J. Fedorowicz Publishing, 2000.

Ziemke, Earl F. *Stalingrad to Berlin: German Defeat in the East*. New York: Dorset Press, 1986.

Ziemke, Earl F., and Magna E. Bauer. *Moscow to Stalingrad: Decision in the East*. Washington, DC: US Army Center of Military History, 1987.

Zinn, Howard. *The Twentieth Century: A People's History*. New York: Harper and Row, 1980.

Index

CPSIA information can be obtained
at www.ICGtesting.com
Printed in the USA
LVHW090313220620
658671LV00008B/78

9 780804 789103